Scattering of Elastic Waves by Obstacles in Solids

Scattering of Elastic Waves by Obstacles in Solids

By

Anders Boström

Cambridge
Scholars
Publishing

Scattering of Elastic Waves by Obstacles in Solids

By Anders Boström

This book first published 2024

Cambridge Scholars Publishing

Lady Stephenson Library, Newcastle upon Tyne, NE6 2PA, UK

British Library Cataloguing in Publication Data
A catalogue record for this book is available from the British Library

ISBN (10): 1-0364-0631-8
ISBN (13): 978-1-0364-0631-8

Contents

Preface

It is almost 50 years since I started as a PhD student in mathematical physics and was introduced to the area of scattering of elastic waves in solids by my supervisor Staffan Ström. This has remained my main area of research, and being professor emeritus since a few years, it now seems appropriate to summarize my views on the subject.

This book is thus concerned with the scattering of elastic waves by various types of obstacles of simple shape, mostly in an infinite solid, but also in a half-space, layered medium, or waveguide. Isotropic media is mostly assumed, but some anisotropic ones are also considered. The focus is on more or less analytical approaches, in most cases leading to "exact" results. Two distinctive features are that the transition matrix (relating the scattered wave coefficients to those of the incident wave coefficients) is calculated when relevant and that this matrix is given to leading order at low frequencies.

There are, of course, useful methods of scattering that are not considered. This includes high frequency asymptotic methods, like ray theory, geometrical theory of diffraction, and Kirchhoff theory. Low frequency asymptotic methods are not considered either, although it should be noted that the methods employed in the book work very well for low frequencies and, as noted above, in a number of cases the leading order T-matrix elements are given for low frequencies. Purely numerical methods, like finite-element methods, are not taken up either, although such methods are of course also useful.

The book is rather unique in its relatively broad coverage of scattering by obstacles in elastic solids, other books being more limited in scope. Pao and Mow [112] use separation of variables for some simply shaped obstacles, mostly in 2D. Zhang and Gross [147] employ integral equation methods, mostly of the same types as here, for scattering by some types of cracks and provide many references. Martin [88] considers multiple scattering for acoustic, electromagnetic, elastic, and surface water waves and this includes a review of scattering by a single obstacle. He also gives an extensive list of references.

The mathematics used in the book is more or less standard, like vector analysis, special functions, separation of variables, analytic function theory, Green functions. Vectors are used as much as possible, although sometimes it is more convenient to use tensors.

I have been influenced by and cooperated with many persons and I thank all these and others that have helped me during the years. I also thank Paul Martin for providing many useful comments on two versions of the manuscript.

1 Elastodynamics

The topic of this book is the scattering of elastic waves (also called acoustic waves [9]) by obstacles in solids. Only linear and time-harmonic problems are considered. In this first chapter the basic equations describing propagation and scattering of linear waves in elastic solids are given. Isotropic media are in particular treated and the relevant potentials are introduced for different coordinate systems. Some of the simpler types of anisotropy are described. As an introduction to the rest of the book the chapter ends with a simple scattering problem.

1.1 Basic equations

In this section the basic equations governing wave propagation in an elastic solid are stated. This includes the equation of motion (Newton's second law), the constitutive relation between stress and strain, and the definition of strain. Throughout the book only linear problems are considered, so all relations are taken in their linear form. The boundary conditions on an elastic body and the interface conditions between two elastic bodies in contact are stated. Energy densities and flow are defined and the energy balance is derived. The Fourier transform between time and frequency is introduced; however, throughout only problems with a fixed frequency are taken up. Reciprocity is a useful concept in elastodynamics and the integral form of reciprocity is derived at fixed frequency. Finally, the equations of motion and the strain are given in cylindrical and spherical coordinates. The material in this section is standard and can be found in many textbooks, for example de Hoop [41], Auld [9], Achenbach [1], Langenberg *et al.* [84], and Aki and Richards [5].

1.1.1 The wave equation

The basic field variable in an elastic solid is the displacement $\boldsymbol{u}(\boldsymbol{r}, t)$, which is a function of position $\boldsymbol{r}$ and time t. The displacement is given components u_x, u_y, u_z in a Cartesian coordinate system (x, y, z) as

$$\boldsymbol{u} = \hat{\boldsymbol{x}} u_x + \hat{\boldsymbol{y}} u_y + \hat{\boldsymbol{z}} u_z$$

As much as possible vectors are used and these are always bold-faced. Unit vectors are denoted with a hat so that $\hat{\boldsymbol{x}}$ is the unit vector in the x-direction. However, sometimes it is convenient to use Cartesian tensors. Then the coordinates are (x_1, x_2, x_3) and the displacement is written

$$\boldsymbol{u} = \hat{\boldsymbol{x}}_1 u_1 + \hat{\boldsymbol{x}}_2 u_2 + \hat{\boldsymbol{x}}_3 u_3 = \hat{\boldsymbol{x}}_i u_i$$

where Einstein's summation convention is used so that a repeated index is summed over $i = 1, 2, 3$.

Both displacements and strains are assumed small and the linear strain tensor is then

$$\varepsilon_{ij} = \frac{1}{2}\left(\partial_j u_i + \partial_i u_j\right)$$

As usual in tensor notation the free indices i and j take on the values 1, 2, 3, so that nine components of strain are defined. Derivatives are sometimes indicated with a short-hand notation

$$\partial_j u_i = \frac{\partial u_i}{\partial x_j}$$

The strain tensor is obviously symmetric. The definition of the strain tensor can also be made in dyadic notation [99].

$$\boldsymbol{\varepsilon} = \frac{1}{2}(\nabla \boldsymbol{u} + \boldsymbol{u}\nabla)$$

where the convention is that $(\nabla \boldsymbol{u})_{ij} = (\boldsymbol{u}\nabla)_{ji} = \partial_i u_j$. According to the identity

$$\boldsymbol{I} \times (\nabla \times \boldsymbol{u}) = \boldsymbol{u}\nabla - \nabla \boldsymbol{u}$$

the strain can be written as

$$\boldsymbol{\varepsilon} = \nabla \boldsymbol{u} + \frac{1}{2}\boldsymbol{I} \times (\nabla \times \boldsymbol{u})$$

where $\boldsymbol{I}$ is the unit dyadic with components δ_{ij}, where δ_{ij} is the Kronecker symbol ($\delta_{ij} = 1$ if $i = j$, $\delta_{ij} = 0$ if $i \neq j$).

The material behaviour is also assumed to be linear and the linear stress tensor σ_{ij} is thus related to the linear strain tensor by the constitutive relation

$$\sigma_{ij} = c_{ijkl}\varepsilon_{kl} \tag{1.1}$$

Here the stiffness tensor c_{ijkl} is a tensor of rank four. The stress tensor is also symmetric and together with the assumption of the existence of a strain energy function this means that the stiffness tensor satisfies the following symmetry properties

$$c_{ijkl} = c_{ijlk} = c_{jikl} = c_{klij}$$

This reduces the number of independent components of the stiffness tensor to 21. Most materials have much fewer independent stiffness components, all the way down to an isotropic material which has only two. Some of the simpler examples of constitutive relations are taken up later in the chapter.

The equation of motion in the linear case is

$$\partial_k \sigma_{ki} + \rho f_i = \rho \ddot{u}_i \tag{1.2}$$

where ρ is the density and f_i is the body force (per unit mass). Time derivatives are denoted with a dot above a variable, thus $\ddot{u}_i$ is the second time derivative of u_i. In vector notation this equation is

$$\nabla \cdot \boldsymbol{\sigma} + \rho \boldsymbol{f} = \rho \ddot{\boldsymbol{u}} \tag{1.3}$$

where $\boldsymbol{\sigma}$ is the stress dyadic. In most cases the body force is abscent, $\boldsymbol{f} = \boldsymbol{0}$. Eliminating the stress and strain the equation of motion becomes

$$\partial_k \left(c_{kilj} \partial_l u_j \right) = \rho \ddot{u}_i \tag{1.4}$$

This is the most general form of the linear elastodynamic wave equation that is valid for general anisotropy and inhomogeneous materials (for which the stiffness tensor is dependent on position).

1.1.2 Boundary conditions

In all scattering problems at least two different domains in space are considered, although one of them may be a domain where no wave propagates, typically vacuum. Then some conditions at the surface of a domain or at an interface between two domains are required. On the other hand no initial conditions are needed as only time harmonic problems are considered.

If a domain where waves are propagating has a boundary to another domain where no waves can propagate then one (vector) boundary condition is required. This can be of different kinds:

- The displacement is specified. Typically the displacement is zero and this corresponds to a boundary to a rigid body where also the tangential components are restrained from moving.
- The traction vector $\boldsymbol{t}^{(n)} = \hat{\boldsymbol{n}} \cdot \boldsymbol{\sigma}$, where $\hat{\boldsymbol{n}}$ is the unit normal to the boundary and $\boldsymbol{\sigma}$ is the stress dyadic, is specified. In tensor notation this is $t_i^{(n)} = n_j \sigma_{ji}$, where n_j are the components of the normal $\hat{\boldsymbol{n}}$. In most cases the traction is zero as this is a good model for a boundary to air (because of the great mismatch in impedance between a solid and air very little energy leaks from a solid into air).
- Normal displacement $u_n = \hat{\boldsymbol{n}} \cdot \boldsymbol{u}$ and tangential traction $\boldsymbol{t}_{tan}^{(n)}$ are specified. If these quantities are both zero this is a model for a boundary to a smooth, rigid body.
- Spring boundary conditions $\boldsymbol{t}^{(n)} = \boldsymbol{\alpha u}$ ($t_j^{(n)} = \alpha_{ji} u_i$), where $\boldsymbol{\alpha}$ is a dyadic, where usually only the diagonal elements are nonzero. This can be used to model a thin surface layer.

There are other possibilities, but there should be one condition in each of three perpendicular directions and they should involve displacement and traction and possibly tangential derivatives thereof.

At the surface between two solids both supporting waves some interface conditions must likewise be satisfied:

- Welded contact where both the displacement $\boldsymbol{u}$ and traction $\boldsymbol{t}^{(n)}$ are continuous. This is by far the most common condition and is usually assumed if two solids are in contact.
- Slip conditions where u_n and $t_n^{(n)} = \hat{\boldsymbol{n}} \cdot \boldsymbol{t}^{(n)}$ are continuous and $\boldsymbol{t}_{tan}^{(n)} = \boldsymbol{0}$ from both sides.
- If a solid has a boundary to a fluid then u_n and $t_n^{(n)} = -p$, where p is the pressure in the fluid, are continuous and $\boldsymbol{t}_{tan}^{(n)} = \boldsymbol{0}$ in the solid.

- Spring interface conditions where the traction $\boldsymbol{t}^{(n)}$ is continuous and where $\boldsymbol{t}^{(n)} = \boldsymbol{\alpha}[\boldsymbol{u}]$, where $[\boldsymbol{u}]$ denotes the jump in displacement and $\boldsymbol{\alpha}$ is a dyadic, where usually only the diagonal elements are nonzero. This condition is widely used to model a thin interface layer or a damaged interface.

Other interface conditions are also used, for instance to obtain a better model of a thin interface layer (for which the spring interface conditions give a somewhat poor model). These conditions can involve tangential derivatives of displacement and traction.

If a solid is of infinite extent, which is the normal situation in scattering problems, then a "boundary condition", called a radiation condition, is needed at infinity. If there is damping in the material this is that the fields decay exponentially towards infinity, and the case without damping can be taken as the limit of this. For a fixed frequency in an isotropic solid that is infinite in all directions this means that the behaviour of the field far away is $\exp(\mathrm{i}kr)/r$ in 3D and $\exp(\mathrm{i}kr)/\sqrt{r}$ in 2D (where r is the radial coordinate in 3D or 2D, respectively) with appropriate wavenumber k and polarization. In other cases without damping, like waves in an infinite plate or for general anisotropy, the situation is more complicated. Generally speaking, the energy flow must always be directed outwards.

1.1.3 Energy densities and flow

It is sometimes of interest to calculate the energy flow or energy densities in waves. To this end the equation of motion is multiplied by $\dot{u}_i$ and summed over i

$$\dot{u}_i \partial_j \sigma_{ji} = \rho\, \ddot{u}_i \dot{u}_i = \frac{\mathrm{d}}{\mathrm{d}t}\left(\frac{1}{2}\rho\, \dot{u}_i \dot{u}_i\right)$$

Manipulate the left-hand side

$$\dot{u}_i \partial_j \sigma_{ji} = \partial_j \left(\dot{u}_i \sigma_{ji}\right) - \left(\partial_j \dot{u}_i\right) \sigma_{ji} = \partial_j \left(\dot{u}_i \sigma_{ji}\right) - \dot{\varepsilon}_{ij} \sigma_{ji}$$

where the symmetry of σ_{ij} is used in the last step. Furthermore,

$$\frac{\mathrm{d}}{\mathrm{d}t}(\varepsilon_{ij}\sigma_{ij}) = \dot{\varepsilon}_{ij}\sigma_{ij} + \varepsilon_{ij} c_{ijkl} \dot{\varepsilon}_{kl} = 2\dot{\varepsilon}_{ij}\sigma_{ij}$$

Introduce the kinetic energy density

$$e_T = \frac{1}{2}\rho\, \dot{u}_i \dot{u}_i$$

the strain energy density

$$e_U = \frac{1}{2}\varepsilon_{ij}\sigma_{ij}$$

and the energy flow density

$$j_i = -\dot{u}_j \sigma_{ji} \tag{1.5}$$

then all the above can be summarized as

$$\frac{\mathrm{d}}{\mathrm{d}t}(e_T + e_U) + \partial_i j_i = 0$$

This is the differential form of energy conservation. The physical meaning becomes clearer if this equation is integrated over a fixed volume V with boundary S and outward unit normal $\hat{\boldsymbol{n}}$ and Gauss' law is employed

$$\frac{\mathrm{d}}{\mathrm{d}t}\int_V (e_T + e_U)\,\mathrm{d}V = -\int_S j_n\,\mathrm{d}S$$

where $j_n = -\dot{u}_i \sigma_{ij} n_j$ and n_j are the components of $\hat{\boldsymbol{n}}$. In words this equation says that "change in energy inside V equals influx of energy through S".

1.1.4 Time harmonic fields

The Fourier transform between time t and angular frequency ω is here defined as

$$\tilde{\boldsymbol{u}}(\omega) = \int_{-\infty}^{\infty} \boldsymbol{u}(t)\mathrm{e}^{\mathrm{i}\omega t}\,\mathrm{d}t \tag{1.6}$$

with the inverse

$$\boldsymbol{u}(t) = \frac{1}{2\pi}\int_{-\infty}^{\infty} \tilde{\boldsymbol{u}}(\omega)\mathrm{e}^{-\mathrm{i}\omega t}\,\mathrm{d}\omega \tag{1.7}$$

where the space dependence has been left out. Switching from the time domain to the frequency domain has the great advantage of removing all time derivatives and replacing them by multiplication by $-\mathrm{i}\omega$, which of course greatly facilitates the solution of the wave equation. As the physical displacement field in the time domain is a real function it follows that $\tilde{\boldsymbol{u}}(-\omega) = \tilde{\boldsymbol{u}}(\omega)^\star$ (where the star denotes the complex conjugate) and

$$\boldsymbol{u}(t) = \frac{1}{\pi}\,\mathrm{Re}\left[\int_0^\infty \tilde{\boldsymbol{u}}(\omega)\mathrm{e}^{-\mathrm{i}\omega t}\,\mathrm{d}\omega\right]$$

Thus the angular frequency ω can always be regarded as a positive real quantity in the following. This is important when propagation direction and damping are discussed.

A note on the definition of the Fourier transform used here is in order. In some areas, particularly in electrical engineering, the sign in the exponential is switched so that instead of the implicit time factor $\exp(-\mathrm{i}\omega t)$ the time factor is $\exp(\mathrm{j}\omega t)$ (where the imaginary unit is called "j"). For wave propagation problems the present definition is more natural as a plane wave like $\exp(+\mathrm{i}kx)$ then propagates in the $+x$ direction. Also the factor $1/2\pi$ in front of the inverse transform can be moved to the transform instead, or a factor $1/\sqrt{2\pi}$ can be put in front of both the transform and the inverse transform (and this makes the Fourier transform a Hermitian operator).

Another approach is to say from the start that a time harmonic problem is considered and put

$$\boldsymbol{u}(t) = \mathrm{Re}\left(\tilde{\boldsymbol{u}}(\omega)\mathrm{e}^{-\mathrm{i}\omega t}\right)$$

In the following only time harmonic problems with a fixed angular frequency ω are considered. The tilde on the displacement field is then skipped as is the dependence on ω. The basic field variable is thus written $\boldsymbol{u}(\boldsymbol{r})$. In principle the field in the time domain can be synthesized by taking an inverse Fourier transform.

It must be remembered that when quadratic quantities are calculated the time factor $\exp(-\mathrm{i}\omega t)$ and the real part must be taken before multiplying together the parts. For the energy flow vector defined in Eq. (1.5) this means that the instantaneous (time dependent) value is

$$\begin{aligned}\boldsymbol{j} &= -\dot{\boldsymbol{u}}\cdot\boldsymbol{\sigma} = -\mathrm{Re}\left(-\mathrm{i}\omega\boldsymbol{u}\,\mathrm{e}^{-\mathrm{i}\omega t}\right)\cdot\mathrm{Re}\left(\boldsymbol{\sigma}\,\mathrm{e}^{-\mathrm{i}\omega t}\right)\\ &= \omega\left(\mathrm{Re}\,\boldsymbol{u}\sin\omega t - \mathrm{Im}\,\boldsymbol{u}\cos\omega t\right)\cdot\left(\mathrm{Re}\,\boldsymbol{\sigma}\cos\omega t + \mathrm{Im}\,\boldsymbol{\sigma}\sin\omega t\right)\end{aligned}$$

where no notational difference is made between the displacement and stress in the time and frequency domain. Taking the mean in time this becomes

$$\langle \boldsymbol{j} \rangle = \frac{\omega}{2} \left(\mathrm{Re}\, \boldsymbol{u} \cdot \mathrm{Im}\, \boldsymbol{\sigma} - \mathrm{Im}\, \boldsymbol{u} \cdot \mathrm{Re}\, \boldsymbol{\sigma} \right) = -\frac{\omega}{2} \mathrm{Im} \left(\boldsymbol{u} \cdot \boldsymbol{\sigma}^{\star} \right) \tag{1.8}$$

Damping in a material is easier to model in the frequency domain by allowing the stiffness constants to be complex. The real part then corresponds to the elasticity of the material while the imaginary part determines the damping. In the time domain damping is more difficult to model, in general there should be memory in the material so that a time integral is needed, considerably complicating the analysis (Carcione [35]). In the frequency domain this corresponds to taking the stiffness constants as functions of frequency, and in practice almost any frequency dependence can be taken. To have damping the imaginary part of the stiffness constants must be negative with the present sign convention for the Fourier transform. It is noted that the negative imaginary part of the stiffness constants gives a positive imaginary part of the wavenumbers.

1.1.5 Reciprocity

A reciprocity relation gives a relation between two elastodynamics states. Here it is assumed that the two states occur in the same body. Consider the two states A and B and write the equation for each state (assuming a fixed frequency)

$$\begin{aligned} \partial_k \sigma_{ki}^{\mathrm{A}} + \rho f_i^{\mathrm{A}} &= -\rho\omega^2 u_i^{\mathrm{A}} \\ \partial_k \sigma_{ki}^{\mathrm{B}} + \rho f_i^{\mathrm{B}} &= -\rho\omega^2 u_i^{\mathrm{B}} \end{aligned}$$

Multiply the first of these with u_i^{B} and the second by u_i^{A} (and sum over i) and subtract the second equation from the first

$$u_i^{\mathrm{B}} \partial_k \sigma_{ki}^{\mathrm{A}} - u_i^{\mathrm{A}} \partial_k \sigma_{ki}^{\mathrm{B}} = u_i^{\mathrm{A}} \rho f_i^{\mathrm{B}} - u_i^{\mathrm{B}} \rho f_i^{\mathrm{A}}$$

The left-hand side is written

$$u_i^{\mathrm{B}} \partial_k \sigma_{ki}^{\mathrm{A}} - u_i^{\mathrm{A}} \partial_k \sigma_{ki}^{\mathrm{B}} = \partial_k \left(u_i^{\mathrm{B}} \sigma_{ki}^{\mathrm{A}} - u_i^{\mathrm{A}} \sigma_{ki}^{\mathrm{B}} \right) - \left(\sigma_{ki}^{\mathrm{A}} \partial_k u_i^{\mathrm{B}} - \sigma_{ki}^{\mathrm{B}} \partial_k u_i^{\mathrm{A}} \right)$$

Using the symmetry of the stress tensor, the stress-strain relation Eq. (1.1), and the symmetry of the stiffness constants the second term on the right-hand side is seen to vanish

$$\sigma_{ki}^{\mathrm{A}} \partial_k u_i^{\mathrm{B}} - \sigma_{ki}^{\mathrm{B}} \partial_k u_i^{\mathrm{A}} = \sigma_{ki}^{\mathrm{A}} \varepsilon_{ki}^{\mathrm{B}} - \sigma_{ki}^{\mathrm{B}} \varepsilon_{ki}^{\mathrm{A}} = c_{kijl} \varepsilon_{jl}^{\mathrm{A}} \varepsilon_{ki}^{\mathrm{B}} - c_{kijl} \varepsilon_{jl}^{\mathrm{B}} \varepsilon_{ki}^{\mathrm{A}} = 0$$

Thus the local form of the reciprocity relation becomes

$$\partial_k \left(\sigma_{ki}^{\mathrm{A}} u_i^{\mathrm{B}} - \sigma_{ki}^{\mathrm{B}} u_i^{\mathrm{A}}\right) = \rho \left(u_i^{\mathrm{A}} f_i^{\mathrm{B}} - u_i^{\mathrm{B}} f_i^{\mathrm{A}}\right)$$

Integrate this equation over a volume V with boundary S and employ Gauss' law to obtain the global form of the reciprocity relation

$$\int_S n_k \left(\sigma_{ki}^{\mathrm{A}} u_i^{\mathrm{B}} - \sigma_{ki}^{\mathrm{B}} u_i^{\mathrm{A}}\right) \mathrm{d}S = \int_V \rho \left(u_i^{\mathrm{A}} f_i^{\mathrm{B}} - u_i^{\mathrm{B}} f_i^{\mathrm{A}}\right) \mathrm{d}V$$

where n_k are the components of the outward pointing normal $\hat{\boldsymbol{n}}$ to S. In vector and dyadic notation this equation is

$$\int_S \hat{\boldsymbol{n}} \cdot \left(\boldsymbol{\sigma}^{\mathrm{A}} \cdot \boldsymbol{u}^{\mathrm{B}} - \boldsymbol{\sigma}^{\mathrm{B}} \cdot \boldsymbol{u}^{\mathrm{A}}\right) \mathrm{d}S = \int_V \rho \left(\boldsymbol{u}^{\mathrm{A}} \cdot \boldsymbol{f}^{\mathrm{B}} - \boldsymbol{u}^{\mathrm{B}} \cdot \boldsymbol{f}^{\mathrm{A}}\right) \mathrm{d}V \quad (1.9)$$

In the derivation it is assumed that the body has a linear stress-strain relation but the body can be both anisotropic and inhomogeneous. Reciprocity relations are very useful in elastodynamics, e.g. in the derivation of the orthogonality properties of the spherical vector wavefunctions and in the derivation of integral representations. For more about reciprocity relations see, for instance, Achenbach [3], Schmerr [119], Aki and Richards [5], de Hoop [41], and Carcione [35].

1.1.6 Cylindrical and spherical coordinates

The formulas above are given in Cartesian tensor notation and are thus directly valid in Cartesian coordinates (x, y, z). In many scattering problems it is natural to use cylindrical (polar in 2D) and spherical coordinates instead. Cylindrical coordinates (r, φ, z) are defined by

$$x = r\cos\varphi, \qquad y = r\sin\varphi, \qquad z = z$$

The equations of motion are

$$\frac{\partial \sigma_{rr}}{\partial r} + \frac{1}{r}\frac{\partial \sigma_{r\varphi}}{\partial \varphi} + \frac{\partial \sigma_{zr}}{\partial z} + \frac{\sigma_{rr} - \sigma_{\varphi\varphi}}{r} = \rho \ddot{u}_r$$

$$\frac{\partial \sigma_{r\varphi}}{\partial r} + \frac{1}{r}\frac{\partial \sigma_{\varphi\varphi}}{\partial \varphi} + \frac{\partial \sigma_{\varphi z}}{\partial z} + \frac{2\sigma_{r\varphi}}{r} = \rho \ddot{u}_\varphi$$

$$\frac{\partial \sigma_{zr}}{\partial r} + \frac{1}{r}\frac{\partial \sigma_{\varphi z}}{\partial \varphi} + \frac{\partial \sigma_{zz}}{\partial z} + \frac{\sigma_{zr}}{r} = \rho \ddot{u}_z \tag{1.10}$$

and the strain-displacement relations are

$$\begin{aligned}
\varepsilon_{rr} &= \frac{\partial u_r}{\partial r} \\
\varepsilon_{\varphi\varphi} &= \frac{1}{r}\frac{\partial u_\varphi}{\partial \varphi} + \frac{u_r}{r} \\
\varepsilon_{zz} &= \frac{\partial u_z}{\partial z} \\
\varepsilon_{\varphi z} &= \frac{1}{2}\left(\frac{1}{r}\frac{\partial u_z}{\partial \varphi} + \frac{\partial u_\varphi}{\partial z}\right) \\
\varepsilon_{zr} &= \frac{1}{2}\left(\frac{\partial u_r}{\partial z} + \frac{\partial u_z}{\partial r}\right) \\
\varepsilon_{r\varphi} &= \frac{1}{2}\left(\frac{\partial u_\varphi}{\partial r} - \frac{u_\varphi}{r} + \frac{1}{r}\frac{\partial u_r}{\partial \varphi}\right)
\end{aligned} \tag{1.11}$$

Spherical coordinates (r, θ, φ) are defined by

$$x = r\sin\theta\cos\varphi, \qquad y = r\sin\theta\sin\varphi, \qquad z = r\cos\theta$$

The radius is denoted r in both the spherical and cylindrical systems, but is of course not the same. This is common practice and there should be no confusion as both coordinate systems are seldom used in the same problem. The azimuthal angle φ, on the other hand, is the same. The equations of motion in spherical coordinates are

$$\frac{\partial \sigma_{rr}}{\partial r} + \frac{1}{r}\frac{\partial \sigma_{r\theta}}{\partial \theta} + \frac{1}{r\sin\theta}\frac{\partial \sigma_{\varphi r}}{\partial \varphi} + \frac{1}{r}\left(2\sigma_{rr} - \sigma_{\theta\theta} - \sigma_{\varphi\varphi} + \cot\theta\,\sigma_{r\theta}\right) = \rho\ddot{u}_r$$

$$\frac{\partial \sigma_{r\theta}}{\partial r} + \frac{1}{r}\frac{\partial \sigma_{\theta\theta}}{\partial \theta} + \frac{1}{r\sin\theta}\frac{\partial \sigma_{\theta\varphi}}{\partial \varphi} + \frac{1}{r}\left(\cot\theta(\sigma_{\theta\theta} - \sigma_{\varphi\varphi}) + 3\sigma_{r\theta}\right) = \rho\ddot{u}_\theta$$

$$\frac{\partial \sigma_{\varphi r}}{\partial r} + \frac{1}{r}\frac{\partial \sigma_{\theta\varphi}}{\partial \theta} + \frac{1}{r\sin\theta}\frac{\partial \sigma_{\varphi\varphi}}{\partial \varphi} + \frac{1}{r}\left(3\sigma_{\varphi r} + 2\cot\theta\,\sigma_{\theta\varphi}\right) = \rho\ddot{u}_\varphi \tag{1.12}$$

and the strain-displacement relations are

$$\varepsilon_{rr} = \frac{\partial u_r}{\partial r}$$

$$\varepsilon_{\theta\theta} = \frac{1}{r}\frac{\partial u_\theta}{\partial \theta} + \frac{u_r}{r}$$

$$\varepsilon_{\varphi\varphi} = \frac{1}{r\sin\theta}\frac{\partial u_\varphi}{\partial \varphi} + \frac{\cot\theta}{r}u_\theta + \frac{u_r}{r}$$

$$\varepsilon_{\theta\varphi} = \frac{1}{2r}\left(\frac{\partial u_\varphi}{\partial \theta} - \cot\theta\, u_\varphi + \frac{1}{\sin\theta}\frac{\partial u_\theta}{\partial \varphi}\right)$$

$$\varepsilon_{\varphi r} = \frac{1}{2}\left(\frac{1}{r\sin\theta}\frac{\partial u_r}{\partial \varphi} + \frac{\partial u_\varphi}{\partial r} - \frac{u_\varphi}{r}\right)$$

$$\varepsilon_{r\theta} = \frac{1}{2}\left(\frac{\partial u_\theta}{\partial r} - \frac{u_\theta}{r} + \frac{1}{r}\frac{\partial u_r}{\partial \theta}\right) \tag{1.13}$$

1.2 Isotropic solids

In many cases it may be assumed that the solid is isotropic and this leads to great simplifications. It becomes, in particular, possible to introduce potentials so that the solution can proceed via scalar Helmholtz equations, thus simplifying the solution procedure. For a comprehensive discussion about potentials and the separability of the elastodynamic wave equation in 3D Morse and Feshbach [99, Chapter 13] should be consulted.

1.2.1 P and S waves

For an isotropic solid the stress-strain relation simplifies in that the number of stiffness constants is reduced from 21 to two, so that this relation now becomes

$$\sigma_{ij} = \lambda\delta_{ij}\varepsilon_{kk} + 2\mu\varepsilon_{ij} = \lambda\delta_{ij}\partial_k u_k + \mu(\partial_i u_j + \partial_j u_i) \tag{1.14}$$

where δ_{ij} is the Kronecker symbol and λ and μ are the Lamé constants. The Lamé constants can be expressed in terms of the engineering constants as

$$\lambda = \frac{\nu E}{(1+\nu)(1-2\nu)}, \qquad \mu = G = \frac{E}{2(1+\nu)}$$

where E is Young's modulus, ν Poisson's number, and G the shear modulus. In vector notation the constitutive equation can be written

$$\boldsymbol{\sigma} = \lambda \boldsymbol{I}\,\nabla\cdot\boldsymbol{u} + \mu[2\nabla\boldsymbol{u} + \boldsymbol{I}\times(\nabla\times\boldsymbol{u})] \tag{1.15}$$

From now on it is assumed that the solid is homogeneous so that λ and μ are constant. It is also assumed that there are no body forces. Inserting the stress from Eq. (1.14) into the equation of motion Eq. (1.2) gives

$$(\lambda + \mu)\partial_i\partial_j u_j + \mu\partial_j\partial_j u_i = -\rho\omega^2 u_i$$

where, as noted above, only time harmonic conditions are considered. In vector notation this can be written

$$k_{\mathrm{p}}^{-2}\nabla(\nabla \cdot \boldsymbol{u}) - k_{\mathrm{s}}^{-2}\nabla \times (\nabla \times \boldsymbol{u}) + \boldsymbol{u} = \mathbf{0} \tag{1.16}$$

Here the following property of the nabla operator is used

$$\nabla^2\boldsymbol{u} = \nabla(\nabla \cdot \boldsymbol{u}) - \nabla \times (\nabla \times \boldsymbol{u})$$

In Cartesian coordinates this is straightforward to show and for other orthogonal coordinates it is the definition of the operator ∇^2 operating on a vector field. Two wavenumbers are introduced

$$k_{\mathrm{p}} = \omega/c_{\mathrm{p}}, \qquad k_{\mathrm{s}} = \omega/c_{\mathrm{s}}$$

where the two wave velocities are

$$c_{\mathrm{p}} = \sqrt{\frac{\lambda + 2\mu}{\rho}}, \qquad c_{\mathrm{s}} = \sqrt{\frac{\mu}{\rho}}$$

When there is damping in the material the wavenumbers are complex, and as noted above the imaginary part must be positive. The wavenumbers are thus located in the first quadrant of the complex plane (or on the positive real axis if there is no damping). Much of this book is concerned with the solution of Eq. (1.16) in various scattering configurations.

To see the significance of the two wavenumbers k_{p} and k_{s} it is useful to take the divergence and curl of Eq. (1.16). First, introduce the scalar potential Φ according to

$$\Phi = -k_{\mathrm{p}}^{-2}\nabla \cdot \boldsymbol{u}$$

and take the divergence of Eq. (1.16) to get (using that $\nabla \cdot (\nabla \times \boldsymbol{v}) = 0$ for arbitrary vector fields $\boldsymbol{v}$)

$$\nabla^2\Phi + k_{\mathrm{p}}^2\Phi = 0$$

which is a scalar Helmholtz equation. Second, introduce the vector potential $\boldsymbol{\Psi}$ according to

$$\boldsymbol{\Psi} = k_{\mathrm{s}}^{-2} \nabla \times \boldsymbol{u}$$

and take the curl of Eq. (1.16) to get (using that $\nabla \times (\nabla s) = \mathbf{0}$ for arbitrary scalar fields s)

$$\nabla \times (\nabla \times \boldsymbol{\Psi}) - k_{\mathrm{s}}^2 \boldsymbol{\Psi} = \mathbf{0}$$

which is a vector Helmholtz equation. As all solutions to this equation has $\nabla \cdot \boldsymbol{\Psi} = \mathbf{0}$ it can alternatively be written

$$\nabla^2 \boldsymbol{\Psi} + k_{\mathrm{s}}^2 \boldsymbol{\Psi} = \mathbf{0}$$

Using the definition of the two potentials in Eq. (1.16) then gives

$$\boldsymbol{u} = \nabla \Phi + \nabla \times \boldsymbol{\Psi}$$

This is just the well known Helmholtz decomposition, which is, in fact, valid for an arbitrary vector field.

There are thus two types of waves in an isotropic, homogeneous solid. Inside the solid they propagate independently, but in general they couple at the boundaries. One is the wave corresponding to the scalar potential Φ which travels with the speed c_{p} and is a compressional wave. It is also longitudinal, i.e. it is irrotational (meaning that $\nabla \times \boldsymbol{u} = 0$), and the displacement vector in a plane wave is in the same direction as the propagation direction (as is also the case for acoustic waves in a fluid). When this wave was first discovered experimentally in seismology it was called the primary wave because it arrives first. Here compressional waves are denoted P waves for short, with P for primary (or pressure). The other wave type is corresponding to the vector potential $\boldsymbol{\Psi}$ and is a shear wave which travels with the slower velocity c_{s}. This wave is also transverse, i.e. it is equivoluminal ($\nabla \cdot \boldsymbol{u} = 0$), and the displacement vector in a plane wave is perpendicular to the propagation direction (as is also the case for electromagnetic waves). This wave was called secondary in seismology. Here shear waves are denoted S waves for short, with S for secondary (or shear). P and S waves are often denoted L (longitudinal) and T (transverse) waves instead.

The potential for the S waves is a vector with three components, but it is only the curl of the potential that has physical significance. This in effect takes away one component, so that an auxiliary restriction can be made, and this is often chosen as

$$\nabla \cdot \boldsymbol{\Psi} = 0$$

However, the way in which potentials are soon introduced this form of the restriction is not always fulfilled.

The P waves are solutions to the scalar Helmholtz equation, for which there exist various ways to obtain analytical solutions depending on the coordinate system chosen. For the S waves the situation is more complex as they satisfy a vector Helmholtz equation.

1.2.2 2D potentials

Looking first at 2D situations (meaning that all fields are independent of the z-coordinate) there is an immediate division into out-of-plane waves and in-plane waves. For the out-of-plane waves the displacement is assumed on the form $\boldsymbol{u}(\boldsymbol{r}) = \hat{\boldsymbol{z}} u_z(x, y)$. Insertion into Eq. (1.16) gives that the scalar displacement must satisfy the Helmholtz equation

$$\nabla^2 u_z + k_\mathrm{s}^2 u_z = 0$$

These waves are apparently S waves without any coupling to P waves. In the following they are denoted 2D SH waves (the designation SH is explained below). As they are the simplest waves of interest in scattering problems, situations involving these are taken up first in many of the later chapters when various types of scattering configurations are investigated. Often it is useful to use polar coordinates (r, φ) instead of Cartesian coordinates (x, y), for instance when the scattering by a circle is studied.

The in-plane waves involve both P and S waves. The displacement vector lies in the xy-plane and can be assumed on the form

$$\boldsymbol{u} = \nabla \Phi + \nabla \times (\hat{\boldsymbol{z}} \Psi) \tag{1.17}$$

where Φ and Ψ are scalar potentials and any convenient coordinates can be used in the xy-plane; in most scattering problems this is polar coordinates (r, φ). Insertion of the first term into Eq. (1.16) gives that the potential Φ must satisfy the 2D Helmholtz equation

$$\nabla^2 \Phi + k_\mathrm{p}^2 \Phi = 0$$

Making a 3D calculation a formula for the nabla operator gives

$$\nabla \times (\hat{\boldsymbol{z}}\Psi) = \Psi(\nabla \times \hat{\boldsymbol{z}}) + (\nabla\Psi) \times \hat{\boldsymbol{z}} = (\nabla\Psi) \times \hat{\boldsymbol{z}}$$

Application of another formula for the nabla operator then gives

$$\begin{aligned}\nabla \times (\nabla \times (\hat{\boldsymbol{z}}\Psi)) &= \nabla\Psi(\nabla \cdot \hat{\boldsymbol{z}}) - \hat{\boldsymbol{z}}\nabla \cdot \nabla\Psi + (\hat{\boldsymbol{z}} \cdot \nabla)\nabla\Psi - (\nabla\Psi \cdot \nabla)\hat{\boldsymbol{z}} \\ &= -\hat{\boldsymbol{z}}\nabla^2\Psi + \frac{\partial}{\partial z}\nabla\Psi\end{aligned}$$

and a final curl thus yields

$$\nabla \times (\nabla \times [\nabla \times (\hat{\boldsymbol{z}}\Psi)]) = -(\nabla\nabla^2\Psi) \times \hat{\boldsymbol{z}}$$

Operating with $\nabla\times$ on Eq.(1.16) it follows that Ψ (or actually $\nabla\Psi$) must satisfy the Helmholtz equation

$$\nabla^2\Psi + k_{\mathrm{s}}^2\Psi = 0$$

It is stressed that the equation for the S potential now is a scalar equation. These S waves are denoted SV waves (explained below). In a homogeneous solid P and SV waves propagate independently, but in general they couple at surfaces of the domain.

If the 2D problem is rotationally symmetric P and SV waves decouple and the displacement can be taken in the form $\boldsymbol{u} = \hat{\boldsymbol{r}}\, u_r(r)$ and $\boldsymbol{u} = \hat{\boldsymbol{\varphi}}\, u_\varphi(r)$, respectively, and no potentials are needed.

1.2.3 3D potentials

Above it is seen that $\hat{\boldsymbol{z}}\Psi$, where Ψ satisfies the scalar Helmholtz equation with the S wavenumber, is a vector potential in 2D. However, as seen by the calculations, it gives a vector potential also in 3D (and it is noticed that its divergence is then nonzero in general). But in 3D there must be another part to the S waves. One way to construct this part is to take a repeated curl so that the displacement has the following representation

$$\boldsymbol{u} = \nabla\Phi + \nabla \times (\hat{\boldsymbol{z}}\Psi_1) + \nabla \times \nabla \times (\hat{\boldsymbol{z}}\Psi_2) \tag{1.18}$$

Here all three potentials satisfy scalar Helmholtz equations. That the last term is a solution for S waves follows directly as the curl of a (S wave) solution is always also a solution. The unit vector $\hat{z}$ that is used in the construction can be substituted with any constant vector and this still yields a solution to the elastodynamic equation of motion. This representation is useful in Cartesian and cylindrical coordinates because $\hat{z}$ is a constant unit vector in these systems.

In spherical coordinates the representation in Eq. (1.18) is not useful. However, there is a similar representation which works in this case [99]

$$\boldsymbol{u} = \nabla\Phi + \nabla \times (\boldsymbol{r}\Psi_1) + \nabla \times \nabla \times (\boldsymbol{r}\Psi_2) \tag{1.19}$$

where $\boldsymbol{r} = \hat{\boldsymbol{r}}r = \hat{\boldsymbol{x}}x + \hat{\boldsymbol{y}}y + \hat{\boldsymbol{z}}z$ is the radius vector, which points in the direction of one of the unit vectors in spherical coordinates. To see that the second term is a solution the following calculations are performed (using rules for the nabla operator)

$$\nabla \times (\boldsymbol{r}\Psi_1) = \Psi_1(\nabla \times \boldsymbol{r}) + (\nabla\Psi_1) \times \boldsymbol{r} = -\boldsymbol{r} \times \nabla\Psi_1$$

$$\begin{aligned}\nabla\times(\nabla\times(\boldsymbol{r}\Psi_1)) &= -\boldsymbol{r}(\nabla\cdot\nabla\Psi_1)+\nabla\Psi_1(\nabla\cdot\boldsymbol{r})-(\nabla\Psi_1\cdot\nabla)\boldsymbol{r}+(\boldsymbol{r}\cdot\nabla)\nabla\Psi_1 \\ &= -\boldsymbol{r}\nabla^2\Psi_1 + 3\nabla\Psi_1 - \nabla\Psi_1 + r\frac{\partial}{\partial r}\nabla\Psi_1\end{aligned}$$

where the following formula is used

$$(\nabla\Psi_1 \cdot \nabla)\boldsymbol{r} = \nabla\Psi_1$$

Taking an additional curl the middle two terms vanish. The last term is written out in spherical coordinates

$$r\frac{\partial}{\partial r}\nabla\Psi_1 = \hat{\boldsymbol{r}}\,r\frac{\partial^2\Psi_1}{\partial r^2} + \hat{\boldsymbol{\theta}}\left(\frac{\partial^2\Psi_1}{\partial r\partial\theta} - \frac{1}{r}\frac{\partial\Psi_1}{\partial\theta}\right) + \hat{\boldsymbol{\varphi}}\,\frac{1}{\sin\theta}\left(\frac{\partial^2\Psi_1}{\partial r\partial\varphi} - \frac{1}{r}\frac{\partial\Psi_1}{\partial\varphi}\right)$$

Making an explicit calculation in spherical coordinates it is seen that the curl of this expression vanishes. Thus there remains

$$\nabla \times [\nabla \times (\nabla \times (\boldsymbol{r}\Psi_1))] = -\nabla \times (\boldsymbol{r}\nabla^2\Psi_1)$$

and this shows that the scalar potential Ψ_1 must satisfy the scalar Helmholtz equation with the S wavenumber. The third term in Eq. (1.19) is a solution in the same way as in Eq. (1.18).

The representation in three terms in Eq. (1.18) and Eq. (1.19) gives a division into one P wave (the first term) and two S waves. The S waves are denoted SH for the second term and SV for the third term. This nomenclature stems from seismology where the "H" stands for "horizontal" and "V" for "vertical", meaning that the second term only has horizontal components (when $\hat{z}$ is vertical), whereas the third term has a vertical component. For spherical problems the original meaning of "horizontal" and "vertical" is lost, but it is convenient to have a way to refer to the three parts. The SH waves often decouple from the SV and P waves in simple scattering problems, whereas the P and SV waves couple, for example in the scattering by a sphere.

For problems in cylindrical or spherical coordinates that are rotationally symmetric the displacement field is independent of the azimuthal angle φ. This type of problem reduces to a simpler SH (torsional) part where $\boldsymbol{u} = \hat{\boldsymbol{\varphi}}\, u_\varphi(r, z)$ in cylindrical coordinates or $\boldsymbol{u} = \hat{\boldsymbol{\varphi}}\, u_\varphi(r, \theta)$ in spherical coordinates. The P and SV waves couple and it is useful to use the representations in Eq. (1.18) or Eq. (1.19) with the middle term omitted.

When using boundary or interface conditions the traction vector is usually involved. On a surface with normal $\hat{\boldsymbol{n}}$ it is

$$t_i^{(n)} = n_j \sigma_{ji} = n_i \lambda \partial_j u_j + \mu n_j (\partial_j u_i + \partial_i u_j)$$

where n_j are the components of $\hat{\boldsymbol{n}}$. In Cartesian coordinates this can be written in vector notation as

$$\boldsymbol{t}^{(n)} = \hat{\boldsymbol{n}}\, \lambda \nabla \cdot \boldsymbol{u} + \mu \left(\frac{\partial \boldsymbol{u}}{\partial n} + \nabla u_n \right)$$

where $u_n = \hat{\boldsymbol{n}} \cdot \boldsymbol{u}$ is the normal component. In cylindrical or spherical coordinates the following form is more useful [which follows from Eq. (1.15)]

$$\boldsymbol{t}^{(n)} = \hat{\boldsymbol{n}}\, \lambda \nabla \cdot \boldsymbol{u} + \mu \left(2\, \frac{\partial \boldsymbol{u}}{\partial n} + \hat{\boldsymbol{n}} \times (\nabla \times \boldsymbol{u}) \right) \tag{1.20}$$

Introducing potentials this expression can be further simplified. Using that $\lambda = \mu(k_\text{s}^2/k_\text{p}^2 - 2)$ the traction for a P wave with potential Φ becomes

$$\boldsymbol{t}^{(n)} = \mu \left(\hat{\boldsymbol{n}} (2k_\text{p}^2 - k_\text{s}^2) \Phi + 2\, \frac{\partial \boldsymbol{u}}{\partial n} \right) \tag{1.21}$$

because the last term vanishes for a P wave. For an S wave the first term vanishes so that

$$\boldsymbol{t}^{(n)} = \mu \left(2\, \frac{\partial \boldsymbol{u}}{\partial n} + \hat{\boldsymbol{n}} \times (\nabla \times \boldsymbol{u}) \right) \tag{1.22}$$

Here the second term can be simplified because the curl of one S wave gives the other S wave multiplied by k_{s}.

The way potentials are introduced and traction vectors are calculated for isotropic solids means that the stress components need not be calculated by the equations given for cylindrical and spherical coordinates in Section 1.1.6, and this leads to simpler calculations.

1.3 Anisotropic solids

Many materials, like pure crystals, composites, and biological materials, are anisotropic. Using analytical approaches, scattering problems for such materials can only be solved in special cases, and some of these are taken up in later chapters. Here a few of the simplest classes of anisotropy are introduced; all different classes of anisotropy are defined by Auld [9] and Ting [129]. van der Hijden [132] and Nayfeh [100] extensively investigates the propagation of waves in stratified anisotropic media.

1.3.1 Orthotropic and transversely isotropic solids

An orthotropic material is characterized by three mutually orthogonal symmetry planes. The stress-strain relations for such a material are

$$\begin{aligned}
\sigma_{xx} &= C_{11}\varepsilon_{xx} + C_{12}\varepsilon_{yy} + C_{13}\varepsilon_{zz} \\
\sigma_{yy} &= C_{12}\varepsilon_{xx} + C_{22}\varepsilon_{yy} + C_{23}\varepsilon_{zz} \\
\sigma_{zz} &= C_{13}\varepsilon_{xx} + C_{23}\varepsilon_{yy} + C_{33}\varepsilon_{zz} \\
\sigma_{yz} &= 2C_{44}\varepsilon_{yz} \\
\sigma_{zx} &= 2C_{55}\varepsilon_{zx} \\
\sigma_{xy} &= 2C_{66}\varepsilon_{xy}
\end{aligned} \tag{1.23}$$

Here a reduced notation for the stiffness constants $C_{IJ}, I, J = 1, 2, 3, 4, 5, 6$, is used, where the stress and strains are collected as vectors with six components (in the order the stresses are given above) and the stiffness tensor is represented by a symmetric six-by-six matrix, see Auld [9] or Ting [129] for more details. An orthotropic material is thus characterized by nine independent stiffness constants. A special case is a cubic material where all the three coordinate directions are equivalent so that $C_{11} = C_{22} = C_{33}$,

$C_{12} = C_{13} = C_{23}$, and $C_{44} = C_{55} = C_{66}$. This material therefore has three independent stiffness constants. The isotropic material is recovered if in addition $2C_{66} = C_{11} - C_{12}$.

Another special case of an orthotropic material is a transversely isotropic material. Taking isotropy in the xy-plane the stress-strain relations are

$$\begin{aligned}
\sigma_{xx} &= C_{11}\varepsilon_{xx} + C_{12}\varepsilon_{yy} + C_{13}\varepsilon_{zz} \\
\sigma_{yy} &= C_{12}\varepsilon_{xx} + C_{11}\varepsilon_{yy} + C_{13}\varepsilon_{zz} \\
\sigma_{zz} &= C_{13}\varepsilon_{xx} + C_{13}\varepsilon_{yy} + C_{33}\varepsilon_{zz} \\
\sigma_{yz} &= 2C_{44}\varepsilon_{yz} \\
\sigma_{zx} &= 2C_{44}\varepsilon_{zx} \\
\sigma_{xy} &= 2C_{66}\varepsilon_{xy}
\end{aligned} \tag{1.24}$$

where $2C_{66} = C_{11} - C_{12}$. This material has five independent stiffness constants. Because of isotropy in the xy-plane it is possible to use cylindrical coordinates (r, φ, z) instead and the stress-strain relations look the same with the change of x and y to r and φ.

For a transversely isotropic material it is possible to introduce potentials in the following way [34, 78] (see also Ding *et al.* [43])

$$\boldsymbol{u} = \nabla \times (\hat{\boldsymbol{z}}\Psi_1) + \hat{\boldsymbol{z}}\frac{\partial \Psi_2}{\partial z} + \nabla_\perp \Psi_3 \tag{1.25}$$

Here $\nabla_\perp$ is the transverse part (in the xy-plane) of the gradient operator. The three scalar potentials are then seen to satisfy

$$\begin{aligned}
&\left(C_{44}\frac{\partial^2}{\partial z^2} + C_{66}\nabla_\perp^2 + \rho\omega^2\right)\Psi_1 = 0 \\
&(C_{13} + C_{44})\frac{\partial^2 \Psi_2}{\partial z^2} + \left(C_{44}\frac{\partial^2}{\partial z^2} + C_{11}\nabla_\perp^2 + \rho\omega^2\right)\Psi_3 = 0 \\
&\left(C_{33}\frac{\partial^2}{\partial z^2} + C_{44}\nabla_\perp^2 + \rho\omega^2\right)\Psi_2 + (C_{13} + C_{44})\nabla_\perp^2\Psi_3 = 0
\end{aligned} \tag{1.26}$$

This decomposition into two coupled equations for Ψ_2 and Ψ_3 and one uncoupled equation for Ψ_1 shows that there exist one independent shear wave for Ψ_1 (called an SH wave) and coupled waves for Ψ_2 and Ψ_3, usually called quasi P-SV waves. In special directions, in particular along the coordinate directions, there still exist pure P and SV waves. Instead of the potentials above an alternative is to use the same potentials as for Cartesian and cylindrical coordinates for isotropic media [64]; the two approaches are essentially equivalent.

1.3.2 Cylindrically and spherically orthotropic solids

Another type of anisotropy that is amenable to analytical solutions is cylindrical and spherical orthotropy. For cylindrical orthotropy the stress-strain relations for an orthotropic material (in Cartesian coordinates) are mimicked in cylindrical coordinates, i.e.

$$\begin{aligned}
\sigma_{rr} &= C_{11}\varepsilon_{rr} + C_{12}\varepsilon_{\varphi\varphi} + C_{13}\varepsilon_{zz}\\
\sigma_{\varphi\varphi} &= C_{12}\varepsilon_{rr} + C_{22}\varepsilon_{\varphi\varphi} + C_{23}\varepsilon_{zz}\\
\sigma_{zz} &= C_{13}\varepsilon_{rr} + C_{23}\varepsilon_{\varphi\varphi} + C_{33}\varepsilon_{zz}\\
\sigma_{\varphi z} &= 2C_{44}\varepsilon_{\varphi z}\\
\sigma_{zr} &= 2C_{55}\varepsilon_{zr}\\
\sigma_{r\varphi} &= 2C_{66}\varepsilon_{r\varphi}
\end{aligned} \tag{1.27}$$

This material is inhomogeneous because, when transformed to the Cartesian coordinates, the stiffness constants depend on the azimuthal angle φ. A cylindrically orthotropic material has a singularity along the z-axis (at $r = 0$) where the fields behave in a nonanalytical way. Cylindrical orthotropy is a plausible material model for wood (Martin and Berger [89]).

For a spherically orthotropic material it is necessary to take it as transversely isotropic in the two angular coordinates θ and φ to be able to have analytical solutions. The stress-strain relations are then

$$\begin{aligned}
\sigma_{rr} &= C_{11}\varepsilon_{rr} + C_{12}\varepsilon_{\theta\theta} + C_{12}\varepsilon_{\varphi\varphi}\\
\sigma_{\theta\theta} &= C_{12}\varepsilon_{rr} + C_{22}\varepsilon_{\theta\theta} + C_{23}\varepsilon_{\varphi\varphi}\\
\sigma_{\varphi\varphi} &= C_{12}\varepsilon_{rr} + C_{23}\varepsilon_{\theta\theta} + C_{22}\varepsilon_{\varphi\varphi}\\
\sigma_{\theta\varphi} &= 2C_{44}\varepsilon_{\theta\varphi}\\
\sigma_{\varphi r} &= 2C_{55}\varepsilon_{\varphi r}
\end{aligned}$$

$$\sigma_{r\theta} = 2C_{55}\varepsilon_{r\theta} \tag{1.28}$$

where $2C_{44} = C_{22} - C_{23}$. This material is inhomogeneous, because transformed to Cartesian coordinates the stiffness constants depend on the spherical angles. A spherically orthotropic material has a singularity at the origin ($r = 0$).

1.4 A scattering problem

In a typical scattering problem a wave is generated somewhere and then propagates through a homogeneous medium and hits a finite obstacle. In the basic case it is assumed that the homogeneous medium fills the whole space outside the obstacle. It is also assumed that the presence of the obstacle has no influence on the generation of the wave. The wave that exists in the absence of the obstacle is called the incident wave and the extra wave that is generated by the presence of the obstacle is called the scattered wave.

As a simple example the scattering of a plane wave by a circular cavity is now briefly studied. To make things as simple as possible only the 2D SH problem is considered. This is a scalar problem for the out-of-plane displacement component u (skipping the index "z" for simplicity), which is treated in much more detail in the next chapter. Outside the circle the field u should thus satisfy the scalar 2D Helmholtz equation

$$\nabla^2 u + k^2 u = 0$$

As always a fixed frequency is assumed and the wavenumber is k, which for this problem is the shear wavenumber, but for simplicity the index "s" is skipped in this section. This equation is to be supplemented with boundary conditions and some type of excitation of the waves. As the cavity is circular it is convenient to use polar coordinates (r, φ) centred at the cavity, the boundary condition on the circle is then that the radial shear stress vanishes

$$\sigma_{zr} = \mu \frac{\partial u}{\partial r} = 0$$

The source of the incident wave is assumed to be situated far away from the obstacle and the incident wave can then be taken as a plane wave. Without loss of generality the plane wave can be assumed to propagate in the x-direction and is thus

$$u^{\text{in}} = A\,\mathrm{e}^{ikx}$$

where A is the amplitude. Remembering the time factor $\exp(-\mathrm{i}\omega t)$, the total phase factor of the incident wave is $\exp(\mathrm{i}(kx-\omega t))$ which gives a wave propagating in the $+x$ direction with the shear wave speed $c_{\mathrm{s}}=\omega/k$. The total wave is then naturally divided into the incident plane wave u^{in} and an additional scattered wave u^{sc} that is due to the cavity

$$u = u^{\mathrm{in}} + u^{\mathrm{sc}}$$

The problem is thus to determine the scattered wave given the incident wave and the properties of the obstacle.

As the incident wave satisfies Helmholtz equation, also the scattered wave must satisfy this equation. The scattered wave must also satisfy a condition at the "boundary" at infinity, this is usually called the radiation condition and it is that energy propagates outwards, as is soon to be explained. To fit the boundary condition on the cavity it is best to express the scattered wave in polar coordinates and it thus satisfies

$$\frac{1}{r}\frac{\partial}{\partial r}\left(r\frac{\partial u^{\mathrm{sc}}}{\partial r}\right)+\frac{1}{r^2}\frac{\partial^2 u^{\mathrm{sc}}}{\partial\varphi^2}+k^2u^{\mathrm{sc}}=0$$

This equation is solved by the standard separation of variables approach. As the problem is even in φ this leads to a sum in $\cos m\varphi$

$$u^{\mathrm{sc}}=\sum_{m=0}^{\infty}R_m(r)\cos m\varphi$$

where R_m satisfies the equation

$$\frac{1}{r}\frac{\mathrm{d}}{\mathrm{d}r}\left(r\frac{\mathrm{d}R_m}{\mathrm{d}r}\right)-\frac{m^2}{r^2}R_m+k^2R_m=0$$

This is Bessel's differential equation, whose solutions are collectively known as Bessel functions. Solutions are the Bessel function $\mathrm{J}_m(kr)$, the Neumann function $\mathrm{Y}_m(kr)$, and the Hankel functions of the first and second kind $\mathrm{H}_m^{(1)}(kr)=\mathrm{J}_m(kr)+\mathrm{i}\mathrm{Y}_m(kr)$ and $\mathrm{H}_m^{(2)}(kr)=\mathrm{J}_m(kr)-\mathrm{i}\mathrm{Y}_m(kr)$. For large arguments the Hankel functions behave as

$$\mathrm{H}_m^{(1,2)}(kr)\simeq\sqrt{\frac{2}{\pi kr}}\,\mathrm{e}^{\pm\mathrm{i}(kr-\pi/4-m\pi/2)}$$

With the implicit time factor this gives a phase factor $\exp(\mathrm{i}(\pm kr - \omega t))$. The Hankel function of the first kind therefore gives waves that propagate in the $+r$ direction and the Hankel function of the second kind waves that propagate in the $-r$ direction, both with the shear wave speed $c_{\mathrm{s}} = \omega/k$. In this simple case energy is propagating with the same velocity in the same direction. Thus the Hankel function of the first kind gives the correct behaviour of the scattered wave far away from the cavity. This "boundary condition" at infinity is called the radiation condition as mentioned earlier. The Bessel and Neumann functions give standing waves far away and are therefore also discarded. This is even more clear if damping is included, because the wavenumber k then has a positive imaginary part and the Hankel function of the first kind decays exponentially far away whereas the other three solutions to Bessel's differential equation grow exponentially. For simplicity of notation (especially for derivatives) the Hankel function of the first kind is simply called H_m. Summing up, the scattered wave admits the representation

$$u^{\mathrm{sc}} = \sum_{m=0}^{\infty} f_m \mathrm{H}_m(kr) \cos m\varphi$$

where the coefficients f_m are to be determined.

When the boundary condition on the cavity is applied it is apparently advantageous if the incident wave can be expressed in the solutions to Helmholtz equation in polar coordinates. This is possible by the following expansion [Eq. (2.17)]

$$\mathrm{e}^{\mathrm{i}kx} = \sum_{m=0}^{\infty} \mathrm{i}^m \epsilon_m \mathrm{J}_m(kr) \cos m\varphi$$

where the Neumann factor $\epsilon_0 = 1$ and $\epsilon_m = 2$, $m = 1, 2, \ldots$. As the $\cos m\varphi$ functions are orthogonal over the circle, the boundary condition gives

$$f_m = -A\,\mathrm{i}^m \epsilon_m \frac{\mathrm{J}'_m(ka)}{\mathrm{H}'_m(ka)}$$

The scattered wave is hereby determined and the solution is formally complete. It is noted that $\mathrm{H}'_m(ka)$ has no real roots (but do have complex ones).

In the solution the expansion coefficients f_m contain information about both the incident wave and the obstacle. However, it is useful to separate these two parts and this is done by noting that the incident plane wave is expanded in the regular solutions (a Bessel function times a trigonometric function) to Helmholtz equation, where the expansion coefficients thus are $A\,\mathrm{i}^m \epsilon_m$. Removing this factor from the expansion coefficients of the scattered wave gives what is called the transition (T) matrix. In this simple case the T-matrix only has diagonal elements

$$T_m = -\frac{\mathrm{J}'_m(ka)}{\mathrm{H}'_m(ka)}$$

Calculation of the transition matrix has the advantage that it completely characterizes the obstacle but is independent of the incident wave.

In some situations the obstacle (or each obstacle in a collection of obstacles) is small compared to relevant wavelengths, this is often called the limit of low frequencies or the Rayleigh limit. This is, for instance, often the case for particles in a particle composite or grains in a metal. In this limit $ka = 2\pi a/\lambda$, where λ is the wavelength, is small. Then the asymptotic form of the Bessel and Hankel functions (see Section 2.1.1) can be employed and this gives that only the first two T-matrix elements are of leading order. They become

$$T_0 = -\mathrm{i}(ka)^2 \frac{\pi}{4}$$

$$T_1 = \mathrm{i}(ka)^2 \frac{\pi}{4}$$

The other T-matrix elements behave as $(ka)^{2m}$ for $m = 2, 3, \ldots$ and can thus be neglected for most purposes.

There are many further aspects of scattering worth investigating, like other incident waves, the scattered energy, and the scattering cross section. Other important factors are of course the shape and material properties of the obstacle. In this connection cracks are of particular interest. Furthermore, there may be more than one obstacle or the obstacle may be located in some sort of structure, such as a half-space or plate. All of the mentioned aspects are more or less taken up in subsequent chapters.

2 Scattering by a circle and cylinder

In this chapter the first scattering problems are taken up. To start as simple as possible the 2D SH scattering by a circle is first investigated. This 2D problem is identical to the corresponding problems in acoustics and electromagnetics, so the solution can be found in many places, see, for instance Pao and Mow [112] and Eringen and Suhubi [48]. Subsequently the corresponding 2D P-SV case is investigated [112], and then the full 3D scattering by an infinite circular cylinder. Being a very similar problem, the scattering by a transversely isotropic cylinder in an isotropic medium is also considered.

2.1 Circle: 2D SH waves

As this is a scalar 2D case, the scattering of SH waves by a circle is first considered. This involves the solution of the scalar Helmholtz equation in polar coordinates and the introduction of the corresponding wavefunctions. Expanding the incident and scattered waves in these wavefunctions the transition (T) matrix of the circle, which is the matrix relating the expansion coefficients of the scattered wave to those of the incident, is determined by applying the boundary conditions on the circle. Both the incident and scattered waves are investigated and the T-matrix is given explicitly for low frequencies. Finally, more general boundary conditions and a layered circle are taken up.

2.1.1 Bessel functions and wavefunctions

The SH waves have only an out-of-plane component and thus satisfy the scalar Helmholtz equation

$$\nabla^2 u + k^2 u = 0$$

where the displacement is only called u (instead of u_z) and the wavenumber similarly is called k (instead of k_s) in this section. The Laplace operator in this case is the 2D one expressed in polar coordinates

$$\nabla^2 u = \frac{1}{r}\frac{\partial}{\partial r}\left(r\frac{\partial u}{\partial r}\right) + \frac{1}{r^2}\frac{\partial^2 u}{\partial \varphi^2}$$

The standard separation of variables approach is to find simple solutions that are a product of a function of r and a function of φ and then add these. Another way to put this is to say that u is expanded in a Fourier series in $\cos m\varphi$ and $\sin m\varphi$, $m = 0, 1, 2, \ldots$, with $m = 0$ excluded for the sine function. This type of dependence on φ is necessary when u is a periodic function of φ with period 2π. Alternatively, a complex Fourier series with expansion functions $\exp(\mathrm{i}m\varphi)$, with $m = 0, \pm 1, \pm 2, \ldots$ can be used. The sine and cosine are odd and even, respectively, and with symmetric obstacles this is an advantage. The Fourier coefficients $R_m(r)$ must be functions of r and insertion into Helmholtz equation gives that they must satisfy (using the orthogonality of the trigonometric functions)

$$\frac{1}{r}\frac{\mathrm{d}}{\mathrm{d}r}\left(r\frac{\mathrm{d}R_m}{\mathrm{d}r}\right) - \frac{m^2}{r^2}R_m + k^2 R_m = 0$$

Putting $z = kr$ this equation becomes

$$\frac{\mathrm{d}^2 R_m}{\mathrm{d}z^2} + \frac{1}{z}\frac{\mathrm{d}R_m}{\mathrm{d}z} + \left(1 - \frac{m^2}{z^2}\right)R_m = 0 \tag{2.1}$$

This is Bessel's differential equation, which can be solved in terms of elementary functions only in exceptional cases. By assuming a solution as z^α times a power series in z it is straightforward to show that one solution is a Bessel function

$$\mathrm{J}_m(z) = \left(\frac{z}{2}\right)^m \sum_{j=0}^{\infty} \frac{(-1)^j}{j!(m+j)!}\left(\frac{z}{2}\right)^{2j} \tag{2.2}$$

This definition is valid also for noninteger (even complex) m if the second factorial is changed to a gamma function as $(m + j)! = \Gamma(m + j + 1)$. The sum is convergent for all (complex) z, but for large z the convergence is very slow with large cancellations.

Bessel's equation is second order and thus has two linearly independent solutions. If m is not an integer $\mathrm{J}_{-m}(z)$ is a second solution, but this function is not defined by Eq. (2.2) when m is a positive integer (and does not give a new solution when $m = 0$). Instead the Neumann function can be taken as the second solution

$$\mathrm{Y}_m(z) = -\frac{1}{\pi}\left(\frac{z}{2}\right)^{-m}\sum_{j=0}^{m-1}\frac{(m-j-1)!}{j!}\left(\frac{z}{2}\right)^{2j} + \left[\frac{2}{\pi}\ln(z/2)+\gamma\right]\mathrm{J}_m(z)$$
$$-\frac{1}{\pi}\sum_{j=0}^{\infty}\frac{(-1)^j}{j!(m+j)!}\left(\frac{z}{2}\right)^{m+2j}\left(\sum_{l=1}^{j}\frac{1}{l}+\sum_{l=1}^{m+j}\frac{1}{l}\right) \tag{2.3}$$

where $\gamma = 0.57721566\ldots$ is Euler's constant and where a sum is defined as zero if the upper summation limit is smaller than the lower one. Both the Bessel and Neumann functions are thus real for real arguments. From the series representation the behaviour when $z \to 0$ follows

$$\mathrm{J}_m(z) \simeq \frac{z^m}{2^m m!}, \qquad m = 0, 1, 2, \ldots$$
$$\mathrm{Y}_0(z) \simeq \frac{2}{\pi}\ln z$$
$$\mathrm{Y}_m(z) \simeq -\frac{2^m(m-1)!}{\pi z^m}, \qquad m = 1, 2, \ldots \tag{2.4}$$

The asymptotic behaviour for large arguments can be shown to be

$$\mathrm{J}_m(z) \simeq \sqrt{\frac{2}{\pi z}}\cos(z-\pi/4-m\pi/2)$$
$$\mathrm{Y}_m(z) \simeq \sqrt{\frac{2}{\pi z}}\sin(z-\pi/4-m\pi/2) \tag{2.5}$$

The normalization of the Bessel and Neumann functions are chosen so that the same factor appears in front of the trigonometric functions. This type of oscillatory behaviour is typical of standing waves, but when considering waves propagating outwards another type of behaviour is needed, which is of exponential type. This is achieved by defining the Hankel functions of the first and second kind

$$\mathrm{H}_m^{(1)}(z) = \mathrm{J}_m(z) + \mathrm{i}\mathrm{Y}_m(z)$$

$$\mathrm{H}_m^{(2)}(z) = \mathrm{J}_m(z) - \mathrm{i}\mathrm{Y}_m(z)$$

It follows that these have the asymptotic behaviour for large arguments

$$\mathrm{H}_m^{(1,2)}(z) \simeq \sqrt{\frac{2}{\pi z}}\, \mathrm{e}^{\pm\mathrm{i}(z-\pi/4-m\pi/2)} \tag{2.6}$$

As the implicit time factor is $\exp(-\mathrm{i}\omega t)$, the Hankel functions of the first kind give waves that are propagating in the $+r$ direction, and thus satisfy a radiation condition. The Hankel functions of the second kind give waves propagating in the $-r$ direction and these are seldom of interest.

There are many relations involving Bessel functions, they can for instance be represented as integrals. They also satisfy recurrence relations; there exist two independent ones, for example

$$\frac{2m}{z}\mathrm{J}_m(z) = \mathrm{J}_{m-1}(z) + \mathrm{J}_{m+1}(z)$$

$$\mathrm{J}_m'(z) = \frac{m}{z}\mathrm{J}_m(z) - \mathrm{J}_{m+1}(z) \tag{2.7}$$

Thus $\mathrm{J}_0'(z) = -\mathrm{J}_1(z)$. These relations are also valid for the Neumann and Hankel functions. Another type of relation is the Wronskian, for example

$$\mathrm{J}_m(z)\mathrm{Y}_m'(z) - \mathrm{J}_m'(z)\mathrm{Y}_m(z) = \frac{2}{\pi z} \tag{2.8}$$

In summary, there are simple solutions to the scalar Helmholtz equation in polar coordinates that are of the form of a product of a trigonometric function and a solution to Bessel's differential equation (these are collectively known as Bessel functions). In most scattering problems only the Bessel function $\mathrm{J}_m(z)$ and the Hankel function of the first kind $\mathrm{H}_m^{(1)}(z)$ are relevant. In the following $\mathrm{H}_m^{(1)}(z)$ is simply called "the Hankel function" with "of the first kind" left out and for ease of notation (especially for derivatives) it is written as $\mathrm{H}_m(z)$.

It is convenient at this stage to introduce wavefunctions according to the following

$$\chi_{\sigma m}^0(\boldsymbol{r}) = \frac{\sqrt{\epsilon_m}}{2}\,\mathrm{J}_m(kr)\left\{\begin{array}{c}\cos m\varphi\\ \sin m\varphi\end{array}\right\} \tag{2.9}$$

Here $m = 0, 1, 2, \ldots$ and $\sigma = \mathrm{e}$ ("e" for even) for the upper row (cosine function) and $\sigma = \mathrm{o}$ ("o" for odd) for the lower row (sine function) with

$m = 0$ excluded for $\sigma = \mathrm{o}$. The Neumann factor $\epsilon_0 = 1$ and $\epsilon_m = 2$ for $m = 1, 2, \ldots$. The factor $(\epsilon_m)^{1/2}/2$ is included to get a simple expansion of the Green function in later chapters. The upper index 0 denotes the regular wavefunctions with a Bessel function; there is a corresponding out-going wavefunction, which is denoted $\chi^+_{\sigma m}(\boldsymbol{r})$, that contains a Hankel function instead of a Bessel function. When needed an index "i" ($= 1, 2$, etc.) is used on the wavefunctions corresponding to an index on the material parameters and wavenumbers, for example $\chi^0_{i,\sigma m}(\boldsymbol{r})$.

2.1.2 The T-matrix

Consider the scattering of an SH wave by a circular obstacle of radius a in an isotropic and homogeneous surrounding. As the circle is a coordinate curve in polar coordinates (r, φ) these coordinates are used. The material outside the circle has shear modulus μ and density ρ. Inside the circle another isotropic and homogeneous material has shear modulus μ_1 and density ρ_1. As special cases the interior can be a cavity (vacuum or air) or a rigid inclusion. The corresponding wavenumbers are $k = \omega(\rho/\mu)^{1/2}$ and $k_1 = \omega(\rho_1/\mu_1)^{1/2}$ (with the index "s" for shear waves omitted in this section as already noted).

The sources of the waves are assumed to lie outside the obstacle. Often they are far away, and a very common assumption is that they are at infinity and the incident wave is a plane wave. At present the incident wave is not specified, but in the next subsection a point force and a plane wave are considered. As there are no sources in the vicinity of the obstacle the incident wave must be finite there and it must be possible to expand it in the regular wavefunctions

$$u^{\mathrm{in}}(\boldsymbol{r}) = \sum_{\sigma,m} b_{\sigma m} \chi^0_{\sigma m}(\boldsymbol{r}) \tag{2.10}$$

where the summation is over $\sigma = \mathrm{e, o}$ and $m = 0, 1, 2, \ldots$ (with $\sigma = \mathrm{o}$ and $m = 0$ excluded). This expansion is valid in a circle centred at the obstacle out to the sources of the incident wave. The expansion coefficients $b_{\sigma m}$ are in principle known.

The wave scattered by the obstacle is carrying energy outwards from the obstacle (the radiation condition) and it is thus possible to expand it in the out-going wavefunctions

$$u^{\mathrm{sc}}(\boldsymbol{r}) = \sum_{\sigma,m} f_{\sigma m} \chi^{+}_{\sigma m}(\boldsymbol{r}) \tag{2.11}$$

Here the expansion coefficients $f_{\sigma m}$ are unknown and the problem is to determine these given the incident wave and the properties of the obstacle. Another way to put this is to say that the T-matrix of the obstacle is to be determined, where this matrix is defined as the linear relation between the expansion coefficients of the incident and scattered waves

$$f_{\sigma m} = \sum_{\sigma',m'} T_{\sigma m,\sigma' m'} b_{\sigma' m'} \tag{2.12}$$

As seen shortly the T-matrix is diagonal and independent of σ in the present case. An advantage with calculating the T-matrix is that it fully characterizes the obstacle, but is independent of the incident wave.

Inside the obstacle there is a different material and a wave is of course generated also there. This wave must be finite and can thus be expanded in the regular wavefunctions

$$u^{1}(\boldsymbol{r}) = \sum_{\sigma,m} c_{\sigma m} \chi^{0}_{1,\sigma m}(\boldsymbol{r}) \tag{2.13}$$

where the properties of the material inside the obstacle must be used, thus the index 1 on the wavefunctions. The expansion coefficients $c_{\sigma m}$ are of course unknown.

Between the obstacle and the surrounding some interface conditions must be employed. The most common one is the welded contact where both displacement and normal stress are continuous and this is assumed for now. In the present 2D SH case this means that u and $\sigma_{zr} = \mu\, \partial u/\partial r$ are continuous. As the trigonometric functions are orthogonal over the circle this leads to the following two equations

$$b_{\sigma m} \mathrm{J}_m(ka) + f_{\sigma m} \mathrm{H}_m(ka) = c_{\sigma m} \mathrm{J}_m(k_1 a)$$

$$\mu k \left(b_{\sigma m} \mathrm{J}'_m(ka) + f_{\sigma m} \mathrm{H}'_m(ka) \right) = \mu_1 k_1 c_{\sigma m} \mathrm{J}'_m(k_1 a)$$

for $m = 0, 1, \ldots$ and $\sigma = \mathrm{e}, \mathrm{o}$. The solution to these equations gives the T-matrix elements

$$T_{\sigma m,\sigma' m'} = \delta_{mm'} \delta_{\sigma\sigma'} T_m$$

where

$$T_m = -\frac{\mu_1 k_1 \mathrm{J}'_m(k_1 a)\mathrm{J}_m(ka) - \mu k \mathrm{J}_m(k_1 a)\mathrm{J}'_m(ka)}{\mu_1 k_1 \mathrm{J}'_m(k_1 a)\mathrm{H}_m(ka) - \mu k \mathrm{J}_m(k_1 a)\mathrm{H}'_m(ka)} \tag{2.14}$$

As mentioned above the T-matrix is diagonal in the m- and σ-indices and is also independent of the σ-indices.

The circular obstacle was assumed to be of another elastic material, but there are other possibilities. The most important case is that of a cavity and this can be obtained by letting the density and shear modulus inside the circle vanish. It is then seen that the T-matrix elements for a cavity are

$$T_m = -\frac{\mathrm{J}'_m(ka)}{\mathrm{H}'_m(ka)}$$

Another case is when the obstacle is regarded as rigid, i.e. $\mu_1 \to \infty$, and this gives

$$T_m = -\frac{\mathrm{J}_m(ka)}{\mathrm{H}_m(ka)}$$

It should be remarked that this corresponds to the boundary condition $u = 0$ and this means that the obstacle, in addition to being rigid, is immovable. For low frequencies this is quite unphysical and leads to a too large T-matrix element for $m = 0$ (which corresponds to rigid body motion). This can be resolved by letting the obstacle be movable and this changes the equations for $m = 0$.

In principle the scattering problem is hereby solved, although it is of interest to further investigate both the incident and scattered waves.

2.1.3 Incident waves

The incident wave was left arbitrary so far. In this subsection a plane wave and an incident wave from a point force are investigated. For a point force at $x = -d$ and $y = 0$ the Helmholtz equation has a source term

$$\nabla^2 u + k^2 u = -A\delta(x + d)\delta(y)$$

where $A = F/\mu$, with the force F (per unit length), gives the strength of the force and the "delta function" $\delta(\cdot)$ appears. This is not an ordinary function at all (it is a distribution), but it is convenient to use and has the following properties

$$\delta(z) = 0, \qquad z \neq 0$$

$$\int \delta(z)\,\mathrm{d}z = 1$$

provided $z = 0$ is an interior point of the integration interval. It follows that

$$\int \delta(z-a) f(z)\,\mathrm{d}z = f(a)$$

provided that $f(z)$ is continuous at $z = a$.

When solving Helmholtz equation with the force it is natural to use polar coordinates centred at the force, call them (r', φ'). In these coordinates the problem is rotationally symmetric so outside $r' = 0$ Helmholtz equation becomes

$$\frac{1}{r'}\frac{\mathrm{d}}{\mathrm{d}r'}\left(r'\frac{\mathrm{d}u}{\mathrm{d}r'}\right) + k^2 u = 0$$

This is Bessel's differential equation [Eq. (2.1)] for $m = 0$ and as the solution must satisfy the radiation condition it is

$$u(r') = C\,\mathrm{H}_0(kr')$$

where C is a constant to be determined. This solution has a singularity at $r' = 0$ which is due to the delta function there. Integrate the differential equation for u over a small circle of radius ϵ around $r' = 0$

$$\int_{r'<\epsilon} \left(\frac{1}{r'}\frac{\mathrm{d}}{\mathrm{d}r'}\left(r'\frac{\mathrm{d}u}{\mathrm{d}r'}\right) + k^2 u\right) r'\mathrm{d}r'\mathrm{d}\varphi' = -A$$

The second term on the left-hand side can be discarded as it vanishes in the limit $\epsilon \to 0$. Using Gauss law in 2D gives

$$\int_0^{2\pi} \left[r'\frac{\mathrm{d}u}{\mathrm{d}r'}\right]_{r'=\epsilon} \mathrm{d}\varphi' = -A$$

As u is independent of φ' this gives

$$2\pi\epsilon \left[\frac{\mathrm{d}u}{\mathrm{d}r'}\right]_{r'=\epsilon} = -A$$

With the asymptotic form of the Hankel function [from Eq. (2.4)]

$$\mathrm{H}_0(kr') \simeq \frac{2\mathrm{i}}{\pi} \ln kr'$$

this gives

$$C = -\frac{\mathrm{i}A}{4}$$

The incident wave from the point force is thus

$$u^{\mathrm{in}}(r') = -\frac{\mathrm{i}A}{4} \mathrm{H}_0(kr') \tag{2.15}$$

The incident wave should be expanded in the polar wavefunctions centred at the obstacle and to accomplish this another approach is taken. With the point force on the x-axis the scattering problem is even in φ and the incident wave can be expanded in a cosine series

$$u^{\mathrm{in}}(r, \varphi) = \sum_{m=0}^{\infty} u_m(r) \cos m\varphi$$

When inserting this in the equation for u^{in} the delta functions are transformed to polar coordinates

$$\delta(x + d)\delta(y) = \frac{1}{d}\delta(r - d)\delta(\varphi - \pi)$$

where the extra factor $1/d$ comes from the surface measure in polar coordinates, i.e. $\mathrm{d}x\,\mathrm{d}y = r\,\mathrm{d}r\,\mathrm{d}\varphi$. Multiplying the equation for u^{in} with $\cos m'\varphi$ and integrating in φ from 0 to 2π then gives

$$\frac{1}{r}\frac{\mathrm{d}}{\mathrm{d}r}\left(r\frac{\mathrm{d}u_m}{\mathrm{d}r}\right) + \left(k^2 - \frac{m^2}{r^2}\right) u_m = -\frac{\epsilon_m(-1)^m}{2\pi d}\delta(r - d)$$

The solution should not satisfy any boundary conditions in the ordinary sense; instead the solution should be out-going for large r and it should be bounded for $r = 0$. Thus it must be of the type

$$u_m(r) = \begin{cases} A_m \mathrm{J}_m(kr), & r < d \\ B_m \mathrm{H}_m(kr), & r > d \end{cases}$$

To determine A_m and B_m some conditions at $r = d$ must be employed. Integrating the equation for u_m over a small interval around $r = d$, it is seen that the derivative of u_m must have a jump $-\epsilon_m(-1)^m/(2\pi d)$ and integrating once more it is seen that u_m must be continuous, thus

$$A_m \mathrm{J}_m(kd) = B_m \mathrm{H}_m(kd)$$

$$k\left(B_m \mathrm{H}_m'(kd) - A_m \mathrm{J}_m'(kd)\right) = -\frac{\epsilon_m(-1)^m}{2\pi d}$$

The determinant of this system can be calculated by a Wronskian relation [essentially Eq. (2.8)]. The solution becomes

$$A_m = \frac{\mathrm{i}A}{4}\,\epsilon_m(-1)^m \mathrm{H}_m(kd)$$

$$B_m = \frac{\mathrm{i}A}{4}\,\epsilon_m(-1)^m \mathrm{J}_m(kd)$$

Collecting everything, the expansion of the incident wave from a point force is

$$u^{\mathrm{in}}(r,\varphi) = \frac{\mathrm{i}A}{4}\sum_{m=0}^{\infty}(-1)^m \epsilon_m \mathrm{J}_m(kr_<)\mathrm{H}_m(kr_>)\cos m\varphi \tag{2.16}$$

where $r_< = \min(r,d)$ and $r_> = \max(r,d)$. This expansion is of the form in Eq. (2.10) with the expansion coefficients

$$b_{em} = \frac{\mathrm{i}\sqrt{\epsilon_m}A}{2}(-1)^m \mathrm{H}_m(kd)$$

and $b_{om} = 0$ for $m = 0, 1, \ldots$. The expansion in Eq. (2.10) is then only valid for $r < d$.

The wave from a point force can be approximated as a plane wave if the force is located sufficiently far away from the obstacle. Using the asymptotic expansion Eq. (2.6) of the Hankel function for large arguments the expansion of the incident wave from a point force becomes

$$u^{\mathrm{in}}(r,\varphi) = \frac{\mathrm{i}A}{4}\sum_{m=0}^{\infty}(-1)^m \epsilon_m \sqrt{\frac{2}{\pi kd}}\,\mathrm{e}^{\mathrm{i}(kd-\pi/4-m\pi/2)}\mathrm{J}_m(kr)\cos m\varphi$$

But the incident wave from the point force is also given by Eq. (2.15). The argument r' in the Hankel function can then be expanded for $r \ll d$ as

$$r' = d + r\cos\varphi = d + x$$

to give

$$u^{\mathrm{in}}(r,\varphi) = \frac{\mathrm{i}A}{4}\sqrt{\frac{2}{\pi kd}}\,\mathrm{e}^{\mathrm{i}(k(d+x)-\pi/4)}$$

where $x \ll d$ is neglected under the square root. Comparing the two expressions for u^{in} it follows that a plane wave propagating in the x-direction has the expansion

$$\mathrm{e}^{\mathrm{i}kx} = \sum_{m=0}^{\infty} \mathrm{i}^m \epsilon_m \mathrm{J}_m(kr)\cos m\varphi \tag{2.17}$$

For an incident plane wave with unit amplitude the expansion coefficients are thus

$$b_{\mathrm{e}m} = 2\sqrt{\epsilon_m}\,\mathrm{i}^m \tag{2.18}$$

and $b_{\mathrm{o}m} = 0$.

2.1.4 The scattered wave

It is of interest to further investigate the scattered wave. To calculate this wave close to the obstacle the full expression for the scattered wave must be used, but if the wave far away from the obstacle is of interest more explicit expressions can be obtained.

If the distance from the obstacle to the field point is large compared to both the wavelength and the radius of the obstacle then the asymptotic formula Eq. (2.6) for the Hankel function can be used so that the scattered wave to leading order can be written

$$u^{\mathrm{sc}}(r,\varphi) = \sqrt{\frac{2}{\pi kr}}\,\mathrm{e}^{\mathrm{i}(kr-\pi/4)}F(\varphi) \tag{2.19}$$

where the far field scattering amplitude is

$$F(\varphi) = \frac{1}{2}\sum_{m=0}^{\infty}\sqrt{\epsilon_m}\,\mathrm{i}^{-m}\left(f_{\mathrm{e}m}\cos m\varphi + f_{\mathrm{o}m}\sin m\varphi\right)$$

Assuming an incident plane wave with unit amplitude propagating in the positive x-direction, this expression can be further simplified

$$F(\varphi) = \sum_{m=0}^{\infty} \epsilon_m T_m \cos m\varphi \tag{2.20}$$

where T_m is the T-matrix element from Eq. (2.14). The scattered wave thus decays as the square root with the distance from the obstacle and the angular dependence is determined by $F(\varphi)$.

The scattered energy is another quantity of interest. The energy flow density of the scattered wave is given by Eq. (1.8). In the present 2D SH wave case the mean in time of the radial energy flow density is

$$\langle j_r \rangle = -\frac{\omega}{2} \operatorname{Im} (\boldsymbol{u} \cdot \boldsymbol{\sigma}^\star) \cdot \hat{\boldsymbol{r}} = -\frac{\omega}{2} \operatorname{Im} \left(u\, \mu \frac{\partial u^\star}{\partial r} \right)$$

The scattered wave in the far field is given in Eq. (2.19). The derivative of the r appearing under the square root sign is neglected in the far field. Using the orthogonality of the trigonometric functions the total radial energy flow then becomes

$$\langle J_r \rangle = \int_0^{2\pi} < j_r > r \, \mathrm{d}\varphi = \frac{\omega\mu}{2} \sum_{\sigma,m} |f_{\sigma m}|^2$$

It is noted that this means that each out-going wavefunction carries the same amount of energy.

In the case of an incident plane wave it is natural to normalize this total scattered energy flow with the energy flow density of the incident plane wave. This flow density is easily calculated and the result is

$$\langle j_x \rangle = -\frac{\omega}{2} \operatorname{Im} \left(u^{\mathrm{in}}\, \mu \frac{\partial u^{\mathrm{in}\star}}{\partial x} \right) = \frac{\omega\mu k}{2} \tag{2.21}$$

Define the total scattering cross section as

$$\gamma = \frac{\langle J_r \rangle}{\langle j_x \rangle} = \frac{4}{k} \sum_{m=0}^{\infty} \epsilon_m |T_m|^2 \tag{2.22}$$

where the coefficients of the scattered wave are expressed in the T-matrix and the expansion coefficients of the incident plane wave. This quantity has the dimension of length in 2D (it is an area in 3D) and is a measure of how large part of the energy incident on the obstacle that is scattered.

2.1.5 Low frequency T-matrix

In some cases the obstacle is small compared to relevant wavelengths, e.g. for a collection of small pores or for grains in a metal. In this limit $ka \ll 1$, where k is any relevant wavenumber, and this is often called the Rayleigh limit. Often it is also called the low frequency limit or the limit of a small obstacle, although this is somewhat inaccurate as both the frequency and the size of an obstacle are quantities with a dimension. For low frequencies it is possible to obtain simple analytical expressions for the T-matrix elements. It is straightforward to use the series representation of the Bessel and Hankel functions (at most two terms are needed) and calculate the first two T- matrix elements to leading order

$$T_0 = \mathrm{i}(ka)^2 \, \frac{\pi}{4} \left(\frac{\rho_1}{\rho} - 1 \right) \tag{2.23}$$

$$T_1 = \mathrm{i}(ka)^2 \, \frac{\pi(\mu - \mu_1)}{4(\mu + \mu_1)} \tag{2.24}$$

The T-matrix elements for $m = 2, 3, \ldots$ behave as $(ka)^{2m}$ and are thus negligible for most purposes. Thus monopole ($m = 0$) and dipole ($m = 1$) scattering is dominant for small obstacles. It is noted that the monopole scattering only depends on the contrast in density. This is natural as the obstacle for $m = 0$ performs a pure translation for low frequencies. In contrast, the dipole scattering only depends on the contrast in stiffness. By putting $\rho_1 = 0$ and $\mu_1 = 0$ the elements for a cavity are obtained.

From these T-matrix elements explicit expressions for the far field scattering amplitude and the total scattering cross section are easily written down. The far field amplitude has a simple dependence on φ with one constant term and one in $\cos\varphi$. The total scattering cross section behaves as k^3a^4.

2.1.6 Other boundary conditions

More general boundary conditions between the circular obstacle and the surrounding are sometimes employed, for example to model a thin interface layer or a damaged interface. The simplest such boundary condition, often called the spring boundary condition, is that the stress is continuous while there is a jump in the displacement proportional to the stress

$$\mu \, \frac{\partial(u^{\mathrm{in}} + u^{\mathrm{sc}})}{\partial r} = \mu_1 \, \frac{\partial u^1}{\partial r} = \alpha(u^{\mathrm{in}} + u^{\mathrm{sc}} - u^1)$$

Here α is a (distributed) spring constant that may depend on the properties of an interface layer or some parameter(s) describing interface damage. It may also depend on frequency. These boundary conditions lead to the following equations

$$\mu k \left(b_{\sigma m} \mathrm{J}'_m(ka) + f_{\sigma m} \mathrm{H}'_m(ka)\right) = \mu_1 k_1 c_{\sigma m} \mathrm{J}'_m(k_1 a)$$

$$\mu_1 k_1 c_{\sigma m} \mathrm{J}'_m(k_1 a) = \alpha \left[b_{\sigma m} \mathrm{J}_m(ka) + f_{\sigma m} \mathrm{H}_m(ka) - c_{\sigma m} \mathrm{J}_m(k_1 a)\right]$$

for $m = 0, 1, \ldots$ and $\sigma = \mathrm{e}, \mathrm{o}$. This system is easily solved for $f_{\sigma m}$ in terms of $b_{\sigma m}$, and thereby for the T-matrix elements.

Even more general boundary conditions are sometimes used, for example to get a better model for a thin interface layer. These may contain a jump also in stress and may, furthermore, contain tangential derivatives of the displacement and/or stress. In a general form they can be given as

$$u - u^1 = \alpha \frac{\partial u}{\partial r} + \alpha_1 \frac{\partial u^1}{\partial r}$$

$$\mu \frac{\partial u}{\partial r} - \mu_1 \frac{\partial u^1}{\partial r} = \beta u + \beta_1 u^1 + \frac{\gamma}{r^2} \frac{\partial^2 u}{\partial \varphi^2} + \frac{\gamma_1}{r^2} \frac{\partial^2 u^1}{\partial \varphi^2} \tag{2.25}$$

where $u = u^{\mathrm{in}} + u^{\mathrm{sc}}$ is the total wave in the surrounding. These boundary conditions between the obstacle and the surrounding can be used as a model for a thin interface layer (meaning that the layer is thin compared to all relevant wavelengths) and in this case the parameters should be chosen according to

$$\alpha = \frac{d}{2} \left(\frac{\mu}{\mu_0} - 1 \right), \qquad \alpha_1 = \frac{d}{2} \left(\frac{\mu_1}{\mu_0} - 1 \right)$$

$$\beta = \frac{\omega^2 d}{2} (\rho - \rho_0), \qquad \beta_1 = \frac{\omega^2 d}{2} (\rho_1 - \rho_0)$$

$$\gamma = \frac{d}{2} (\mu - \mu_0), \qquad \gamma_1 = \frac{d}{2} (\mu_1 - \mu_0)$$

Here ρ_0, μ_0, and d are the density, shear modulus, and thickness, respectively, of the thin interface layer. These interface conditions are derived and discussed by Bövik and Olsson [32] and it is in particular noted that they are exact to first order in the layer thickness. These interface conditions lead to the following equations

$$\begin{aligned} & b_{\sigma m}\mathrm{J}_m(ka) + f_{\sigma m}\mathrm{H}_m(ka) - c_{\sigma m}\mathrm{J}_m(k_1 a) \\ & \qquad = \alpha k \left(b_{\sigma m}\mathrm{J}'_m(ka) + f_{\sigma m}\mathrm{H}'_m(ka)\right) + \alpha_1 k_1 c_{\sigma m}\mathrm{J}'_m(k_1 a) \end{aligned}$$

$$\begin{aligned} & \mu k \left(b_{\sigma m}\mathrm{J}'_m(ka) + f_{\sigma m}\mathrm{H}'_m(ka)\right) - \mu_1 k_1 c_{\sigma m}\mathrm{J}'_m(k_1 a) \\ & \qquad = \left(\beta - \frac{\gamma m^2}{a^2}\right)\left(b_{\sigma m}\mathrm{J}_m(ka) + f_{\sigma m}\mathrm{H}_m(ka)\right) \\ & \qquad\qquad + \left(\beta_1 - \frac{\gamma_1 m^2}{a^2}\right) c_{\sigma m}\mathrm{J}_m(k_1 a) \end{aligned}$$

This system is straightforward to solve to obtain the T-matrix. In the present case with a circular obstacle it is as easy to consider a layered circle, but for other cases these interface conditions may be a better alternative.

2.1.7 Layered circle

Another generalization of some interest is to consider a layered circular obstacle. Consider the simplest such case with a circular core with radius a_1, density ρ_1, and shear modulus μ_1 surrounded by a concentric layer with outer radius a_2, density ρ_2, and shear modulus μ_2. The waves in the core are expanded as previously

$$u^1(\boldsymbol{r}) = \sum_{\sigma,m} c_{1\sigma m}\chi^0_{1,\sigma m}(\boldsymbol{r})$$

The waves in the layer must contain two linearly independent solutions to Bessel's differential equation. As these wavefunctions are used elsewhere, the waves in the layer are expanded as a sum of regular and out-going wavefunctions

$$u^2(\boldsymbol{r}) = \sum_{\sigma,m} \left[c_{2\sigma m}\chi^0_{2,\sigma m}(\boldsymbol{r}) + d_{2\sigma m}\chi^+_{2,\sigma m}(\boldsymbol{r})\right]$$

Another choice is to make the expansion in terms of the out-going and ingoing waves (containing Hankel functions of the second kind) and this would more correspond to the usual way in a plane layer where the expansion usually is in terms of up-going and down-going waves. If the usual welded boundary conditions are assumed at both interfaces, the following equations are obtained

$$b_{\sigma m}\mathrm{J}_m(ka_2) + f_{\sigma m}\mathrm{H}_m(ka_2) = c_{2\sigma m}\mathrm{J}_m(k_2a_2) + d_{2\sigma m}\mathrm{H}_m(k_2a_2)$$

$$\mu k\left(b_{\sigma m}\mathrm{J}'_m(ka_2) + f_{\sigma m}\mathrm{H}'_m(ka_2)\right) = \mu_2 k_2\left(c_{2\sigma m}\mathrm{J}'_m(k_2a_2) + d_{2\sigma m}\mathrm{H}'_m(k_2a_2)\right)$$

$$c_{2\sigma m}\mathrm{J}_m(k_2a_1) + d_{2\sigma m}\mathrm{H}_m(k_2a_1) = c_{1\sigma m}\mathrm{J}_m(k_1a_1)$$

$$\mu_2 k_2\left(c_{2\sigma m}\mathrm{J}'_m(k_2a_1) + d_{2\sigma m}\mathrm{H}'_m(k_2a_1)\right) = \mu_1 k_1 c_{1\sigma m}\mathrm{J}'_m(k_1a_1)$$

Solving these equations gives the T-matrix elements.

It should be apparent how cases with more than one layer are treated.

2.2 Circle: 2D P-SV waves

With the simplest 2D scalar scattering problem solved, the next step is to consider the corresponding 2D in-plane vector problem that involves both P and SV waves. The development then relies very much on the scalar case.

2.2.1 Wavefunctions

The potentials for the 2D vector problem are given in Section 1.2.2 with the representation of the displacement field as [Eq. (1.17)]

$$\boldsymbol{u} = \nabla\Phi + \nabla\times(\hat{\boldsymbol{z}}\Psi)$$

where the potentials Φ and Ψ satisfy scalar Helmholtz equations with wavenumber k_p and k_s, respectively. Wavefunctions can be constructed in the same way from the corresponding scalar ones of the preceding section. Thus the following two polar vector wavefunctions are defined

$$\boldsymbol{\chi}^0_{1\sigma m}(\boldsymbol{r}) = \frac{\sqrt{\epsilon_m}}{2k_\mathrm{s}}\nabla\times\left[\hat{\boldsymbol{z}}\mathrm{J}_m(k_\mathrm{s}r)\left\{\begin{array}{c}\cos m\varphi\\ \sin m\varphi\end{array}\right\}\right]$$

$$= \frac{\sqrt{\epsilon_m}}{2}\left[\hat{\boldsymbol{r}}\frac{m}{k_\mathrm{s}r}\mathrm{J}_m(k_\mathrm{s}r)\left\{\begin{array}{c}-\sin m\varphi\\ \cos m\varphi\end{array}\right\} - \hat{\boldsymbol{\varphi}}\mathrm{J}'_m(k_\mathrm{s}r)\left\{\begin{array}{c}\cos m\varphi\\ \sin m\varphi\end{array}\right\}\right]$$

$$\chi^0_{2\sigma m}(\boldsymbol{r}) = \frac{\sqrt{\epsilon_m}}{2k_\mathrm{s}} \nabla \left[\mathrm{J}_m(k_\mathrm{p} r) \left\{ \begin{array}{c} \cos m\varphi \\ \sin m\varphi \end{array} \right\} \right]$$

$$= \frac{\sqrt{\epsilon_m}\, k_\mathrm{p}}{2k_\mathrm{s}} \left[\hat{\boldsymbol{r}} \mathrm{J}'_m(k_\mathrm{p} r) \left\{ \begin{array}{c} \cos m\varphi \\ \sin m\varphi \end{array} \right\} + \hat{\boldsymbol{\varphi}} \frac{m}{k_\mathrm{p} r} \mathrm{J}_m(k_\mathrm{p} r) \left\{ \begin{array}{c} -\sin m\varphi \\ \cos m\varphi \end{array} \right\} \right] \tag{2.26}$$

Here the first index $\tau = 1$ or $\tau = 2$ on the wavefunctions denotes SV or P waves, respectively, $m = 0, 1, 2, \ldots$, and $\sigma = \mathrm{e}$ (even) for the upper row or $\sigma = \mathrm{o}$ (odd) for the lower row, with $\sigma = \mathrm{o}$ excluded for $m = 0$. The Neumann factor is $\epsilon_0 = 1$ and $\epsilon_m = 2$ for $m = 1, 2, \ldots$. The upper index 0 on the wavefunctions means that they are regular containing Bessel functions; the corresponding out-going wavefunctions contain Hankel functions and are denoted by an upper index $+$. The wavefunctions are dimensionless and the factor $k_\mathrm{p}/k_\mathrm{s}$ in front of the second wavefunction is chosen so as to give a symmetric T-matrix. When needed an extra index is put on the wavefunctions to denote what material parameters that are used, for example $\chi^0_{i,\tau\sigma m}(\boldsymbol{r})$.

As defined, the wavefunctions for $\tau = 1$ and $\tau = 2$ have different dependence on φ in that the cosine and sine functions have switched place in the r- and φ-components. For the scattering by a circle this leads to a coupling between the wavefunctions for $\tau = 1$, $\sigma = \mathrm{e}$ and $\tau = 2$, $\sigma = \mathrm{o}$ and vice versa. To avoid this coupling between different σ-values, the definition of the wavefunctions could be changed so that the trigonometric functions in one of them change place with each other. However, this leads to complications in other situations, in particular when transformation to plane wavefunctions are involved.

The corresponding radial tractions on $r = a$ are needed for the boundary conditions. The traction operator is given in Eqs. (1.21) and (1.22) for P and SV waves, respectively. Calculation of the traction of the wavefunctions gives

$$\boldsymbol{t}^{(r)}(\chi^0_{1\sigma m}) = \frac{\mu\sqrt{\epsilon_m}}{2} \left[\hat{\boldsymbol{r}}\, 2m \frac{\mathrm{d}}{\mathrm{d}r} \left(\frac{\mathrm{J}_m(k_\mathrm{s} r)}{k_\mathrm{s} r} \right) \left\{ \begin{array}{c} -\sin m\varphi \\ \cos m\varphi \end{array} \right\} \right.$$

$$\left. -\hat{\boldsymbol{\varphi}} k_\mathrm{s} \left(2\mathrm{J}''_m(k_\mathrm{s} r) + \mathrm{J}_m(k_\mathrm{s} r) \right) \left\{ \begin{array}{c} \cos m\varphi \\ \sin m\varphi \end{array} \right\} \right]$$

$$
\begin{aligned}
\boldsymbol{t}^{(r)}(\boldsymbol{\chi}^0_{2\sigma m}) = \frac{\mu\sqrt{\epsilon_m}}{2k_\mathrm{s}} \Bigg[\hat{\boldsymbol{r}} \Big(2k_\mathrm{p}^2 \mathrm{J}''_m(k_\mathrm{p}r) \\
+ (2k_\mathrm{p}^2 - k_\mathrm{s}^2)\mathrm{J}_m(k_\mathrm{p}r) \Big) \left\{ \begin{array}{c} \cos m\varphi \\ \sin m\varphi \end{array} \right\} \\
+ \hat{\boldsymbol{\varphi}}\, 2mk_\mathrm{p} \frac{\mathrm{d}}{\mathrm{d}r} \left(\frac{\mathrm{J}_m(k_\mathrm{p}r)}{k_\mathrm{p}r} \right) \left\{ \begin{array}{c} -\sin m\varphi \\ \cos m\varphi \end{array} \right\} \Bigg]
\end{aligned} \tag{2.27}
$$

If desired derivatives can be eliminated by using recurrence relations and Bessel's differential equation

$$
\begin{aligned}
\mathrm{J}''_m(z) &= -\mathrm{J}_m(z) + \frac{1}{z}\mathrm{J}_{m+1}(z) \\
\frac{\mathrm{d}}{\mathrm{d}z}\left(\frac{\mathrm{J}_m(z)}{z} \right) &= \frac{m-1}{z^2}\mathrm{J}_m(z) - \frac{1}{z}\mathrm{J}_{m+1}(z)
\end{aligned}
$$

When needed an index is put on the traction operator, thus $\boldsymbol{t}_i^{(r)}(\boldsymbol{\chi}^0_{i,\tau\sigma m})$. The traction of the out-going wavefunctions is of course obtained by changing all Bessel functions to Hankel functions.

For convenience when stating boundary conditions the radial dependence of the wavefunction components are now introduced

$$
\begin{aligned}
V^0_{r,1m}(r) &= \frac{\sqrt{\epsilon_m}\, m}{2k_\mathrm{s}r}\mathrm{J}_m(k_\mathrm{s}r) \\
V^0_{\varphi,1m}(r) &= -\frac{\sqrt{\epsilon_m}}{2}\mathrm{J}'_m(k_\mathrm{s}r) \\
V^0_{r,2m}(r) &= \frac{\sqrt{\epsilon_m}\, k_\mathrm{p}}{2k_\mathrm{s}}\mathrm{J}'_m(k_\mathrm{p}r) \\
V^0_{\varphi,2m}(r) &= -\frac{\sqrt{\epsilon_m}\, m}{2k_\mathrm{s}r}\mathrm{J}_m(k_\mathrm{p}r)
\end{aligned} \tag{2.28}
$$

Note that these are defined without the σ-index, but are taken for the lower row for $\tau = 1$ and the upper row for $\tau = 2$, thus with a $\cos m\varphi$ behaviour for the r-components and a $\sin m\varphi$ behaviour for the φ-components. There is an exception to this and that is the $\tau = 1$ and $m = 0$ case where the upper row is used (because the lower row does not exist). The other group (the $\tau = 1$ and $m = 0$ case actually belongs to this group) has some sign changes but this does not affect the T-matrix, but must otherwise be taken care of. The corresponding functions with Bessel functions changed to Hankel functions are denoted $V^{+}_{j,\tau m}$, $j = r, \varphi$, $\tau = 1, 2$, and $m = 0, 1, 2, \ldots$. Similarly the radial dependence of the traction components are introduced

$$\begin{aligned}
S^{0}_{r,1m}(r) &= \mu m \sqrt{\epsilon_m} \frac{\mathrm{d}}{\mathrm{d}r}\left(\frac{\mathrm{J}_m(k_\mathrm{s} r)}{k_\mathrm{s} r}\right), \\
S^{0}_{\varphi,1m}(r) &= -\frac{\mu k_\mathrm{s}\sqrt{\epsilon_m}}{2}\left(2\mathrm{J}''_m(k_\mathrm{s} r) + \mathrm{J}_m(k_\mathrm{s} r)\right) \\
S^{0}_{r,2m}(r) &= \frac{\mu\sqrt{\epsilon_m}}{2k_\mathrm{s}}\left(2k_\mathrm{p}^2\mathrm{J}''_m(k_\mathrm{p} r) + (2k_\mathrm{p}^2 - k_\mathrm{s}^2)\mathrm{J}_m(k_\mathrm{p} r)\right) \\
S^{0}_{\varphi,2m}(r) &= -\frac{\mu m k_\mathrm{p}\sqrt{\epsilon_m}}{k_\mathrm{s}}\frac{\mathrm{d}}{\mathrm{d}r}\left(\frac{\mathrm{J}_m(k_\mathrm{p} r)}{k_\mathrm{p} r}\right)
\end{aligned} \tag{2.29}$$

The corresponding functions with Bessel functions changed to Hankel functions are denoted $S^{+}_{j,\tau m}$, $j = r, \varphi$, $\tau = 1, 2$, and $m = 0, 1, 2, \ldots$. If needed an extra index "i" is put on these quantities to denote the domain, for example $V^{+}_{ji,\tau m}$ and $S^{+}_{ji,\tau m}$.

2.2.2 The T-matrix

Consider the scattering of an incident wave by a circular obstacle of radius a in an isotropic and homogeneous surrounding. The material outside the circle has Lamé constants λ and μ and density ρ. Inside the circle another isotropic and homogeneous material has Lamé constants λ_1 and μ_1 and density ρ_1. As special cases the interior can be a cavity, a fluid, or a rigid inclusion. The corresponding wavenumbers are $k_\mathrm{p} = \omega(\rho/(\lambda + 2\mu))^{1/2}$ and $k_\mathrm{s} = \omega(\rho/\mu)^{1/2}$ outside the circle and $k_{\mathrm{p}1} = \omega(\rho_1/(\lambda_1 + 2\mu_1))^{1/2}$ and $k_{\mathrm{s}1} = \omega(\rho_1/\mu_1)^{1/2}$ inside the circle.

As for the SH case the sources of the incident wave are assumed to lie outside the obstacle, often a plane wave is assumed. At present the incident wave is not specified, but in the next subsection plane waves are considered. As there are no sources in the vicinity of the obstacle the incident wave must be finite there and close enough to the obstacle it must then be possible to expand it in the regular wavefunctions

$$\boldsymbol{u}^{\mathrm{in}}(\boldsymbol{r}) = \sum_{\tau,\sigma,m} b_{\tau\sigma m}\boldsymbol{\chi}^{0}_{\tau\sigma m}(\boldsymbol{r}) \tag{2.30}$$

$$
\begin{aligned}
V^0_{\varphi,1m}(a)b_{1om} + V^0_{\varphi,2m}(a)b_{2em} + V^+_{\varphi,1m}(a)f_{1om} + V^+_{\varphi,2m}(a)f_{2em} \\
= V^0_{\varphi 1,1m}(a)c_{1om} + V^0_{\varphi 1,2m}(a)c_{2em} \\
S^0_{r,1m}(a)b_{1om} + S^0_{r,2m}(a)b_{2em} + S^+_{r,1m}(a)f_{1om} + S^+_{r,2m}(a)f_{2em} \\
= S^0_{r1,1m}(a)c_{1om} + S^0_{r1,2m}(a)c_{2em} \\
S^0_{\varphi,1m}(a)b_{1om} + S^0_{\varphi,2m}(a)b_{2em} + S^+_{\varphi,1m}(a)f_{1om} + S^+_{\varphi,2m}(a)f_{2em} \\
= S^0_{\varphi 1,1m}(a)c_{1om} + S^0_{\varphi 1,2m}(a)c_{2em}
\end{aligned}
$$

Solving these equations by expressing the coefficients of the scattered wave $f_{\tau\sigma m}$ in terms of those of the incident wave $b_{\tau\sigma m}$, the T-matrix elements can be read off.

As for the SH case it can be of interest to investigate other boundary conditions, in particular spring boundary conditions. This is straightforward given the expressions for the wavefunctions and their tractions. Also a layered circle is straightforward to consider.

2.2.3 Incident plane waves

An incident plane wave can be of either SV or P type. A plane SV wave propagating in the x-direction is

$$
\boldsymbol{u}^{\text{in}} = \hat{\boldsymbol{y}}\, \mathrm{e}^{\mathrm{i}k_{\text{s}}x} = \frac{\mathrm{i}}{k_{\text{s}}} \nabla \times \left(\hat{\boldsymbol{z}}\, \mathrm{e}^{\mathrm{i}k_{\text{s}}x} \right)
$$

Taking the curl of the scalar plane wave expansion Eq. (2.17) multiplied by $\hat{\boldsymbol{z}}$, it follows that the only nonzero expansion coefficients are

$$
b_{1em} = 2\sqrt{\epsilon_m}\, \mathrm{i}^{m+1}
$$

for $m = 0, 1, 2, \ldots$. Similarly a plane P wave is

$$
\boldsymbol{u}^{\text{in}} = \hat{\boldsymbol{x}}\, \mathrm{e}^{\mathrm{i}k_{\text{p}}x} = -\frac{\mathrm{i}}{k_{\text{p}}} \nabla \left(\mathrm{e}^{\mathrm{i}k_{\text{p}}x} \right)
$$

so that the only nonzero expansion coefficients are

$$
b_{2em} = \frac{2k_{\text{s}}}{k_{\text{p}}} \sqrt{\epsilon_m}\, \mathrm{i}^{m-1}
$$

for $m = 0, 1, 2, \ldots$.

corresponding functions with Bessel functions changed to Hankel functions are denoted $V^+_{j,\tau m}$, $j = r, \varphi$, $\tau = 1, 2$, and $m = 0, 1, 2, \ldots$. Similarly the radial dependence of the traction components are introduced

$$
\begin{aligned}
S^0_{r,1m}(r) &= \mu m\sqrt{\epsilon_m}\frac{\mathrm{d}}{\mathrm{d}r}\left(\frac{\mathrm{J}_m(k_\mathrm{s}r)}{k_\mathrm{s}r}\right),\\
S^0_{\varphi,1m}(r) &= -\frac{\mu k_\mathrm{s}\sqrt{\epsilon_m}}{2}\left(2\mathrm{J}''_m(k_\mathrm{s}r) + \mathrm{J}_m(k_\mathrm{s}r)\right)\\
S^0_{r,2m}(r) &= \frac{\mu\sqrt{\epsilon_m}}{2k_\mathrm{s}}\left(2k_\mathrm{p}^2\mathrm{J}''_m(k_\mathrm{p}r) + (2k_\mathrm{p}^2 - k_\mathrm{s}^2)\mathrm{J}_m(k_\mathrm{p}r)\right)\\
S^0_{\varphi,2m}(r) &= -\frac{\mu m k_\mathrm{p}\sqrt{\epsilon_m}}{k_\mathrm{s}}\frac{\mathrm{d}}{\mathrm{d}r}\left(\frac{\mathrm{J}_m(k_\mathrm{p}r)}{k_\mathrm{p}r}\right)
\end{aligned}
\tag{2.29}
$$

The corresponding functions with Bessel functions changed to Hankel functions are denoted $S^+_{j,\tau m}$, $j = r, \varphi$, $\tau = 1, 2$, and $m = 0, 1, 2, \ldots$. If needed an extra index "i" is put on these quantities to denote the domain, for example $V^+_{ji,\tau m}$ and $S^+_{ji,\tau m}$.

2.2.2 The T-matrix

Consider the scattering of an incident wave by a circular obstacle of radius a in an isotropic and homogeneous surrounding. The material outside the circle has Lamé constants λ and μ and density ρ. Inside the circle another isotropic and homogeneous material has Lamé constants λ_1 and μ_1 and density ρ_1. As special cases the interior can be a cavity, a fluid, or a rigid inclusion. The corresponding wavenumbers are $k_\mathrm{p} = \omega(\rho/(\lambda + 2\mu))^{1/2}$ and $k_\mathrm{s} = \omega(\rho/\mu)^{1/2}$ outside the circle and $k_\mathrm{p1} = \omega(\rho_1/(\lambda_1 + 2\mu_1))^{1/2}$ and $k_\mathrm{s1} = \omega(\rho_1/\mu_1)^{1/2}$ inside the circle.

As for the SH case the sources of the incident wave are assumed to lie outside the obstacle, often a plane wave is assumed. At present the incident wave is not specified, but in the next subsection plane waves are considered. As there are no sources in the vicinity of the obstacle the incident wave must be finite there and close enough to the obstacle it must then be possible to expand it in the regular wavefunctions

$$
\boldsymbol{u}^{\mathrm{in}}(\boldsymbol{r}) = \sum_{\tau,\sigma,m} b_{\tau\sigma m}\boldsymbol{\chi}^0_{\tau\sigma m}(\boldsymbol{r}) \tag{2.30}
$$

where the summation is over $\tau = 1, 2$, $\sigma = \mathrm{e}, \mathrm{o}$, and $m = 0, 1, 2, \ldots$ (with $\sigma = \mathrm{o}$ and $m = 0$ excluded). The expansion coefficients $b_{\tau\sigma m}$ are in principle known. The wave scattered by the obstacle is carrying energy outwards from the obstacle (the radiation condition) and is thus possible to expand in the out-going wavefunctions

$$\boldsymbol{u}^{\mathrm{sc}}(\boldsymbol{r}) = \sum_{\tau,\sigma,m} f_{\tau\sigma m} \boldsymbol{\chi}^{+}_{\tau\sigma m}(\boldsymbol{r}) \tag{2.31}$$

Here the expansion coefficients $f_{\tau\sigma m}$ are unknown and the problem is thus to determine these given the incident wave and the properties of the obstacle. As for the SH case this is done in full generality by determining the T-matrix, which is defined as the linear relation between the expansion coefficients of the incident and scattered waves

$$f_{\tau\sigma m} = \sum_{\tau',\sigma',m'} T_{\tau\sigma m,\tau'\sigma' m'} b_{\tau'\sigma' m'} \tag{2.32}$$

For the circular obstacle the T-matrix has the following structure

$$T_{\tau\sigma m,\tau'\sigma' m'} = \delta_{mm'} \xi_{\tau\sigma,\tau'\sigma'} T^{m}_{\tau\tau'} \tag{2.33}$$

where

$$\xi_{\tau\sigma,\tau'\sigma'} = \begin{cases} 1, & \text{if } \tau = \tau' \text{ and } \sigma = \sigma' \text{ or } \tau \neq \tau' \text{ and } \sigma \neq \sigma' \\ 0, & \text{otherwise} \end{cases}$$

In words this means that the T-matrix is diagonal except that mode coupling elements between P and SV waves are nonzero if the m-values are the same and the σ-values different. For $m = 0$ these mode coupling elements vanish. With the chosen normalization of the wavefunctions the T-matrix is, furthermore, symmetric, so that $T^{m}_{\tau\tau'} = T^{m}_{\tau'\tau}$.

To start simple, and yet with a case of interest, the circular cavity is first considered. The boundary condition is that both traction components vanish. For $m = 0$ there is only an azimuthal traction component for $\tau = 1$ and only a radial traction component for $\tau = 2$. There is thus no coupling between $\tau = 1$ and $\tau = 2$. For $\tau = 1$ the boundary condition gives

$$S^{0}_{\varphi,10}(a) b_{1\mathrm{e}0} + S^{+}_{\varphi,10}(a) f_{1\mathrm{e}0} = 0$$

and after simplification the corresponding T-matrix element becomes

$$T_{11}^{0} = -\frac{k_{\mathrm{s}}a\mathrm{J}_0(k_{\mathrm{s}}a) - 2\mathrm{J}_1(k_{\mathrm{s}}a)}{k_{\mathrm{s}}a\mathrm{H}_0(k_{\mathrm{s}}a) - 2\mathrm{H}_1(k_{\mathrm{s}}a)}$$

For $\tau = 2$ the boundary condition similarly gives

$$S_{r,20}^{0}(a)b_{2\mathrm{e}0} + S_{r,20}^{+}(a)f_{2\mathrm{e}0} = 0$$

and after simplification the corresponding T-matrix element becomes

$$T_{22}^{0} = -\frac{k_{\mathrm{s}}^2 a\mathrm{J}_0(k_{\mathrm{p}}a) - 2k_{\mathrm{p}}\mathrm{J}_1(k_{\mathrm{p}}a)}{k_{\mathrm{s}}^2 a\mathrm{H}_0(k_{\mathrm{p}}a) - 2k_{\mathrm{p}}\mathrm{H}_1(k_{\mathrm{p}}a)}$$

For $m = 1, 2, \ldots$ the vanishing traction leads to the following two equations

$$S_{r,1m}^{0}(a)b_{1\mathrm{o}m} + S_{r,2m}^{0}(a)b_{2\mathrm{e}m} + S_{r,1m}^{+}(a)f_{1\mathrm{o}m} + S_{r,2m}^{+}(a)f_{2\mathrm{e}m} = 0$$

$$S_{\varphi,1m}^{0}(a)b_{1\mathrm{o}m} + S_{\varphi,2m}^{0}(a)b_{2\mathrm{e}m} + S_{\varphi,1m}^{+}(a)f_{1\mathrm{o}m} + S_{\varphi,2m}^{+}(a)f_{2\mathrm{e}m} = 0$$

Solving these equations by expressing the coefficients of the scattered wave $f_{\tau\sigma m}$ in terms of those of the incident wave $b_{\tau\sigma m}$, the T-matrix elements can be read off. It is noted that the two off-diagonal elements T_{12}^{m} and T_{21}^{m} become equal. To show this analytically recurrence relations and a Wronskian for Bessel functions are used.

When there is a different material inside the obstacle a wave is of course generated also there. This wave must be finite and can thus be expanded in the regular wavefunctions

$$\boldsymbol{u}^1(\boldsymbol{r}) = \sum_{\tau,\sigma,m} c_{\tau\sigma m}\boldsymbol{\chi}_{1,\tau\sigma m}^{0}(\boldsymbol{r}) \tag{2.34}$$

where the properties of the material inside the obstacle must be used, thus the index 1 on the wavefunctions. The expansion coefficients $c_{\tau\sigma m}$ are of course also unknown. Assuming welded contact the following system of equations is obtained

$$V_{r,1m}^{0}(a)b_{1\mathrm{o}m} + V_{r,2m}^{0}(a)b_{2\mathrm{e}m} + V_{r,1m}^{+}(a)f_{1\mathrm{o}m} + V_{r,2m}^{+}(a)f_{2\mathrm{e}m}$$
$$= V_{r1,1m}^{0}(a)c_{1\mathrm{o}m} + V_{r1,2m}^{0}(a)c_{2\mathrm{e}m}$$

$$V^0_{\varphi,1m}(a)b_{1om} + V^0_{\varphi,2m}(a)b_{2em} + V^+_{\varphi,1m}(a)f_{1om} + V^+_{\varphi,2m}(a)f_{2em} = V^0_{\varphi 1,1m}(a)c_{1om} + V^0_{\varphi 1,2m}(a)c_{2em}$$

$$S^0_{r,1m}(a)b_{1om} + S^0_{r,2m}(a)b_{2em} + S^+_{r,1m}(a)f_{1om} + S^+_{r,2m}(a)f_{2em} = S^0_{r1,1m}(a)c_{1om} + S^0_{r1,2m}(a)c_{2em}$$

$$S^0_{\varphi,1m}(a)b_{1om} + S^0_{\varphi,2m}(a)b_{2em} + S^+_{\varphi,1m}(a)f_{1om} + S^+_{\varphi,2m}(a)f_{2em} = S^0_{\varphi 1,1m}(a)c_{1om} + S^0_{\varphi 1,2m}(a)c_{2em}$$

Solving these equations by expressing the coefficients of the scattered wave $f_{\tau\sigma m}$ in terms of those of the incident wave $b_{\tau\sigma m}$, the T-matrix elements can be read off.

As for the SH case it can be of interest to investigate other boundary conditions, in particular spring boundary conditions. This is straightforward given the expressions for the wavefunctions and their tractions. Also a layered circle is straightforward to consider.

2.2.3 Incident plane waves

An incident plane wave can be of either SV or P type. A plane SV wave propagating in the x-direction is

$$\boldsymbol{u}^{\text{in}} = \hat{\boldsymbol{y}}\, \mathrm{e}^{\mathrm{i}k_{\mathrm{s}}x} = \frac{\mathrm{i}}{k_{\mathrm{s}}} \nabla \times \left(\hat{\boldsymbol{z}}\, \mathrm{e}^{\mathrm{i}k_{\mathrm{s}}x}\right)$$

Taking the curl of the scalar plane wave expansion Eq. (2.17) multiplied by $\hat{\boldsymbol{z}}$, it follows that the only nonzero expansion coefficients are

$$b_{1em} = 2\sqrt{\epsilon_m}\, \mathrm{i}^{m+1}$$

for $m = 0, 1, 2, \ldots$. Similarly a plane P wave is

$$\boldsymbol{u}^{\text{in}} = \hat{\boldsymbol{x}}\, \mathrm{e}^{\mathrm{i}k_{\mathrm{p}}x} = -\frac{\mathrm{i}}{k_{\mathrm{p}}} \nabla \left(\mathrm{e}^{\mathrm{i}k_{\mathrm{p}}x}\right)$$

so that the only nonzero expansion coefficients are

$$b_{2em} = \frac{2k_{\mathrm{s}}}{k_{\mathrm{p}}}\sqrt{\epsilon_m}\, \mathrm{i}^{m-1}$$

for $m = 0, 1, 2, \ldots$.

2.2.4 The scattered wave

When the distance from the obstacle to the field point is large compared to both the wavelength and the radius of the obstacle, the asymptotic formula Eq. (2.6) for the Hankel function can be used so that the scattered wave to leading order becomes

$$\boldsymbol{u}^{\mathrm{sc}}(r,\varphi) = \hat{\boldsymbol{\varphi}}\sqrt{\frac{2}{\pi k_{\mathrm{s}} r}}\,\mathrm{e}^{\mathrm{i}(k_{\mathrm{s}} r-\pi/4)} F_{\mathrm{s}}(\varphi) + \hat{\boldsymbol{r}}\sqrt{\frac{2}{\pi k_{\mathrm{p}} r}}\,\mathrm{e}^{\mathrm{i}(k_{\mathrm{p}} r-\pi/4)} F_{\mathrm{p}}(\varphi)$$

where the far field scattering amplitudes are

$$F_{\mathrm{s}}(\varphi) = -\frac{\mathrm{i}}{2}\sum_{m=0}^{\infty}\sqrt{\epsilon_m}\,\mathrm{i}^{-m}\left(f_{1\mathrm{e}m}\cos m\varphi + f_{1\mathrm{o}m}\sin m\varphi\right)$$

$$F_{\mathrm{p}}(\varphi) = \frac{\mathrm{i}k_{\mathrm{p}}}{2k_{\mathrm{s}}}\sum_{m=0}^{\infty}\sqrt{\epsilon_m}\,\mathrm{i}^{-m}\left(f_{2\mathrm{e}m}\cos m\varphi + f_{2\mathrm{o}m}\sin m\varphi\right)$$

It is noted that the scattered S wave is polarized in the azimuthal direction and the scattered P wave in the radial direction. However, this is only true in the far field. For an incident plane wave the far field amplitudes can be expressed in terms of the T-matrix as in the SH case.

The scattered energy is calculated as for the SH case and this gives the total radial energy flow

$$\langle J_r\rangle = \frac{\omega\mu}{2}\sum_{\tau,\sigma,m}|f_{\tau\sigma m}|^2$$

For an incident plane SV wave the total scattering cross section then becomes

$$\gamma_{\mathrm{s}} = \frac{4}{k_{\mathrm{s}}}\sum_{m=0}^{\infty}\epsilon_m\left(|T_{11}^m|^2 + |T_{21}^m|^2\right) \tag{2.35}$$

For an incident plane P wave it becomes

$$\gamma_{\mathrm{p}} = \frac{4}{k_{\mathrm{p}}}\sum_{m=0}^{\infty}\epsilon_m\left(|T_{12}^m|^2 + |T_{22}^m|^2\right) \tag{2.36}$$

In these expressions the first term is due to the scattered S wave and the second to the scattered P wave.

2.2.5 Low frequency T-matrix

For low frequencies the largest T-matrix elements behave as $(ka)^2$, where k is any of the wavenumbers. These elements appear for $m = 0$, 1, and 2, corresponding to monopole, dipole, and quadrupole scattering. The leading order elements for an elastic obstacle are given below. From these special cases are easily deduced, thus a fluid-filled circle is obtained by putting $\mu_1 = 0$ and a cavity by further putting $\lambda_1 = 0$ and $\rho_1 = 0$. However, the rigid (immovable) circle can not be obtained in this way.

Starting with the $m = 0$ (monopole) elements, one element that is not of leading order is given, namely the SV ($\tau = 1$) and even $m = 0$ element which is

$$T^0_{11} = \mathrm{i}(k_\mathrm{s}a)^4 \, \frac{\pi}{32} \left(\frac{\rho_1}{\rho} - 1 \right) \tag{2.37}$$

This element thus goes as the frequency to the fourth power and gives the scattering of the shear (torsional) wave which is thus seen to depend on the difference in density, not on the stiffness properties of the circle. This is quite natural as the torsional motion is a pure rigid body rotation of the circle without any deformation. Thus, it is to be expected that the scattering is proportional to the difference in mass moment of inertia which for a circle is $\pi a^4(\rho_1 - \rho)/2$.

The purely radial motion is a P wave ($\tau = 2$) and is even with $m = 0$, the corresponding monopole T-matrix element is

$$T^0_{22} = \mathrm{i}(k_\mathrm{p}a)^2 \, \frac{\pi(\lambda + \mu - (\lambda_1 + \mu_1))}{4(\mu + \lambda_1 + \mu_1)} \tag{2.38}$$

This element only depends on the stiffness properties of the circle, not on its density. That there is no dependence on density is natural as there is no net translation or rotation of the circle, this argument also holds for the $m = 2$ (quadrupole) elements below.

For the dipole waves for $m = 1$ the T-matrix elements are

$$\begin{aligned} T^1_{11} &= \mathrm{i}(k_\mathrm{s}a)^2 C_1 \\ T^1_{12} &= T^1_{21} = \mathrm{i}k_\mathrm{p}k_\mathrm{s}a^2 C_1 \\ T^1_{22} &= \mathrm{i}(k_\mathrm{p}a)^2 C_1 \end{aligned} \tag{2.39}$$

where

$$C_1 = \frac{\pi}{8}\left(\frac{\rho_1}{\rho} - 1\right)$$

As the $m = 1$ motion corresponds to a rigid body translation of the circle for low frequencies, it is natural that these elements only depend on the difference in density, more precisely on the mass difference $\pi a^2(\rho_1 - \rho)$, not on the stiffness properties of the circle.

For the quadrupole waves for $m = 2$ the T-matrix elements are

$$\begin{aligned} T_{11}^2 &= \mathrm{i}\left(\frac{k_\mathrm{s}^2 a}{k_\mathrm{p}}\right)^2 C_2 \\ T_{12}^2 &= T_{21}^2 = \mathrm{i}(k_\mathrm{s}a)^2 C_2 \\ T_{22}^2 &= \mathrm{i}(k_\mathrm{p}a)^2 C_2 \end{aligned} \tag{2.40}$$

where

$$C_2 = \frac{\pi(\mu - \mu_1)}{4(\lambda + \mu + \mu_1(3 + \lambda/\mu))}$$

These elements depend almost linearly on the difference in shear modulus, but are independent of λ_1 (and ρ_1). They have no counterpart in acoustics or electromagnetics where quadrupole elements do not contribute to leading order for low frequencies.

2.3 Isotropic cylinder

Having considered the two decoupled 2D scattering problems for a circle it is time to turn to the 3D scattering by an infinite circular cylinder. Wait [142] considers a similar electromagnetic problem.

2.3.1 Wavefunctions

In the 3D case the representation of the displacement field in cylindrical coordinates is [Eq. (1.18)]

$$\boldsymbol{u} = \nabla\Phi + \nabla \times (\hat{\boldsymbol{z}}\Psi_1) + \nabla \times \nabla \times (\hat{\boldsymbol{z}}\Psi_2)$$

where the potentials satisfy scalar Helmholtz equations. In 3D the scalar Helmholtz equation in cylindrical coordinates is

$$\frac{1}{r}\frac{\partial}{\partial r}\left(r\frac{\partial\Phi}{\partial r}\right)+\frac{1}{r^2}\frac{\partial^2\Phi}{\partial\varphi^2}+\frac{\partial^2\Phi}{\partial z^2}+k^2\Phi=0$$

The separation of this equation is performed in a way very similar to the 2D case. Assume a dependence on z as an exponential $\exp(ihz)$, where h is the wavenumber in the z-direction. For a plane wave h has a fixed value, but in general an integration is performed over h (corresponding to a Fourier transform). The remaining equation is then the 2D Helmholtz equation with wavenumber $q=(k^2-h^2)^{1/2}$, where it is noted that q^2 may be negative.

Thus three vector wavefunctions are introduced

$$\begin{aligned}
\boldsymbol{\chi}^0_{1\sigma m}(h,\boldsymbol{r}) &= \sqrt{\frac{\epsilon_m}{8\pi}}\,\frac{1}{q_\mathrm{s}}\,\nabla\times\left[\hat{\boldsymbol{z}}\mathrm{J}_m(q_\mathrm{s}r)\left\{\begin{array}{c}\cos m\varphi\\ \sin m\varphi\end{array}\right\}\mathrm{e}^{\mathrm{i}hz}\right]\\
&= \sqrt{\frac{\epsilon_m}{8\pi}}\,\mathrm{e}^{\mathrm{i}hz}\left[\hat{\boldsymbol{r}}\,\frac{m}{q_\mathrm{s}r}\mathrm{J}_m(q_\mathrm{s}r)\left\{\begin{array}{c}-\sin m\varphi\\ \cos m\varphi\end{array}\right\}\right.\\
&\qquad\left.-\hat{\boldsymbol{\varphi}}\mathrm{J}'_m(q_\mathrm{s}r)\left\{\begin{array}{c}\cos m\varphi\\ \sin m\varphi\end{array}\right\}\right]\\
\boldsymbol{\chi}^0_{2\sigma m}(h,\boldsymbol{r}) &= \sqrt{\frac{\epsilon_m}{8\pi}}\,\frac{1}{k_\mathrm{s}q_\mathrm{s}}\,\nabla\times\nabla\times\left[\hat{\boldsymbol{z}}\mathrm{J}_m(q_\mathrm{s}r)\left\{\begin{array}{c}\cos m\varphi\\ \sin m\varphi\end{array}\right\}\mathrm{e}^{\mathrm{i}hz}\right]\\
&= \sqrt{\frac{\epsilon_m}{8\pi}}\,\mathrm{e}^{\mathrm{i}hz}\left[\hat{\boldsymbol{r}}\,\frac{\mathrm{i}h}{k_\mathrm{s}}\mathrm{J}'_m(q_\mathrm{s}r)\left\{\begin{array}{c}\cos m\varphi\\ \sin m\varphi\end{array}\right\}\right.\\
&\left.+\hat{\boldsymbol{\varphi}}\,\frac{\mathrm{i}mh}{k_\mathrm{s}q_\mathrm{s}r}\mathrm{J}_m(q_\mathrm{s}r)\left\{\begin{array}{c}-\sin m\varphi\\ \cos m\varphi\end{array}\right\}+\hat{\boldsymbol{z}}\,\frac{q_\mathrm{s}}{k_\mathrm{s}}\mathrm{J}_m(q_\mathrm{s}r)\left\{\begin{array}{c}\cos m\varphi\\ \sin m\varphi\end{array}\right\}\right]\\
\boldsymbol{\chi}^0_{3\sigma m}(h,\boldsymbol{r}) &= \sqrt{\frac{\epsilon_m}{8\pi}}\,\frac{k_\mathrm{p}^{1/2}}{k_\mathrm{s}^{3/2}}\,\nabla\left[\mathrm{J}_m(q_\mathrm{p}r)\left\{\begin{array}{c}\cos m\varphi\\ \sin m\varphi\end{array}\right\}\mathrm{e}^{\mathrm{i}hz}\right]\\
&= \sqrt{\frac{\epsilon_m}{8\pi}}\,\frac{k_\mathrm{p}^{1/2}}{k_\mathrm{s}^{3/2}}\,\mathrm{e}^{\mathrm{i}hz}\left[\hat{\boldsymbol{r}}q_\mathrm{p}\mathrm{J}'_m(q_\mathrm{p}r)\left\{\begin{array}{c}\cos m\varphi\\ \sin m\varphi\end{array}\right\}\right.\\
&\left.+\hat{\boldsymbol{\varphi}}\,\frac{m}{r}\mathrm{J}_m(q_\mathrm{p}r)\left\{\begin{array}{c}-\sin m\varphi\\ \cos m\varphi\end{array}\right\}+\hat{\boldsymbol{z}}\,\mathrm{i}h\mathrm{J}_m(q_\mathrm{p}r)\left\{\begin{array}{c}\cos m\varphi\\ \sin m\varphi\end{array}\right\}\right]
\end{aligned}\qquad(2.41)$$

The first index $\tau = 1, 2, 3$ corresponds to SH, SV, and P waves, respectively. It is noted that $\tau = 1, 3$ corresponds to 2D SV and P waves, and there is, therefore, the little strange situation that what is called a 3D SH wave goes into an SV wave when specializing to 2D. The axial wavenumber is h and as in 2D $m = 0, 1, 2, \dots$ and $\sigma = \mathrm{e}, \mathrm{o}$ (with $m = 0$ and $\sigma = \mathrm{o}$ excluded). The upper index 0 means that the wavefunctions are regular containing Bessel functions. The corresponding out-going wavefunctions are denoted with an upper index $+$. The normalization is a little different than in 2D, but the factor $(k_\mathrm{p}/k_\mathrm{s})^{3/2}$ is chosen to give a symmetric T-matrix, and the factor $(\epsilon_m/8\pi)^{1/2}$ is to get a simple expansion of the Green function. An extra index "i" is added to denote the domain when needed, for example $\boldsymbol{\chi}^0_{i,\tau\sigma m}(h, \boldsymbol{r})$.

In the definition of the wavefunctions the wavenumber h in the z-direction is chosen to be the same in all three wavefunctions. With an integration over all real h this is a Fourier transform in the z-coordinate, which is appropriate when considering the scattering by an infinite cylinder. There are then two different radial wavenumbers $q_\mathrm{s} = (k_\mathrm{s}^2 - h^2)^{1/2}$ and $q_\mathrm{p} = (k_\mathrm{p}^2 - h^2)^{1/2}$. The branch of the square root is chosen so that $\operatorname{Im} q_\mathrm{s} \geq 0$ and $\operatorname{Im} q_\mathrm{p} \geq 0$ to ensure the correct behaviour for the out-going waves. In other problems with cylindrical wavefunctions it may be appropriate to take a common radial wavenumber, corresponding to a Hankel transform, and correspondingly different wavenumbers in the z-direction (this is used in Section 10.5).

It is seen that the first two wavefunctions are connected by

$$\nabla \times \boldsymbol{\chi}^0_{1\sigma m}(\boldsymbol{r}) = k_\mathrm{s} \boldsymbol{\chi}^0_{2\sigma m}(\boldsymbol{r})$$

$$\nabla \times \boldsymbol{\chi}^0_{2\sigma m}(\boldsymbol{r}) = k_\mathrm{s} \boldsymbol{\chi}^0_{1\sigma m}(\boldsymbol{r})$$

The radial tractions of the wavefunctions are needed in the boundary conditions and they are [Eqs. (1.21) and (1.22)]

$$\boldsymbol{t}^{(r)}\left(\boldsymbol{\chi}^0_{1\sigma m}(h, \boldsymbol{r})\right) = 2\mu \frac{\partial}{\partial r} \boldsymbol{\chi}^0_{1\sigma m}(h, \boldsymbol{r}) + \mu k_\mathrm{s} \hat{\boldsymbol{r}} \times \boldsymbol{\chi}^0_{2\sigma m}(h, \boldsymbol{r})$$

$$= \mu \sqrt{\frac{\epsilon_m}{8\pi}}\, \mathrm{e}^{\mathrm{i}hz} \left[\hat{\boldsymbol{r}}\, 2m \frac{\mathrm{d}}{\mathrm{d}r} \left(\frac{\mathrm{J}_m(q_\mathrm{s} r)}{q_\mathrm{s} r} \right) \left\{ \begin{array}{c} -\sin m\varphi \\ \cos m\varphi \end{array} \right\} - \hat{\boldsymbol{\varphi}}\, q_\mathrm{s} (2\mathrm{J}''_m(q_\mathrm{s} r) \right.$$

$$\left. + \mathrm{J}_m(q_\mathrm{s} r)) \left\{ \begin{array}{c} \cos m\varphi \\ \sin m\varphi \end{array} \right\} + \hat{\boldsymbol{z}}\, \frac{\mathrm{i}mh}{q_\mathrm{s} r} \mathrm{J}_m(q_\mathrm{s} r) \left\{ \begin{array}{c} -\sin m\varphi \\ \cos m\varphi \end{array} \right\} \right]$$

$$\boldsymbol{t}^{(r)}\left(\boldsymbol{\chi}^0_{2\sigma m}(h,\boldsymbol{r})\right) = 2\mu\frac{\partial}{\partial r}\boldsymbol{\chi}^0_{2\sigma m}(h,\boldsymbol{r}) + \mu k_{\mathrm{s}}\hat{\boldsymbol{r}}\times\boldsymbol{\chi}^0_{1\sigma m}(h,\boldsymbol{r})$$

$$= \mu\sqrt{\frac{\epsilon_m}{8\pi}}\,\mathrm{e}^{\mathrm{i}hz}\left[\hat{\boldsymbol{r}}\,\frac{2\mathrm{i}q_{\mathrm{s}}h}{k_{\mathrm{s}}}\mathrm{J}''_m(q_{\mathrm{s}}r)\left\{\begin{array}{c}\cos m\varphi\\ \sin m\varphi\end{array}\right\}\right.$$

$$+\hat{\boldsymbol{\varphi}}\,\frac{2\mathrm{i}mh}{k_{\mathrm{s}}}\frac{\mathrm{d}}{\mathrm{d}r}\left(\frac{\mathrm{J}_m(q_{\mathrm{s}}r)}{q_{\mathrm{s}}r}\right)\left\{\begin{array}{c}-\sin m\varphi\\ \cos m\varphi\end{array}\right\}$$

$$\left.+\hat{\boldsymbol{z}}\,\frac{2q_{\mathrm{s}}^2-k_{\mathrm{s}}^2}{k_{\mathrm{s}}}\mathrm{J}'_m(q_{\mathrm{s}}r)\left\{\begin{array}{c}\cos m\varphi\\ \sin m\varphi\end{array}\right\}\right]$$

$$\boldsymbol{t}^{(r)}\left(\boldsymbol{\chi}^0_{3\sigma m}(h,\boldsymbol{r})\right) = \hat{\boldsymbol{r}}\mu\left(2k_{\mathrm{p}}^2-k_{\mathrm{s}}^2\right)\Phi^0_{\sigma m}(h,\boldsymbol{r}) + 2\mu\frac{\partial}{\partial r}\boldsymbol{\chi}^0_{3\sigma m}(h,\boldsymbol{r})$$

$$= \mu\sqrt{\frac{\epsilon_m}{8\pi}\,\frac{k_{\mathrm{p}}^{1/2}}{k_{\mathrm{s}}^{3/2}}}\,\mathrm{e}^{\mathrm{i}hz}\left[\hat{\boldsymbol{r}}\left((2k_{\mathrm{p}}^2-k_{\mathrm{s}}^2)\mathrm{J}_m(q_{\mathrm{p}}r)+2q_{\mathrm{p}}^2\mathrm{J}''_m(q_{\mathrm{p}}r)\right)\left\{\begin{array}{c}\cos m\varphi\\ \sin m\varphi\end{array}\right\}\right.$$

$$+\hat{\boldsymbol{\varphi}}\,2mq_{\mathrm{p}}\frac{\mathrm{d}}{\mathrm{d}r}\left(\frac{\mathrm{J}_m(q_{\mathrm{p}}r)}{q_{\mathrm{p}}r}\right)\left\{\begin{array}{c}-\sin m\varphi\\ \cos m\varphi\end{array}\right\}$$

$$\left.+\hat{\boldsymbol{z}}\,2\mathrm{i}q_{\mathrm{p}}h\mathrm{J}'_m(q_{\mathrm{p}}r)\left\{\begin{array}{c}\cos m\varphi\\ \sin m\varphi\end{array}\right\}\right] \tag{2.42}$$

where $\Phi^0_{\sigma m}(h,\boldsymbol{r})$ is the potential of the P wavefunction. The traction of the out-going wavefunctions are of course obtained by changing all Bessel functions to Hankel functions.

As for the 2D P-SV waves the radial dependence of the wavefunction components and their traction components is introduced. They are called $V^0_{j,\tau m}(h,r)$ and $S^0_{j,\tau m}(h,r)$, respectively, where $j=r,\varphi,z$ gives the component and $\tau=1,2,3$ the mode. Note that these are defined for the lower row for $\tau=1$ and the upper row for $\tau=2,3$, thus with a $\cos m\varphi$ behaviour for the r- and z-components and a $\sin m\varphi$ behaviour for the φ-components. Comparing with the 2D case it should be apparent how they are defined, so they are not listed here (however, note that the exponential factor $\exp(\mathrm{i}hz)$ is not included). The corresponding functions for the out-going wavefunctions are similarly defined and are denoted by an upper index $+$ instead of 0.

2.3.2 The T-matrix

Consider now a wave incident on an infinite cylinder of radius a in a homogeneous and isotropic surrounding. The material outside the cylinder has Lamé constants λ and μ and density ρ. Inside the cylinder another isotropic and homogeneous material has Lamé constants λ_1 and μ_1 and density ρ_1. As special cases the interior can be a cavity, a fluid, or a rigid inclusion. The corresponding wavenumbers are $k_\mathrm{p} = \omega(\rho/(\lambda + 2\mu))^{1/2}$ and $k_\mathrm{s} = \omega(\rho/\mu)^{1/2}$ outside the cylinder and $k_{\mathrm{p}1} = \omega(\rho_1/(\lambda_1+2\mu_1))^{1/2}$ and $k_{\mathrm{s}1} = \omega(\rho_1/\mu_1)^{1/2}$ inside the cylinder.

At present the incident wave is not specified, it is only assumed that its sources lie outside the cylinder. Close enough to the cylinder the incident wave can then be expanded in the regular wavefunctions

$$\boldsymbol{u}^{\mathrm{in}}(\boldsymbol{r}) = \sum_{\tau,\sigma,m} \int_{-\infty}^{\infty} b_{\tau\sigma m}(h)\boldsymbol{\chi}^{0}_{\tau\sigma m}(h,\boldsymbol{r})\,\mathrm{d}h \tag{2.43}$$

where the summation is over $\tau = 1, 2, 3$, $\sigma = \mathrm{e}, \mathrm{o}$, and $m = 0, 1, 2, \ldots$ (with $\sigma = \mathrm{o}$ and $m = 0$ excluded). The expansion coefficients $b_{\tau\sigma m}(h)$ are in principle known. The integration in h is a Fourier transform in the z-coordinate. For an incident plane wave there is a single value of the axial wavenumber h, so in this case no integral is needed. The wave scattered by the cylinder is carrying energy outwards (the radiation condition) and is thus possible to expand in the out-going wavefunctions

$$\boldsymbol{u}^{\mathrm{sc}}(\boldsymbol{r}) = \sum_{\tau,\sigma,m} \int_{-\infty}^{\infty} f_{\tau\sigma m}(h)\boldsymbol{\chi}^{+}_{\tau\sigma m}(h,\boldsymbol{r})\,\mathrm{d}h \tag{2.44}$$

Here the expansion coefficients $f_{\tau\sigma m}(h)$ are unknown and the problem is thus to determine these given the incident wave and the properties of the obstacle. This is done in full generality by determining the T-matrix, which is defined as the linear relation between the expansion coefficients of the incident and scattered waves

$$f_{\tau\sigma m}(h) = \sum_{\tau',\sigma',m'} T_{\tau\sigma m,\tau'\sigma' m'}(h) b_{\tau'\sigma' m'}(h) \tag{2.45}$$

The axial wavenumber h acts as a parameter because the cylinder is translationally invariant. For a circular cylindrical obstacle the T-matrix has the following structure

$$T_{\tau\sigma m,\tau'\sigma' m'}(h) = \delta_{mm'}\xi_{\tau\sigma,\tau'\sigma'}T^{m}_{\tau\tau'}(h) \tag{2.46}$$

The T-matrix is thus diagonal in the azimuthal m-index, but there is mode conversion between all three modes $\tau = 1, 2, 3$, so for a fixed m the T-matrix is a three-by-three matrix. With the chosen normalization of the wavefunctions the T-matrix is, furthermore, symmetric, so that $T^{m}_{\tau\tau'}(h) = T^{m}_{\tau'\tau}(h)$. The factor $\xi_{\tau\sigma,\tau'\sigma'}$ is 0, 1, or -1 as follows. Due to symmetry the scattering is divided into two groups. The symmetric group (with respect to the xz-plane) corresponds to the lower row for the azimuthal factor in the SH wavefunction and the upper row for the SV and P wavefunctions and vice versa for the antisymmetric group. Thus $\xi_{\tau\sigma,\tau'\sigma'} = 0$ for values of the indices that couple these two groups. For indices belonging to the symmetric group $\xi_{\tau\sigma,\tau'\sigma'} = 1$ and for the antisymmetric group $\xi_{\tau\sigma,\tau'\sigma'} = 1$, except for the elements coupling SH wavefunctions to SV and P wavefunctions ($\tau = 1$, $\tau' = 2, 3$ and $\tau = 2, 3$, $\tau' = 1$) for which $\xi_{\tau\sigma,\tau'\sigma'} = -1$. In the following the T-matrix for the symmetric group is determined.

To determine the T-matrix the boundary conditions between the surrounding and the obstacle is set up. For simplicity the case with a cavity is first considered. The vanishing of the traction on the cylinder $r = a$ gives the following three equations

$$\sum_{\tau=1}^{3}\left[S^{0}_{j,\tau m}(h,a)b_{\tau\sigma m}(h) + S^{+}_{j,\tau m}(h,a)f_{\tau\sigma m}(h)\right] = 0, \qquad j = r, \varphi, z$$

where $\sigma = \mathrm{o}$ for $\tau = 1$ and $\sigma = \mathrm{e}$ for $\tau = 2, 3$. Solving these equations by expressing the coefficients of the scattered wave $f_{\tau\sigma m}(h)$ in terms of those of the incident wave $b_{\tau\sigma m}(h)$, the T-matrix elements can be read off. For $m = 0$ the SH waves do not couple to the SV and P waves.

If there is a different material inside the cylinder a wave exists there and can be expanded in the regular wavefunctions

$$\boldsymbol{u}^{1}(\boldsymbol{r}) = \sum_{\tau,\sigma,m}\int_{-\infty}^{\infty} c_{\tau\sigma m}(h)\boldsymbol{\chi}^{0}_{1,\tau\sigma m}(h,\boldsymbol{r})\,\mathrm{d}h \tag{2.47}$$

where the first index 1 on the wavefunctions means that the material properties of the cylinder are used. The boundary conditions are continuity of displacement and traction, leading to the following six equations:

$$\sum_{\tau=1}^{3} \left[V^{0}_{j,\tau m}(h,a) b_{\tau\sigma m}(h) + V^{+}_{j,\tau m}(h,a) f_{\tau\sigma m}(h)\right.$$
$$\left. - V^{0}_{j1,\tau m}(h,a) c_{\tau\sigma m}(h)\right] = 0, \qquad j = r, \varphi, z$$

$$\sum_{\tau=1}^{3} \left[S^{0}_{j,\tau m}(h,a) b_{\tau\sigma m}(h) + S^{+}_{j,\tau m}(h,a) f_{\tau\sigma m}(h)\right.$$
$$\left. - S^{0}_{j1,\tau m}(h,a) c_{\tau\sigma m}(h)\right] = 0, \qquad j = r, \varphi, z$$

where $\sigma = \mathrm{o}$ for $\tau = 1$ and $\sigma = \mathrm{e}$ for $\tau = 2, 3$. Solving the equations by expressing the coefficients of the scattered wave $f_{\tau\sigma m}(h)$ in terms of those of the incident wave $b_{\tau\sigma m}(h)$, the T-matrix elements can be read off.

2.3.3 Incident plane waves

An incident plane wave can be of SH, SV, or P type. A plane SH wave propagating in the xz-plane is

$$\boldsymbol{u}^{\mathrm{in}} = \hat{\boldsymbol{y}}\, \mathrm{e}^{\mathrm{i}(q_{\mathrm{s}0} x + h_0 z)} = \frac{\mathrm{i}}{q_{\mathrm{s}0}} \nabla \times \left(\hat{\boldsymbol{z}}\, \mathrm{e}^{\mathrm{i}(q_{\mathrm{s}0} x + h_0 z)}\right)$$

The two wavenumbers h_0 and $q_{\mathrm{s}0}$ are related to the direction of propagation as $h_0 = k_{\mathrm{s}} \cos\theta_0$ and $q_{\mathrm{s}0} = k_{\mathrm{s}} \sin\theta_0$, where θ_0 is the angle between the z-axis and the direction of propagation of the plane wave. Inserting the scalar plane wave expansion Eq. (2.17) it follows that the only nonzero expansion coefficients are

$$b_{1\mathrm{e}m}(h) = \sqrt{8\pi\epsilon_m}\, \mathrm{i}^{m+1} \delta(h - h_0)$$

for $m = 0, 1, 2, \ldots$. Note that the delta function leads to a single value of h_0 so that no integral in h is needed in Eqs. (2.43) and (2.44).

Similarly a plane SV wave is

$$\boldsymbol{u}^{\mathrm{in}} = \frac{1}{k_{\mathrm{s}}} \left(-\hat{\boldsymbol{x}} h_0 + \hat{\boldsymbol{z}} q_{\mathrm{s}0}\right) \mathrm{e}^{\mathrm{i}(q_{\mathrm{s}0} x + h_0 z)} = \frac{1}{k_{\mathrm{s}} q_{\mathrm{s}0}} \nabla \times \left[\nabla \times \left(\hat{\boldsymbol{z}}\, \mathrm{e}^{\mathrm{i}(q_{\mathrm{s}0} x + h_0 z)}\right)\right]$$

so that the only nonzero expansion coefficients are

$$b_{2\mathrm{e}m}(h) = \sqrt{8\pi\epsilon_m}\, \mathrm{i}^{m} \delta(h - h_0)$$

for $m = 0, 1, 2, \ldots$.

An incident P wave in the xz-plane is

$$\boldsymbol{u}^{\text{in}} = \frac{1}{k_{\text{p}}} \left(\hat{\boldsymbol{x}} q_{\text{p}0} + \hat{\boldsymbol{z}} h_0 \right) \mathrm{e}^{\mathrm{i}(q_{\text{p}0} x + h_0 z)} = -\frac{\mathrm{i}}{k_{\text{p}}} \nabla \left(\mathrm{e}^{\mathrm{i}(q_{\text{p}0} x + h_0 z)} \right)$$

so that the only nonzero expansion coefficients are

$$b_{3\text{e}m}(h) = \left(\frac{k_{\text{s}}}{k_{\text{p}}} \right)^{3/2} \sqrt{8\pi\epsilon_m}\, \mathrm{i}^{m-1} \delta(h - h_0)$$

for $m = 0, 1, 2, \ldots$.

Similarly to the 2D cases, far field amplitudes and scattering cross sections can be introduced.

2.4 Transversely isotropic cylinder

The scattering by an isotropic cylindrical obstacle is investigated in the previous section. A very similar approach is possible if the material inside the cylinder is instead transversely isotropic with the symmetry axis along the cylinder axis and this case is now briefly described. This configuration is a special case of the situation where the surrounding medium is also transversely isotropic, see Niklasson and Datta [102]. The scattering by the transversely isotropic cylinder can also be considered by using a somewhat different representation in the transversely isotropic cylinder [64, 49].

For a transversely isotropic material potentials are introduced so that the displacement field is represented as [Eq. (1.25)]

$$\boldsymbol{u} = \nabla \times (\hat{\boldsymbol{z}} \Psi_1) + \hat{\boldsymbol{z}} \frac{\partial \Psi_2}{\partial z} + \nabla_{\perp} \Psi_3$$

where the potentials satisfy Eqs. (1.26). The first term represents pure SH waves which are decoupled from the quasi P-SV waves represented by the last two terms. Assume now that each potential can be taken as an elementary cylindrical wave

$$\mathrm{J}_m(qr) \mathrm{e}^{\mathrm{i}hz} \left\{ \begin{array}{c} \cos m\varphi \\ \sin m\varphi \end{array} \right\}$$

where h is the axial wavenumber and q the radial wavenumber. Insertion into Eqs. (1.26) gives the dispersion relations

$$C_{66}q^2 + C_{44}h^2 - \rho\omega^2 = 0$$

$$C_{11}C_{44}q^4 + \left[\left(C_{11}C_{33} + C_{13}^2 - 2C_{13}C_{44}\right)h^2 - \left(C_{11} + C_{44}\right)\rho\omega^2\right]q^2 + \left[C_{33}C_{44}h^4 - \left(C_{33} + C_{44}\right)\rho\omega^2 h^2 + \rho^2\omega^4\right] = 0$$

This gives three values of q^2. To be definite (and to suite the situation when Hankel functions are used for out-going waves) the three q_j, $j = 1, 2, 3$ (where q_1 is the solution of the first equation) are chosen so that $\operatorname{Im} q_j \geq 0$. For lossless materials this gives q_j that are positive real or positive purely imaginary. Three wavefunctions are introduced

$$\boldsymbol{\zeta}^0_{1\sigma m}(h, \boldsymbol{r}) = \frac{1}{q_1}\nabla \times \left[\hat{\boldsymbol{z}}\mathrm{J}_m(q_1 r)\left\{\begin{array}{c}\cos m\varphi\\ \sin m\varphi\end{array}\right\}\mathrm{e}^{\mathrm{i}hz}\right]$$

$$= \mathrm{e}^{\mathrm{i}hz}\left[\hat{\boldsymbol{r}}\,\frac{m}{q_1 r}\mathrm{J}_m(q_1 r)\left\{\begin{array}{c}-\sin m\varphi\\ \cos m\varphi\end{array}\right\} - \hat{\boldsymbol{\varphi}}\mathrm{J}'_m(q_1 r)\left\{\begin{array}{c}\cos m\varphi\\ \sin m\varphi\end{array}\right\}\right]$$

$$\boldsymbol{\zeta}^0_{\tau\sigma m}(h, \boldsymbol{r}) = \left(\hat{\boldsymbol{z}}\,\frac{1}{h}\frac{\partial}{\partial z}\left[\mathrm{J}_m(q_\tau r)\left\{\begin{array}{c}\cos m\varphi\\ \sin m\varphi\end{array}\right\}\mathrm{e}^{\mathrm{i}hz}\right] - \frac{s_\tau}{q_\tau}\nabla_\perp\left[\mathrm{J}_m(q_\tau r)\left\{\begin{array}{c}\cos m\varphi\\ \sin m\varphi\end{array}\right\}\mathrm{e}^{\mathrm{i}hz}\right]\right)$$

$$= \mathrm{e}^{\mathrm{i}hz}\left[-\hat{\boldsymbol{r}}s_\tau\mathrm{J}'_m(q_\tau r)\left\{\begin{array}{c}\cos m\varphi\\ \sin m\varphi\end{array}\right\} - \hat{\boldsymbol{\varphi}}\,\frac{m s_\tau}{q_\tau r}\mathrm{J}_m(q_\tau r)\left\{\begin{array}{c}-\sin m\varphi\\ \cos m\varphi\end{array}\right\} + \hat{\boldsymbol{z}}\,\mathrm{i}\mathrm{J}_m(q_\tau r)\left\{\begin{array}{c}\cos m\varphi\\ \sin m\varphi\end{array}\right\}\right] \tag{2.48}$$

where

$$s_\tau = \frac{C_{44}q_\tau^2 + C_{33}h^2 - \rho\omega^2}{\left(C_{13} + C_{44}\right)q_\tau h}$$

and $\tau = 2, 3$. As previously $m = 0, 1, 2, \ldots$ and $\sigma = \mathrm{e}, \mathrm{o}$ for the upper and lower row, respectively. The upper index 0 denotes wavefunctions that are regular containing Bessel functions. Corresponding out-going wavefunctions containing Hankel functions can be introduced, but are not needed here. As these wavefunctions are only used in one domain no material index is used for simplicity. They are dimensionless but are otherwise not normalized in any particular way (as the normalization does not affect the T-matrix).

The stresses corresponding to these wavefunctions are calculated from Eqs. (1.24) and (1.11). Collecting these into a radial traction vector the result is

$$\begin{aligned}
\boldsymbol{t}^{(r)}\left(\boldsymbol{\zeta}^{0}_{1\sigma m}(h,\boldsymbol{r})\right) = \mathrm{e}^{ihz}\Bigg[&\hat{\boldsymbol{r}}\,2C_{66}m\frac{\mathrm{d}}{\mathrm{d}r}\left(\frac{\mathrm{J}_m(q_1 r)}{q_1 r}\right)\left\{\begin{array}{c}-\sin m\varphi\\ \cos m\varphi\end{array}\right\}\\
&-\hat{\boldsymbol{\varphi}}\,C_{66}q_1\left[2J''_m(q_1 r)+\mathrm{J}_m(q_1 r)\right]\left\{\begin{array}{c}\cos m\varphi\\ \sin m\varphi\end{array}\right\}\\
&+\hat{\boldsymbol{z}}C_{44}\frac{imh}{q_1 r}\mathrm{J}_m(q_1 r)\left\{\begin{array}{c}-\sin m\varphi\\ \cos m\varphi\end{array}\right\}\Bigg]\\
\boldsymbol{t}^{(r)}\left(\boldsymbol{\zeta}^{0}_{\tau\sigma m}(h,\boldsymbol{r})\right) = \mathrm{e}^{ihz}\Bigg[&\hat{\boldsymbol{r}}\Bigg(-C_{11}q_\tau s_\tau \mathrm{J}''_m(q_\tau r) - C_{12}\frac{s_\tau}{r}\mathrm{J}'_m(q_\tau r)\\
&+\left(C_{12}\frac{m^2 s_\tau}{q_\tau r^2} - C_{13}h\right)\mathrm{J}_m(q_\tau r)\Bigg)\left\{\begin{array}{c}\cos m\varphi\\ \sin m\varphi\end{array}\right\}\\
&+\hat{\boldsymbol{\varphi}}\,2C_{66}ms_\tau\frac{\mathrm{d}}{\mathrm{d}r}\left(\frac{\mathrm{J}_m(q_\tau r)}{q_\tau r}\right)\left\{\begin{array}{c}-\sin m\varphi\\ \cos m\varphi\end{array}\right\}\\
&+\hat{\boldsymbol{z}}\,\mathrm{i}C_{44}(q_\tau - hs_\tau)\mathrm{J}'_m(q_\tau r)\left\{\begin{array}{c}\cos m\varphi\\ \sin m\varphi\end{array}\right\}\Bigg]
\end{aligned} \tag{2.49}$$

where $\tau = 2, 3$.

From here on the scattering problem is treated exactly as for an isotropic cylinder. It is simply to exchange the displacement and traction components from those of the isotropic cylinder to those of the anisotropic cylinder. Therefore no more details are given.

3 Scattering by a sphere

The scattering by a spherical obstacle is a classical problem and it has a long history which is reviewed by Pao and Mow [112], Eringen and Suhubi [48], and Martin [88], where many references to the literature are given. It is remarked that many investigations, even relatively recent ones, do not use the vector spherical harmonics to full advantage, as noticed by Waterman [139], Boström [13], and Norris [103].

To solve the scattering by an isotropic sphere it is best to use the spherical vector wavefunctions, and these are thus introduced in the first section. This involves the solution of the scalar Helmholtz equation in spherical coordinates in terms of spherical Bessel functions and spherical harmonics. Then the vector spherical harmonics are defined and the spherical vector wavefunctions are expressed in terms of these. The scattering problem is solved in much the same way as in the previous chapter with the determination of the transition (T) matrix. The elements of the T-matrix are given explicitly for low frequencies. Finally, more general boundary conditions between the sphere and the surrounding are taken up.

3.1 Vector wavefunctions

In this section the spherical vector wavefunctions are introduced and some of their properties are investigated, see Morse and Feshbach [99, Chapter 13]. To construct the spherical vector wavefunctions the recipe in Eq. (1.19) is used:

$$\boldsymbol{u} = \nabla\Phi + \nabla \times (\boldsymbol{r}\Psi_1) + \nabla \times \nabla \times (\boldsymbol{r}\Psi_2)$$

where Φ satisfies a scalar Helmholtz equation with the P wavenumber and Ψ_1 and Ψ_2 satisfy scalar Helmholtz equations with the S wavenumber. To this end the wavefunctions for the scalar Helmholtz equation in spherical coordinates are first needed.

3.1.1 Scalar wavefunctions

Written out in spherical coordinates the scalar Helmholtz equation is

$$\frac{1}{r^2}\frac{\partial}{\partial r}\left(r^2\frac{\partial\Phi}{\partial r}\right)+\frac{1}{r^2\sin\theta}\frac{\partial}{\partial\theta}\left(\sin\theta\frac{\partial\Phi}{\partial\theta}\right)+\frac{1}{r^2\sin^2\theta}\frac{\partial^2\Phi}{\partial\varphi^2}+k^2\Phi=0$$

Using separation of variables, it is again realized that the azimuthal factor must be a trigonometric function $\cos m\varphi$ or $\sin m\varphi$ with $m = 0, 1, 2, \ldots$. Alternatively, the exponential $\exp(im\varphi)$, $m = 0, \pm 1, \pm 2, \ldots$, can be used, but the trigonometric functions have the advantage of being even or odd. The polar factor must then satisfy

$$\frac{1}{\sin\theta}\frac{\mathrm{d}}{\mathrm{d}\theta}\left(\sin\theta\frac{\mathrm{d}T}{\mathrm{d}\theta}\right)+\left(\lambda-\frac{m^2}{\sin^2\theta}\right)T=0$$

where λ is a separation constant to be determined. Changing variable to $x = \cos\theta$, with $-1 \leq x \leq 1$, this equation becomes

$$\frac{\mathrm{d}}{\mathrm{d}x}\left(\left(1-x^2\right)\frac{\mathrm{d}T}{\mathrm{d}x}\right)+\left(\lambda-\frac{m^2}{(1-x^2)}\right)T=0 \tag{3.1}$$

With $m = 0$ this is Legendre's differential equation. The only solutions to this equation that are finite at the singular points $x = \pm 1$ are the Legendre polynomials $\mathrm{P}_l(x)$, for which the separation constant $\lambda = l(l+1)$. One way to define the Legendre polynomials is by Rodrigues' formula

$$\mathrm{P}_l(x)=\frac{1}{2^l l!}\frac{\mathrm{d}^l}{\mathrm{d}x^l}\left(x^2-1\right)^l \tag{3.2}$$

The Legendre polynomial $\mathrm{P}_l(x)$ is a polynomial of degree l and it is even or odd if l is even or odd, respectively. The first four Legendre polynomials are

$$\mathrm{P}_0(x)=1 \qquad \mathrm{P}_1(x)=x \qquad \mathrm{P}_2(x)=\frac{3}{2}x^2-\frac{1}{2} \qquad \mathrm{P}_3(x)=\frac{5}{2}x^3-\frac{3}{2}x$$

The Legendre polynomials are normalized so that $\mathrm{P}_l(1) = 1$. They constitute a complete orthogonal system on the interval $-1 \leq x \leq 1$ with the orthogonality relation

$$\int_{-1}^{1} \mathrm{P}_l(x)\mathrm{P}_{l'}(x)\,\mathrm{d}x = \frac{2}{2l+1}\,\delta_{ll'} \tag{3.3}$$

They satisfy the following recurrence relations

$$\begin{aligned} &(l+1)\mathrm{P}_{l+1}(x) = (2l+1)x\mathrm{P}_l(x) - l\mathrm{P}_{l-1}(x) \\ &\left(1-x^2\right)\mathrm{P}_l'(x) = l\mathrm{P}_{l-1}(x) - lx\mathrm{P}_l(x) \end{aligned} \tag{3.4}$$

Turning to $m \neq 0$, the solutions to the polar equation are the associated Legendre functions which are defined by

$$\mathrm{P}_l^m(x) = \left(1-x^2\right)^{m/2} \frac{\mathrm{d}^m}{\mathrm{d}x^m}\mathrm{P}_l(x) \tag{3.5}$$

for real $-1 \leq x \leq 1$, $m = 0, 1, 2, \ldots$, and $l = m, m+1, \ldots$. The separation constant is still $\lambda = l(l+1)$. It is seen that $\mathrm{P}_l^0(x) = \mathrm{P}_l(x)$. For fixed m and $l = m, m+1, m+2, \ldots$ the associated Legendre functions constitute a complete orthogonal system on $-1 \leq x \leq 1$ with orthogonality relation

$$\int_{-1}^{1} \mathrm{P}_l^m(x)\mathrm{P}_{l'}^m(x)\,\mathrm{d}x = \frac{2(l+m)!}{(2l+1)(l-m)!}\,\delta_{ll'} \tag{3.6}$$

With fixed order m the associated Legendre functions satisfy the following recurrence relations

$$\begin{aligned} &(l-m+1)\mathrm{P}_{l+1}^m(x) = (2l+1)x\mathrm{P}_l^m(x) - (l+m)\mathrm{P}_{l-1}^m(x) \\ &\left(1-x^2\right)\mathrm{P}_l^{m\prime}(x) = (l+m)\mathrm{P}_{l-1}^m(x) - lx\mathrm{P}_l^m(x) \end{aligned} \tag{3.7}$$

With fixed degree l the following recurrence relations are satisfied (for $m = 1, 2, \ldots$)

$$\begin{aligned} &\frac{2mx}{\sqrt{1-x^2}}\mathrm{P}_l^m(x) = \mathrm{P}_l^{m+1}(x) + (l+m)(l-m+1)\mathrm{P}_l^{m-1}(x) \\ &2\sqrt{1-x^2}\mathrm{P}_l^{m\prime}(x) = \mathrm{P}_l^{m+1}(x) - (l+m)(l-m+1)\mathrm{P}_l^{m-1}(x) \end{aligned} \tag{3.8}$$

A recurrence relation involving both degree and order is

$$(2l+1)\sqrt{1-x^2}\mathrm{P}_l^m(x) = \mathrm{P}_{l+1}^{m+1}(x) - \mathrm{P}_{l-1}^{m+1}(x) \tag{3.9}$$

At this point it is convenient to introduce the real spherical harmonics

$$\mathrm{Y}_{\sigma ml}(\theta,\varphi) = \gamma_{ml}\mathrm{P}_l^m(\cos\theta)\left\{\begin{array}{c}\cos m\varphi\\ \sin m\varphi\end{array}\right\} \tag{3.10}$$

with $l = 0, 1, 2, \ldots, m = 0, 1, \ldots, l$, and $\sigma = \mathrm{e}$ (upper row) or $\sigma = \mathrm{o}$ (lower row). The normalization constant of the spherical harmonics is

$$\gamma_{ml} = \sqrt{\frac{\epsilon_m(2l+1)(l-m)!}{4\pi(l+m)!}} \tag{3.11}$$

The spherical harmonics constitute a complete orthonormal set on the unit sphere $0 \le \theta \le \pi$ and $0 \le \varphi < 2\pi$

$$\int_0^{2\pi}\int_0^{\pi} \mathrm{Y}_{\sigma ml}(\theta,\varphi)\mathrm{Y}_{\sigma' m' l'}(\theta,\varphi)\,\sin\theta\,\mathrm{d}\theta\,\mathrm{d}\varphi = \delta_{\sigma\sigma'}\delta_{mm'}\delta_{ll'} \tag{3.12}$$

A useful equality is the addition theorem for the Legendre polynomials

$$\mathrm{P}_l(\cos\gamma) = \frac{4\pi}{2l+1}\sum_{m=0}^{l}\sum_{\sigma=\mathrm{e,o}} \mathrm{Y}_{\sigma ml}(\theta,\varphi)\mathrm{Y}_{\sigma ml}(\theta',\varphi') \tag{3.13}$$

where γ is the angle between the two directions specified by the spherical angles (θ, φ) and (θ', φ'). This means that

$$\cos\gamma = \sin\theta\sin\theta'\cos(\varphi-\varphi') + \cos\theta\cos\theta'$$

With the separation constant $\lambda = l(l+1)$ the equation for the radial factor in the separation of variables approach becomes

$$\frac{1}{r^2}\frac{\mathrm{d}}{\mathrm{d}r}\left(r^2\frac{\mathrm{d}R}{\mathrm{d}r}\right) + \left(k^2 - \frac{l(l+1)}{r^2}\right)R = 0 \tag{3.14}$$

By the substitution $R(r) = S(r)/\sqrt{r}$ and $z = kr$ this equation is transformed to Bessel's differential equation for $S(z)$ with half-integer order $l + 1/2$. It is convenient to introduce the spherical Bessel functions

$$\mathrm{j}_l(z) = \sqrt{\frac{\pi}{2z}}\mathrm{J}_{l+1/2}(z) \tag{3.15}$$

Spherical Neumann functions $\mathrm{y}_l(z)$ and Hankel functions of the first and second kind $\mathrm{h}_l^{(1)}(z)$ and $\mathrm{h}_l^{(2)}(z)$, respectively, are similarly defined. As for the ordinary Hankel function, the spherical Hankel function of the first kind is henceforth denoted $\mathrm{h}_l(z)$ for ease of notation (especially for derivatives). All these functions can be expressed as a finite sum of elementary functions. The first few are

$$\begin{aligned}
\mathrm{j}_0(z) &= \frac{\sin z}{z} \qquad & \mathrm{j}_1(z) &= \frac{\sin z}{z^2} - \frac{\cos z}{z} \\
\mathrm{y}_0(z) &= -\frac{\cos z}{z} \qquad & \mathrm{y}_1(z) &= -\frac{\cos z}{z^2} - \frac{\sin z}{z} \\
\mathrm{h}_0(z) &= -\frac{\mathrm{i}\mathrm{e}^{\mathrm{i}z}}{z} \qquad & \mathrm{h}_1(z) &= -\frac{\mathrm{i}\mathrm{e}^{\mathrm{i}z}}{z^2} - \frac{\mathrm{e}^{\mathrm{i}z}}{z}
\end{aligned} \tag{3.16}$$

This also directly gives the asymptotic behaviour for large arguments, which for general order is

$$\begin{aligned}
\mathrm{j}_l(z) &\simeq \frac{1}{z}\sin(z - l\pi/2) \\
\mathrm{h}_l(z) &\simeq -\frac{\mathrm{i}}{z}\,\mathrm{e}^{\mathrm{i}(z-l\pi/2)}
\end{aligned} \tag{3.17}$$

The series representations follow from those for the Bessel functions

$$\begin{aligned}
\mathrm{j}_l(z) &= z^l \sum_{j=0}^{\infty} \frac{(-1)^j}{j!(2l+2j+1)!!}\left(\frac{z^2}{2}\right)^j \\
\mathrm{y}_l(z) &= -\frac{1}{z^{l+1}} \sum_{j=0}^{\infty} \frac{(2l-2j-1)!!}{j!}\left(\frac{z^2}{2}\right)^j
\end{aligned}$$

where the semifactorial is $k!! = k(k-2)(k-4)\ldots$. From these the behaviour when $z \to 0$ follows

$$\begin{aligned}
\mathrm{j}_l(z) &\simeq \frac{z^l}{(2l+1)!!} \\
\mathrm{y}_l(z) &\simeq -\frac{(2l-1)!!}{z^{l+1}}
\end{aligned} \tag{3.18}$$

(where $(2l-1)!! = 1$ if $l = 0$). Recurrence relations for the spherical Bessel functions are

$$\frac{2l+1}{z}\mathrm{j}_l(z) = \mathrm{j}_{l-1}(z) + \mathrm{j}_{l+1}(z)$$

$$(2l+1)\mathrm{j}_l'(z) = l\mathrm{j}_{l-1}(z) - (l+1)\mathrm{j}_{l+1}(z) \tag{3.19}$$

which are also valid for the spherical Neumann and Hankel functions. Thus $\mathrm{j}_0'(z) = -\mathrm{j}_1(z)$. A Wronskian relation is

$$\mathrm{j}_l(z)\mathrm{y}_l'(z) - \mathrm{j}_l'(z)\mathrm{y}_l(z) = \frac{1}{z^2} \tag{3.20}$$

Summarizing, the solutions to the scalar Helmholtz equation in spherical coordinates are a product of a spherical Bessel or Hankel function and a spherical harmonic. The spherical harmonic is a product of an associated Legendre function in the polar coordinate θ and a trigonometric function in the azimuthal coordinate φ and the system of spherical harmonics constitute a complete orthonormal system on the unit sphere.

3.1.2 Vector spherical harmonics

When the spherical vector wavefunctions are constructed from the scalar ones, differentiations of the spherical harmonics appear. It is then natural to define the vector spherical harmonics

$$\begin{aligned}
\boldsymbol{A}_{1\sigma ml}(\theta,\varphi) &= \frac{1}{\sqrt{l(l+1)}}\nabla\times(\boldsymbol{r}\mathrm{Y}_{\sigma ml}(\theta,\varphi)) \\
&= \frac{1}{\sqrt{l(l+1)}}\nabla\mathrm{Y}_{\sigma ml}(\theta,\varphi)\times\boldsymbol{r} \\
&= \frac{1}{\sqrt{l(l+1)}}\left(\hat{\boldsymbol{\theta}}\,\frac{1}{\sin\theta}\frac{\partial}{\partial\varphi}\mathrm{Y}_{\sigma ml}(\theta,\varphi) - \hat{\boldsymbol{\varphi}}\,\frac{\partial}{\partial\theta}\mathrm{Y}_{\sigma ml}(\theta,\varphi)\right) \\
\boldsymbol{A}_{2\sigma ml}(\theta,\varphi) &= \frac{1}{\sqrt{l(l+1)}}\,r\nabla\mathrm{Y}_{\sigma ml}(\theta,\varphi) \\
&= \frac{1}{\sqrt{l(l+1)}}\left(\hat{\boldsymbol{\theta}}\,\frac{\partial}{\partial\theta}\mathrm{Y}_{\sigma ml}(\theta,\varphi) + \hat{\boldsymbol{\varphi}}\,\frac{1}{\sin\theta}\frac{\partial}{\partial\varphi}\mathrm{Y}_{\sigma ml}(\theta,\varphi)\right)
\end{aligned}$$

$$\boldsymbol{A}_{3\sigma ml}(\theta,\varphi) = \hat{\boldsymbol{r}}\,\mathrm{Y}_{\sigma ml}(\theta,\varphi) \tag{3.21}$$

Here $\boldsymbol{r} = \hat{\boldsymbol{r}}r$ is the radius vector. The indices are $l = 0, 1, 2, \ldots$, $m = 0, 1, \ldots, l$, $\sigma = \mathrm{e}, \mathrm{o}$ (with $m = 0$ and $\sigma = \mathrm{o}$ excluded as usual). For $l = 0$ only the third vector spherical harmonic is defined. The first index is denoted $\tau\ (= 1, 2, 3)$. The vector spherical harmonics constitute a complete orthonormal vector basis system for vector-valued functions on the unit sphere and thus

$$\int_0^{2\pi}\int_0^{\pi} \boldsymbol{A}_{\tau\sigma ml}(\theta,\varphi)\cdot\boldsymbol{A}_{\tau'\sigma'm'l'}(\theta,\varphi)\,\sin\theta\,\mathrm{d}\theta\,\mathrm{d}\varphi = \delta_{\tau\tau'}\delta_{\sigma\sigma'}\delta_{mm'}\delta_{ll'} \tag{3.22}$$

The first two vector spherical harmonics are both tangent to the unit sphere and

$$\boldsymbol{A}_{2\sigma ml}(\theta,\varphi) = \hat{\boldsymbol{r}}\times\boldsymbol{A}_{1\sigma ml}(\theta,\varphi)$$

$$\boldsymbol{A}_{1\sigma ml}(\theta,\varphi) = -\hat{\boldsymbol{r}}\times\boldsymbol{A}_{2\sigma ml}(\theta,\varphi) \tag{3.23}$$

The divergence and curl of the vector spherical harmonics are obtained as

$$\nabla\cdot\boldsymbol{A}_{1\sigma ml}(\theta,\varphi) = 0$$

$$\nabla\cdot\boldsymbol{A}_{2\sigma ml}(\theta,\varphi) = -\frac{1}{r}\sqrt{l(l+1)}\,\mathrm{Y}_{\sigma ml}(\theta,\varphi)$$

$$\nabla\cdot\boldsymbol{A}_{3\sigma ml}(\theta,\varphi) = \frac{2}{r}\,\mathrm{Y}_{\sigma ml}(\theta,\varphi)$$

$$\nabla\times\boldsymbol{A}_{1\sigma ml}(\theta,\varphi) = \frac{1}{r}\left(\boldsymbol{A}_{2\sigma ml}(\theta,\varphi) + \sqrt{l(l+1)}\boldsymbol{A}_{3\sigma ml}(\theta,\varphi)\right)$$

$$\nabla\times\boldsymbol{A}_{2\sigma ml}(\theta,\varphi) = -\frac{1}{r}\boldsymbol{A}_{1\sigma ml}(\theta,\varphi)$$

$$\nabla\times\boldsymbol{A}_{3\sigma ml}(\theta,\varphi) = \frac{1}{r}\sqrt{l(l+1)}\boldsymbol{A}_{1\sigma ml}(\theta,\varphi) \tag{3.24}$$

3.1.3 Spherical vector wavefunctions

Following the recipe in Eq. (1.19) the following three spherical vector wavefunctions are defined

$$\begin{aligned}
\boldsymbol{\psi}^0_{1\sigma ml}(\boldsymbol{r}) &= \frac{1}{\sqrt{l(l+1)}}\nabla\times(\boldsymbol{r}\mathrm{j}_l(k_\mathrm{s}r)\mathrm{Y}_{\sigma ml}(\theta,\varphi)) \\
&= \mathrm{j}_l(k_\mathrm{s}r)\boldsymbol{A}_{1\sigma ml}(\theta,\varphi) \\
\boldsymbol{\psi}^0_{2\sigma ml}(\boldsymbol{r}) &= \frac{1}{\sqrt{l(l+1)}}\frac{1}{k_\mathrm{s}}\nabla\times\nabla\times(\boldsymbol{r}\mathrm{j}_l(k_\mathrm{s}r)\mathrm{Y}_{\sigma ml}(\theta,\varphi)) \\
&= \left(\mathrm{j}_l'(k_\mathrm{s}r)+\frac{\mathrm{j}_l(k_\mathrm{s}r)}{k_\mathrm{s}r}\right)\boldsymbol{A}_{2\sigma ml}(\theta,\varphi)+\sqrt{l(l+1)}\,\frac{\mathrm{j}_l(k_\mathrm{s}r)}{k_\mathrm{s}r}\boldsymbol{A}_{3\sigma ml}(\theta,\varphi) \\
\boldsymbol{\psi}^0_{3\sigma ml}(\boldsymbol{r}) &= \left(\frac{k_\mathrm{p}}{k_\mathrm{s}}\right)^{3/2}\frac{1}{k_\mathrm{p}}\nabla\left(\mathrm{j}_l(k_\mathrm{p}r)\mathrm{Y}_{\sigma ml}(\theta,\varphi)\right) \\
&= \left(\frac{k_\mathrm{p}}{k_\mathrm{s}}\right)^{3/2}\left(\mathrm{j}_l'(k_\mathrm{p}r)\boldsymbol{A}_{3\sigma ml}(\theta,\varphi)+\sqrt{l(l+1)}\,\frac{\mathrm{j}_l(k_\mathrm{p}r)}{k_\mathrm{p}r}\boldsymbol{A}_{2\sigma ml}(\theta,\varphi)\right)
\end{aligned} \tag{3.25}$$

As for the vector spherical harmonics the indices here run through $l = 0,1,2,\ldots$, $m = 0,1,\ldots,l$, $\sigma = \mathrm{e},\mathrm{o}$. The first index is denoted $\tau = 1,2,3$ for the SH, SV, and P wavefunctions, respectively. For $l = 0$ only the P wavefunction exists, the others are not defined. The normalization is chosen so the wavefunctions are dimensionless and the factor $(k_\mathrm{p}/k_\mathrm{s})^{3/2}$ is inserted to get a symmetric T-matrix (and to later give a simple expansion of the Green tensor). The upper index 0 indicates that the wavefunctions are regular, containing spherical Bessel functions. The corresponding out-going wavefunctions are denoted with the upper index $+$ and contain spherical Hankel functions. If the wavefunctions are needed for a domain where the material properties and the wavenumbers have an index "i", an index "i" is put on the wavefunctions, for example $\boldsymbol{\psi}^0_{i,\tau\sigma ml}(\boldsymbol{r})$.

The wavefunctions have simple properties under the divergence and curl operators

$$\begin{aligned}
\nabla\cdot\boldsymbol{\psi}^0_{1\sigma ml} &= 0, \qquad & \nabla\times\boldsymbol{\psi}^0_{1\sigma ml} &= k_\mathrm{s}\boldsymbol{\psi}^0_{2\sigma ml} \\
\nabla\cdot\boldsymbol{\psi}^0_{2\sigma ml} &= 0, \qquad & \nabla\times\boldsymbol{\psi}^0_{2\sigma ml} &= k_\mathrm{s}\boldsymbol{\psi}^0_{1\sigma ml}
\end{aligned}$$

$$\nabla \cdot \boldsymbol{\psi}^0_{3\sigma ml} = -\left(k_\mathrm{p}/k_\mathrm{s}\right)^{3/2} k_\mathrm{p}\mathrm{j}_l(k_\mathrm{p}r)\mathrm{Y}_{\sigma ml}, \qquad \nabla \times \boldsymbol{\psi}^0_{3\sigma ml} = 0 \quad (3.26)$$

These properties facilitate the calculation of the traction of the wavefunctions.

The radial tractions of the wavefunctions are needed in the boundary conditions and they become [using Eqs. (1.21)–(1.22)]

$$\begin{aligned}
\boldsymbol{t}^{(r)}\left(\boldsymbol{\psi}^0_{1\sigma ml}(\boldsymbol{r})\right) &= 2\mu\frac{\partial}{\partial r}\boldsymbol{\psi}^0_{1\sigma ml}(\boldsymbol{r}) + \mu k_\mathrm{s}\hat{\boldsymbol{r}} \times \boldsymbol{\psi}^0_{2\sigma ml}(\boldsymbol{r}) \\
&= \mu r\frac{\mathrm{d}}{\mathrm{d}r}\left(\frac{\mathrm{j}_l(k_\mathrm{s}r)}{r}\right)\boldsymbol{A}_{1\sigma ml}(\theta,\varphi) \\
\boldsymbol{t}^{(r)}\left(\boldsymbol{\psi}^0_{2\sigma ml}(\boldsymbol{r})\right) &= 2\mu\frac{\partial}{\partial r}\boldsymbol{\psi}^0_{2\sigma ml}(\boldsymbol{r}) + \mu k_\mathrm{s}\hat{\boldsymbol{r}} \times \boldsymbol{\psi}^0_{1\sigma ml}(\boldsymbol{r}) \\
= \mu\Bigg[\bigg(2k_\mathrm{s}\mathrm{j}''_l(k_\mathrm{s}r) &+ \frac{2\mathrm{j}'_l(k_\mathrm{s}r)}{r} - \frac{2\mathrm{j}_l(k_\mathrm{s}r)}{k_\mathrm{s}r^2} + k_\mathrm{s}\mathrm{j}_l(k_\mathrm{s}r)\bigg)\boldsymbol{A}_{2\sigma ml}(\theta,\varphi) \\
&+2\sqrt{l(l+1)}\frac{\mathrm{d}}{\mathrm{d}r}\left(\frac{\mathrm{j}_l(k_\mathrm{s}r)}{k_\mathrm{s}r}\right)\boldsymbol{A}_{3\sigma ml}(\theta,\varphi)\Bigg] \\
\boldsymbol{t}^{(r)}\left(\boldsymbol{\psi}^0_{3\sigma ml}(\boldsymbol{r})\right) &= \hat{\boldsymbol{r}}\mu\left(k_\mathrm{s}^2/k_\mathrm{p}^2 - 2\right)\nabla \cdot \boldsymbol{\psi}^0_{3\sigma ml}(\boldsymbol{r}) + 2\mu\frac{\partial}{\partial r}\boldsymbol{\psi}^0_{3\sigma ml}(\boldsymbol{r}) \\
= \mu\left(\frac{k_\mathrm{p}}{k_\mathrm{s}}\right)^{3/2}\Bigg[\bigg(2k_\mathrm{p}\mathrm{j}''_l(k_\mathrm{p}r) &+ \frac{2k_\mathrm{p}^2 - k_\mathrm{s}^2}{k_\mathrm{p}}\mathrm{j}_l(k_\mathrm{p}r)\bigg)\boldsymbol{A}_{3\sigma ml}(\theta,\varphi) \\
&+2\sqrt{l(l+1)}\frac{\mathrm{d}}{\mathrm{d}r}\left(\frac{\mathrm{j}_l(k_\mathrm{p}r)}{k_\mathrm{p}r}\right)\boldsymbol{A}_{2\sigma ml}(\theta,\varphi)\Bigg] \quad (3.27)
\end{aligned}$$

If the tractions are needed for a domain where the material properties and the wavenumbers have an index "i", an index "i" is put on the traction operator, for example $\boldsymbol{t}_i^{(r)}\left(\boldsymbol{\psi}^0_{i,\tau\sigma ml}(\boldsymbol{r})\right)$. The tractions of the out-going wavefunctions are of course obtained by changing all Bessel functions to Hankel functions.

It is convenient to introduce the nonzero (vector spherical harmonics) components $V^0_{\tau',\tau l}$ of the wavefunctions (where τ' specifies the vector spherical harmonic component and τ the vector wavefunction)

$$V^0_{1,1l}(r) = \mathrm{j}_l(k_\mathrm{s}r)$$

$$V^0_{2,2l}(r) = \mathrm{j}'_l(k_\mathrm{s}r) + \frac{\mathrm{j}_l(k_\mathrm{s}r)}{k_\mathrm{s}r}$$

$$V^0_{3,2l}(r) = \sqrt{l(l+1)}\,\frac{\mathrm{j}_l(k_\mathrm{s}r)}{k_\mathrm{s}r}$$

$$V^0_{2,3l}(r) = \left(\frac{k_\mathrm{p}}{k_\mathrm{s}}\right)^{3/2}\sqrt{l(l+1)}\,\frac{\mathrm{j}_l(k_\mathrm{p}r)}{k_\mathrm{p}r}$$

$$V^0_{3,3l}(r) = \left(\frac{k_\mathrm{p}}{k_\mathrm{s}}\right)^{3/2}\mathrm{j}_l'(k_\mathrm{p}r) \tag{3.28}$$

Likewise, the nonzero components of the tractions of the wavefunctions are introduced

$$S^0_{1,1l}(r) = \mu r\frac{\mathrm{d}}{\mathrm{d}r}\left(\frac{\mathrm{j}_l(k_\mathrm{s}r)}{r}\right)$$

$$S^0_{2,2l}(r) = \mu\left(2k_\mathrm{s}\mathrm{j}_l''(k_\mathrm{s}r) + \frac{2\mathrm{j}_l'(k_\mathrm{s}r)}{r} - \frac{2\mathrm{j}_l(k_\mathrm{s}r)}{k_\mathrm{s}r^2} + k_\mathrm{s}\mathrm{j}_l(k_\mathrm{s}r)\right)$$

$$S^0_{3,2l}(r) = 2\mu\sqrt{l(l+1)}\,\frac{\mathrm{d}}{\mathrm{d}r}\left(\frac{\mathrm{j}_l(k_\mathrm{s}r)}{k_\mathrm{s}r}\right)$$

$$S^0_{2,3l}(r) = 2\mu\sqrt{l(l+1)}\left(\frac{k_\mathrm{p}}{k_\mathrm{s}}\right)^{3/2}\frac{\mathrm{d}}{\mathrm{d}r}\left(\frac{\mathrm{j}_l(k_\mathrm{p}r)}{k_\mathrm{p}r}\right)$$

$$S^0_{3,3l}(r) = \mu\left(\frac{k_\mathrm{p}}{k_\mathrm{s}}\right)^{3/2}\left(2k_\mathrm{p}\mathrm{j}_l''(k_\mathrm{p}r) + \frac{2k_\mathrm{p}^2 - k_\mathrm{s}^2}{k_\mathrm{p}}\mathrm{j}_l(k_\mathrm{p}r)\right) \tag{3.29}$$

It is noted that components of the wavefunctions and their tractions are independent of the order m of the spherical harmonics. If needed, a domain index "i" is added, for example $V^0_{\tau' i,\tau l}(r)$. The corresponding functions with the spherical Bessel functions replaced by spherical Hankel functions are denoted $V^+_{\tau',\tau l}(r)$ and $S^+_{\tau',\tau l}(r)$.

It is also useful to have the asymptotic form of the out-going wavefunctions and their radial tractions for large r. Using the asymptotic form of the Hankel functions for large arguments the wavefunction components become to leading order (including only terms that behave as r^{-1})

$$V^+_{1,1l}(r) \simeq -\frac{\mathrm{i}}{k_\mathrm{s}r}\,\mathrm{e}^{\mathrm{i}(k_\mathrm{s}r - l\pi/2)}$$

$$V^+_{2,2l}(r) \simeq \frac{1}{k_\mathrm{s}r}\,\mathrm{e}^{\mathrm{i}(k_\mathrm{s}r - l\pi/2)}$$

$$
\begin{aligned}
V^{+}_{3,2l}(r) &\simeq 0 \\
V^{+}_{2,3l}(r) &\simeq 0 \\
V^{+}_{3,3l}(r) &\simeq \left(\frac{k_{\mathrm{p}}}{k_{\mathrm{s}}}\right)^{3/2} \frac{1}{k_{\mathrm{p}} r}\, \mathrm{e}^{\mathrm{i}(k_{\mathrm{p}} r - l\pi/2)}
\end{aligned} \tag{3.30}
$$

The tractions are likewise

$$
\begin{aligned}
S^{+}_{1,1l}(r) &\simeq \frac{\mu}{r}\, \mathrm{e}^{\mathrm{i}(k_{\mathrm{s}} r - l\pi/2)} \\
S^{+}_{2,2l}(r) &\simeq \frac{\mathrm{i}\mu}{r}\, \mathrm{e}^{\mathrm{i}(k_{\mathrm{s}} r - l\pi/2)} \\
S^{+}_{3,2l}(r) &\simeq 0 \\
S^{+}_{2,3l}(r) &\simeq 0 \\
S^{+}_{3,3l}(r) &\simeq \left(\frac{k_{\mathrm{p}}}{k_{\mathrm{s}}}\right)^{3/2} \frac{\mathrm{i}\mu k_{\mathrm{s}}^2}{k_{\mathrm{p}}^2 r}\, \mathrm{e}^{\mathrm{i}(k_{\mathrm{p}} r - l\pi/2)}
\end{aligned} \tag{3.31}
$$

3.1.4 Orthogonality of the vector wavefunctions

The spherical vector wavefunctions satisfy certain orthogonality relations, see Pao [111]. The starting point is the reciprocity relation derived in Eq. (1.9)

$$
\int_S \hat{\boldsymbol{n}} \cdot \left(\boldsymbol{\sigma}^{\mathrm{A}} \cdot \boldsymbol{u}^{\mathrm{B}} - \boldsymbol{\sigma}^{\mathrm{B}} \cdot \boldsymbol{u}^{\mathrm{A}}\right) \mathrm{d}S = \int_V \rho \left(\boldsymbol{u}^{\mathrm{A}} \cdot \boldsymbol{f}^{\mathrm{B}} - \boldsymbol{u}^{\mathrm{B}} \cdot \boldsymbol{f}^{\mathrm{A}}\right) \mathrm{d}V
$$

The two states A and B are now chosen as a regular or out-going vector wavefunction. These wavefunctions have no sources except at the origin for the out-going wavefunctions and at infinity for the regular ones. Let the surface S be composed of two surfaces, one that may be any closed surface with the origin as an interior point, and one that is a large sphere with radius tending to infinity, both with outward-pointing normal. The volume V is then the domain between these two surfaces. As there are no sources in V the integrals over both surfaces are the same. On the large sphere the asymptotic forms of the wavefunctions and their tractions [Eqs. (3.30) and (3.31)] can be used. As the vector spherical harmonics are orthonormal the integral over the large sphere vanishes except when the two wavefunctions have identical indices (except that $\tau = 2$ and $\tau' = 3$ couple; but then the radial dependence gives a vanishing result). The radial dependence can be calculated from the asymptotic form and the final result is the three orthogonality relations

$$\int_S \Big[\boldsymbol{t}^{(n)} \left(\boldsymbol{\psi}^0_{\tau\sigma ml}(\boldsymbol{r}) \right) \cdot \boldsymbol{\psi}^0_{\tau'\sigma' m' l'}(\boldsymbol{r}) \\ - \boldsymbol{t}^{(n)} \left(\boldsymbol{\psi}^0_{\tau'\sigma' m' l'}(\boldsymbol{r}) \right) \cdot \boldsymbol{\psi}^0_{\tau\sigma ml}(\boldsymbol{r}) \Big] \, \mathrm{d}S = 0$$

$$\int_S \Big[\boldsymbol{t}^{(n)} \left(\boldsymbol{\psi}^+_{\tau\sigma ml}(\boldsymbol{r}) \right) \cdot \boldsymbol{\psi}^+_{\tau'\sigma' m' l'}(\boldsymbol{r}) \\ - \boldsymbol{t}^{(n)} \left(\boldsymbol{\psi}^+_{\tau'\sigma' m' l'}(\boldsymbol{r}) \right) \cdot \boldsymbol{\psi}^+_{\tau\sigma ml}(\boldsymbol{r}) \Big] \, \mathrm{d}S = 0$$

$$\int_S \Big[\boldsymbol{t}^{(n)} \left(\boldsymbol{\psi}^+_{\tau\sigma ml}(\boldsymbol{r}) \right) \cdot \boldsymbol{\psi}^0_{\tau'\sigma' m' l'}(\boldsymbol{r}) \\ - \boldsymbol{t}^{(n)} \left(\boldsymbol{\psi}^0_{\tau'\sigma' m' l'}(\boldsymbol{r}) \right) \cdot \boldsymbol{\psi}^+_{\tau\sigma ml}(\boldsymbol{r}) \Big] \, \mathrm{d}S = \frac{\mathrm{i}\mu}{k_\mathrm{s}} \delta_{\tau\tau'} \delta_{\sigma\sigma'} \delta_{mm'} \delta_{ll'} \tag{3.32}$$

The final result is of course independent of the radius of the large sphere. The surface S can now be any closed surface surrounding the origin. These orthogonality relations are often called Betti identities.

3.2 Scattering problem

With the spherical vector wavefunctions defined it is time to turn to the scattering problem for a sphere, and in particular to determine the T-matrix. Also incident and scattered waves are further investigated and the T-matrix is given explicitly for low frequencies.

3.2.1 The T-matrix

Consider the scattering of an incident wave by a spherical obstacle of radius a. The surrounding medium has density ρ and Lamé constants λ and μ and the corresponding wavenumbers are k_s and k_p. The material inside the sphere is another elastic material with corresponding constants ρ_1, λ_1, μ_1, $k_{\mathrm{s}1}$, and $k_{\mathrm{p}1}$. As special cases the inside may be a fluid or a cavity.

The incident wave is not specified at present, it is only required that its sources are located outside the spherical obstacle. Close enough to the obstacle the incident wave admits an expansion in the regular wavefunctions

$$\boldsymbol{u}^{\mathrm{in}}(\boldsymbol{r}) = \sum_{\tau,\sigma,m,l} b_{\tau\sigma ml}\boldsymbol{\psi}^{0}_{\tau\sigma ml}(\boldsymbol{r}) \tag{3.33}$$

where the summation is over $\tau = 1,2,3$, $\sigma = \mathrm{e},\mathrm{o}$, $m = 0,1,\ldots,l$, and $l = 0,1,2,\ldots$ (with $l = 0$ excluded for $\tau = 1,2$). The expansion coefficients $b_{\tau\sigma ml}$ are in principle known. The scattered wave is carrying energy outwards from the obstacle (the radiation condition) and is thus possible to expand in the out-going wavefunctions

$$\boldsymbol{u}^{\mathrm{sc}}(\boldsymbol{r}) = \sum_{\tau,\sigma,m,l} f_{\tau\sigma ml}\boldsymbol{\psi}^{+}_{\tau\sigma ml}(\boldsymbol{r}) \tag{3.34}$$

The expansion coefficients $f_{\tau\sigma ml}$ are to be determined. This is done in full generality by determining the T-matrix, which is defined as the linear relation between the expansion coefficients of the incident and scattered waves

$$f_{\tau\sigma ml} = \sum_{\tau',\sigma',m',l'} T_{\tau\sigma ml,\tau'\sigma'm'l'}b_{\tau'\sigma'm'l'} \tag{3.35}$$

For a spherical obstacle the T-matrix has the structure

$$T_{\tau\sigma ml,\tau'\sigma'm'l'} = \delta_{ll'}\delta_{mm'}\delta_{\sigma\sigma'}T^{l}_{\tau\tau'} \tag{3.36}$$

The T-matrix is thus diagonal in all indices except the mode index τ and is, furthermore, independent of the order m (and the parity σ). As the first wavefunction and its traction only depend on the first vector spherical harmonic and the second and third wavefunctions and their tractions only depend on the second and third vector spherical harmonics, it follows that there is no coupling between the first (SH) wavefunction and the other two and thus $T^{l}_{12} = T^{l}_{21} = T^{l}_{13} = T^{l}_{31} = 0$. With the chosen normalization the T-matrix is symmetric so that $T^{l}_{32} = T^{l}_{23}$. All these properties of the T-matrix are, in fact, true for all spherically symmetric obstacles (even for inhomogeneous ones).

Inside the obstacle waves are also set up and these can be expanded in the regular wavefunctions

$$\boldsymbol{u}^{1}(\boldsymbol{r}) = \sum_{\tau,\sigma,m,l} c_{\tau\sigma ml}\boldsymbol{\psi}^{0}_{1,\tau\sigma ml}(\boldsymbol{r}) \tag{3.37}$$

where the first index 1 on the wavefunctions means that the material properties of the obstacle are used.

The boundary conditions between the obstacle and the surrounding are continuity of displacement and traction. For the decoupled SH waves ($\tau = 1$) this gives

$$V^0_{1,1l}(a)b_{1\sigma ml} + V^+_{1,1l}(a)f_{1\sigma ml} - V^0_{11,1l}(a)c_{1\sigma ml} = 0$$

$$S^0_{1,1l}(a)b_{1\sigma ml} + S^+_{1,1l}(a)f_{1\sigma ml} - S^0_{11,1l}(a)c_{1\sigma ml} = 0$$

Solving these two equations for $f_{1\sigma ml}$ in terms of $b_{1\sigma ml}$ the following T-matrix element is obtained

$$T^l_{11} = \frac{-\mu \mathrm{j}_l(k_{\mathrm{s}1}a)\left[k_{\mathrm{s}}a\mathrm{j}_l(k_{\mathrm{s}}a) - \mathrm{j}'_l(k_{\mathrm{s}}a)\right] + \mu_1 \mathrm{j}_l(k_{\mathrm{s}}a)\left[k_{\mathrm{s}1}a\mathrm{j}_l(k_{\mathrm{s}1}a) - \mathrm{j}'_l(k_{\mathrm{s}1}a)\right]}{\mu \mathrm{j}_l(k_{\mathrm{s}1}a)\left[k_{\mathrm{s}}a\mathrm{h}_l(k_{\mathrm{s}}a) - \mathrm{h}'_l(k_{\mathrm{s}}a)\right] - \mu_1 \mathrm{h}_l(k_{\mathrm{s}}a)\left[k_{\mathrm{s}1}a\mathrm{j}_l(k_{\mathrm{s}1}a) - \mathrm{j}'_l(k_{\mathrm{s}1}a)\right]}$$

For the coupled P-SV waves the boundary conditions give

$$\sum_{\tau'=2,3}\left[V^0_{\tau,\tau' l}(a)b_{\tau'\sigma ml} + V^+_{\tau,\tau' l}(a)f_{\tau'\sigma ml} - V^0_{\tau 1,\tau' l}(a)c_{\tau'\sigma ml}\right] = 0$$

$$\sum_{\tau'=2,3}\left[S^0_{\tau,\tau' l}(a)b_{\tau'\sigma ml} + S^+_{\tau,\tau' l}(a)f_{\tau'\sigma ml} - S^0_{\tau 1,\tau' l}(a)c_{\tau'\sigma ml}\right] = 0$$

for $\tau = 2, 3$. The expressions for the T-matrix elements T^l_{22}, $T^l_{23} = T^l_{32}$, and T^l_{33} are obtained from these four equations (solving for $f_{2\sigma ml}$ and $f_{3\sigma ml}$ in terms of $b_{2\sigma ml}$ and $b_{3\sigma ml}$).

The scattering by a spherical cavity and a fluid-filled sphere can be obtained as special cases, taking the appropriate limits, but more straightforward is to use the correct boundary conditions directly, thereby making the system of equations smaller. It should be clear how this is done so no more details are given.

3.2.2 Incident waves

To get expressions for incident plane waves in an elastic solid it is first necessary to consider the corresponding scalar problem. This starts with the problem of waves in a fluid that are excited by a point source. With the source at the z-axis at $z = -d$ the scalar Helmholtz equation becomes

$$\nabla^2 u + k^2 u = -A\delta(x)\delta(y)\delta(z+d)$$

where A is a measure of the strength of the source and a product of three delta functions appears. To solve this equation it is convenient to use spherical coordinates centred at the source, call them (r', θ', φ'). The problem is then spherically symmetric so outside $r' = 0$ the Helmholtz equation becomes

$$\frac{1}{(r')^2}\frac{\mathrm{d}}{\mathrm{d}r'}\left((r')^2\frac{\mathrm{d}u}{\mathrm{d}r'}\right) + k^2 u = 0$$

This is the differential equation for spherical Bessel functions of order 0 and as the solution must satisfy a radiation condition it is (essentially the zero order spherical Hankel function)

$$u = C\,\frac{\mathrm{e}^{ikr'}}{kr'}$$

where C is a constant to be determined. Integrate the differential equation over a small spherical volume of radius ϵ around $r' = 0$

$$\int_{r'<\epsilon}\left(\frac{1}{(r')^2}\frac{\mathrm{d}}{\mathrm{d}r'}\left((r')^2\frac{\mathrm{d}u}{\mathrm{d}r'}\right) + k^2 u\right)(r')^2 \sin\theta'\,\mathrm{d}r'\,\mathrm{d}\theta'\,\mathrm{d}\varphi' = -A$$

The second term vanishes in the limit $\epsilon \to 0$. Using Gauss law to convert the volume integral into a surface integral over the sphere with radius ϵ gives

$$\int_0^{2\pi}\int_0^{\pi}\left[(r')^2\frac{\mathrm{d}u}{\mathrm{d}r'}\right]_{r'=\epsilon}\sin\theta'\,\mathrm{d}\theta'\,\mathrm{d}\varphi' = -A$$

As u is independent of θ' and φ' this gives

$$4\pi\epsilon^2\left[\frac{\mathrm{d}u}{\mathrm{d}r'}\right]_{r'=\epsilon} = -A$$

Inserting the explicit form of u leads to

$$C = \frac{kA}{4\pi}$$

The wave from a point source in an acoustic medium thus has the very simple form

$$u(r') = \frac{A\mathrm{e}^{\mathrm{i}kr'}}{4\pi r'} \tag{3.38}$$

To be useful in the present context the wave from the point source should be expanded in the wavefunctions centred at the obstacle. As the source is situated on the z-axis the problem is rotationally symmetric about this axis and is thus independent of φ. The incident wave can then be expanded in the Legendre polynomials

$$u(r,\theta) = \sum_{l=0}^{\infty} u_l(r)\mathrm{P}_l(\cos\theta)$$

When inserting into the Helmholtz equation the delta functions are transformed to spherical coordinates

$$\delta(x)\delta(y)\delta(z+d) = \frac{1}{r^2\sin\theta}\delta(r-d)\delta(\theta-\pi)\delta(\varphi-\varphi_0)$$

where $\theta = \pi$ must be regarded as an interior point for the θ interval and the value of φ_0 does not matter. Multiplying the Helmholtz equation with $\mathrm{P}_{l'}(\cos\theta)$ and integrating in θ and φ over the unit sphere then gives

$$\frac{1}{r^2}\frac{\mathrm{d}}{\mathrm{d}r}\left(r^2\frac{\mathrm{d}u_l}{\mathrm{d}r}\right) + \left(k^2 - \frac{l(l+1)}{r^2}\right)u_l = -\frac{A(2l+1)}{4\pi d^2}\delta(r+d)$$

The solution should satisfy no boundary condition in the ordinary sense; instead the solution must be bounded for $r = 0$ and it must be out-going for large r. Thus it has the form

$$u_l(r) = \begin{cases} A_l\mathrm{j}_l(kr), & r < d \\ B_l\mathrm{h}_l(kr), & r > d \end{cases}$$

The constants A_l and B_l must be determined from conditions at $r = d$. Integrating the equation for u_l over a small interval around $r = d$ it is seen that the derivative of u_l must have a jump $-A(2l+1)/(4\pi d^2)$ and that u_l must be continuous

$$A_l\mathrm{j}_l(kd) = B_l\mathrm{h}_l(kd)$$

$$k\left(B_l \mathrm{h}_l'(kd) - A_l \mathrm{j}_l'(kd)\right) = -\frac{A(2l+1)}{4\pi d^2}$$

The determinant of this system can be calculated using a Wronskian relation [essentially Eq. (3.20)] and this gives

$$A_l = \frac{\mathrm{i}Ak(2l+1)}{4\pi}\mathrm{h}_l(kd)$$

$$B_l = \frac{\mathrm{i}Ak(2l+1)}{4\pi}\mathrm{j}_l(kd)$$

For an incident wave from a point source the expansion thus is

$$u(r,\theta) = \mathrm{i}Ak\sum_{l=0}^{\infty}(-1)^l\frac{2l+1}{4\pi}\mathrm{j}_l(kr_<)\mathrm{h}_l(kr_>)\mathrm{P}_l(\cos\theta) \tag{3.39}$$

where $r_< = \min(r,d)$ and $r_> = \max(r,d)$.

To get the expansion of a plane wave, the two expressions Eqs. (3.38) and (3.39) are put equal, the distance d to the point source is pushed far away, and the asymptotic form of the Hankel function is used to get

$$\mathrm{e}^{\mathrm{i}kz} = \sum_{l=0}^{\infty}\mathrm{i}^l(2l+1)\mathrm{j}_l(kr)\mathrm{P}_l(\cos\theta) \tag{3.40}$$

With the scalar problem finished, it is straightforward to obtain the expansion of a plane P wave. It is just to take the gradient of Eq. (3.40)

$$\hat{\boldsymbol{z}}\,\mathrm{e}^{\mathrm{i}k_\mathrm{p}z} = -\frac{\mathrm{i}}{k_\mathrm{p}}\nabla\left(\mathrm{e}^{\mathrm{i}k_\mathrm{p}z}\right) = -\frac{\mathrm{i}}{k_\mathrm{p}}\nabla\left(\sum_{l=0}^{\infty}\mathrm{i}^l(2l+1)\mathrm{j}_l(kr)\mathrm{P}_l(\cos\theta)\right)$$

from which it follows that the only nonzero expansion coefficients for a plane P wave propagating in the z-direction are

$$b_{3\mathrm{e}0l} = \mathrm{i}^{l-1}\sqrt{4\pi(2l+1)}\left(\frac{k_\mathrm{s}}{k_\mathrm{p}}\right)^{3/2} \tag{3.41}$$

for $l = 0, 1, 2, \ldots$.

For a plane S wave the situation is a little more complicated. Without loss of generality the plane S wave can be taken as

$$\boldsymbol{u}^{\mathrm{in}} = \hat{\boldsymbol{x}}\,\mathrm{e}^{\mathrm{i}k_\mathrm{s}z}$$

As the unit vector in the x-direction is

$$\hat{\boldsymbol{x}} = \hat{\boldsymbol{r}} \sin\theta \cos\varphi + \hat{\boldsymbol{\theta}} \cos\theta \cos\varphi - \hat{\boldsymbol{\varphi}} \sin\varphi$$

it follows that the expansion only includes the azimuthal order $m = 1$ and must be of the form

$$\hat{\boldsymbol{x}}\, \mathrm{e}^{\mathrm{i}k_\mathrm{s}z} = \sum_{l=1}^{\infty} \left(b_{1\mathrm{o}1l} \boldsymbol{\psi}^0_{1\mathrm{o}1l} + b_{2\mathrm{e}1l} \boldsymbol{\psi}^0_{2\mathrm{e}1l} \right)$$

$$= \sum_{l=1}^{\infty} \Bigg(b_{1\mathrm{o}1l} \mathrm{j}_l(k_\mathrm{s}r) \boldsymbol{A}_{1\mathrm{o}1l}(\theta,\varphi) + b_{2\mathrm{e}1l} \Bigg[\left(\mathrm{j}'_l(k_\mathrm{s}r) + \frac{\mathrm{j}_l(k_\mathrm{s}r)}{k_\mathrm{s}r} \right) \boldsymbol{A}_{2\mathrm{e}1l}(\theta,\varphi)$$
$$+ \sqrt{l(l+1)}\, \frac{\mathrm{j}_l(k_\mathrm{s}r)}{k_\mathrm{s}r} \boldsymbol{A}_{3\mathrm{e}1l}(\theta,\varphi) \Bigg] \Bigg)$$

Taking the scalar product with the third vector spherical harmonic and integrating over the unit sphere this gives [inserting also the plane wave expansion Eq. (3.40)]

$$b_{2\mathrm{e}1l} \sqrt{l(l+1)}\, \frac{\mathrm{j}_l(k_\mathrm{s}r)}{k_\mathrm{s}r} = \int_0^{2\pi} \int_0^{\pi} \hat{\boldsymbol{x}} \cdot \boldsymbol{A}_{3\mathrm{e}1l}(\theta,\varphi) \mathrm{e}^{\mathrm{i}k_\mathrm{s}z}\, \sin\theta\, \mathrm{d}\theta\, \mathrm{d}\varphi$$

$$= \pi \int_0^{\pi} \sqrt{\frac{(2l+1)(l-1)!}{2\pi(l+1)!}} \mathrm{P}_l^1(\cos\theta)$$
$$\times \sum_{n=0}^{\infty} \mathrm{i}^n (2n+1) \mathrm{P}_n(\cos\theta) \mathrm{j}_n(k_\mathrm{s}r) \sin^2\theta\, \mathrm{d}\theta$$

$$= \pi \int_0^{\pi} \sqrt{\frac{(2l+1)(l-1)!}{2\pi(l+1)!}} \mathrm{P}_l^1(\cos\theta)$$
$$\times \sum_{n=0}^{\infty} \mathrm{i}^n \left(\mathrm{P}^1_{n+1}(\cos\theta) - \mathrm{P}^1_{n-1}(\cos\theta) \right) \mathrm{j}_n(k_\mathrm{s}r) \sin\theta\, \mathrm{d}\theta$$

$$= \sqrt{\frac{\pi(2l+1)(l-1)!}{2(l+1)!}} \frac{2(l+1)!}{(2l+1)(l-1)!} \left[\mathrm{i}^{l-1} \mathrm{j}_{l-1}(k_\mathrm{s}r) - \mathrm{i}^{l+1} \mathrm{j}_{l+1}(k_\mathrm{s}r) \right]$$

$$= \mathrm{i}^{l-1} \sqrt{2\pi l(l+1)(2l+1)}\, \frac{\mathrm{j}_l(k_\mathrm{s}r)}{k_\mathrm{s}r}$$

In the third equality a recurrence relation for Legendre functions [Eq. (3.8)] is used, in the following equality the orthogonality of the Legendre functions [Eq. (3.3)] is used, and in the last equality a recurrence relation for spherical Bessel functions [Eq. (3.19)] is used. It follows that

$$b_{2\mathrm{e}1l} = \mathrm{i}^{l-1}\sqrt{2\pi(2l+1)} \tag{3.42}$$

for $l = 1, 2, 3, \ldots$. Likewise, a projection on the first vector spherical harmonic gives

$$\begin{aligned}
b_{1\mathrm{o}1l}\mathrm{j}_l(k_\mathrm{s}r) &= \int_0^{2\pi}\int_0^{\pi} \hat{\boldsymbol{x}} \cdot \boldsymbol{A}_{1\mathrm{o}1l}(\theta,\varphi)\mathrm{e}^{\mathrm{i}k_\mathrm{s}z}\,\sin\theta\,\mathrm{d}\theta\,\mathrm{d}\varphi \\
&= \int_0^{\pi}\sqrt{\frac{\pi(2l+1)(l-1)!}{2l(l+1)(l+1)!}} \\
&\qquad \times \left(\frac{\cos\theta}{\sin\theta}\,\mathrm{P}_l^1(\cos\theta) - \sin\theta\,\mathrm{P}_l^{1\prime}(\cos\theta)\right)\mathrm{e}^{\mathrm{i}k_\mathrm{s}z}\,\sin\theta\,\mathrm{d}\theta \\
&= \int_0^{\pi}\sqrt{\frac{\pi l(l+1)(2l+1)(l-1)!}{2(l+1)!}}\,\mathrm{P}_l(\cos\theta) \\
&\qquad \times \sum_{n=0}^{\infty}\mathrm{i}^n(2n+1)\mathrm{P}_n(\cos\theta)\mathrm{j}_n(k_\mathrm{s}r)\,\sin\theta\,\mathrm{d}\theta \\
&= \mathrm{i}^l\sqrt{2\pi(2l+1)}\,\mathrm{j}_l(k_\mathrm{s}r)
\end{aligned}$$

A recurrence relation for Legendre functions [Eq. (3.7)] and the orthogonality of the Legendre polynomials [Eq. (3.3)] have been used. It follows that

$$b_{1\mathrm{o}1l} = \mathrm{i}^l\sqrt{2\pi(2l+1)} \tag{3.43}$$

for $l = 1, 2, 3, \ldots$. All the nonzero expansion coefficients for a plane S wave are hereby determined.

It is noted that a more general and elegant derivation of the plane wave expansion coefficients are given in Section 5.1.3.

3.2.3 The scattered wave

In the far field (when the distance to the field point is large compared to both the wavelength and the sphere radius; the distance must, in fact, also be large compared to "the near field length", see Section 13.1.1) the asymptotic formula for the Hankel function [Eq. (3.17)] can be used so that the scattered wave becomes

$$\boldsymbol{u}^{\mathrm{sc}}(\boldsymbol{r}) = \frac{\mathrm{e}^{\mathrm{i}k_{\mathrm{s}}r}}{k_{\mathrm{s}}r}\boldsymbol{F}_{\mathrm{s}}(\theta,\varphi) + \hat{\boldsymbol{r}}\,\frac{\mathrm{e}^{\mathrm{i}k_{\mathrm{p}}r}}{k_{\mathrm{p}}r}F_{\mathrm{p}}(\theta,\varphi) \tag{3.44}$$

The S wave far field scattering amplitude is

$$\boldsymbol{F}_{\mathrm{s}}(\theta,\varphi) = \sum_{\sigma,m,l} \mathrm{i}^{-l}\left[-\mathrm{i}f_{1\sigma ml}\boldsymbol{A}_{1\sigma ml}(\theta,\varphi) + f_{2\sigma ml}\boldsymbol{A}_{2\sigma ml}(\theta,\varphi)\right] \tag{3.45}$$

where it is noted that the summation over l starts with $l = 1$. The P wave far field scattering amplitude is

$$F_{\mathrm{p}}(\theta,\varphi) = \left(\frac{k_{\mathrm{p}}}{k_{\mathrm{s}}}\right)^{3/2} \sum_{\sigma,m,l} \mathrm{i}^{-l} f_{3\sigma ml}\mathrm{Y}_{\sigma ml}(\theta,\varphi) \tag{3.46}$$

The S wave far field scattering amplitude is a vector which have θ- and φ-components, whereas the P wave far field scattering amplitude can be written as a scalar as the P wave only has an r-component.

For an incident plane wave the far field scattering amplitudes can be expressed in the T-matrix elements. For an incident S wave the S wave far field scattering amplitude is

$$\boldsymbol{F}_{\mathrm{ss}}(\theta,\varphi) = -\mathrm{i}\sum_{l=1}^{\infty}\sqrt{2\pi(2l+1)}\left[T_{11}^{l}\boldsymbol{A}_{1\mathrm{o}1l}(\theta,\varphi) + T_{22}^{l}\boldsymbol{A}_{2\mathrm{e}1l}(\theta,\varphi)\right] \tag{3.47}$$

and the P wave far field scattering amplitude is

$$F_{\mathrm{ps}}(\theta,\varphi) = \mathrm{i}\left(\frac{k_{\mathrm{p}}}{k_{\mathrm{s}}}\right)^{3/2}\sum_{l=1}^{\infty}\sqrt{2\pi(2l+1)}\,T_{32}^{l}\mathrm{Y}_{\mathrm{e}1l}(\theta,\varphi) \tag{3.48}$$

Similarly, for an incident P wave the P wave far field scattering amplitude is

$$F_{\mathrm{pp}}(\theta) = \sum_{l=0}^{\infty}\sqrt{4\pi(2l+1)}\,T_{33}^{l}\mathrm{Y}_{\mathrm{e}0l}(\theta) \tag{3.49}$$

and the S wave far field scattering amplitude is

$$\boldsymbol{F}_{\rm sp}(\theta) = \left(\frac{k_{\rm s}}{k_{\rm p}}\right)^{3/2} \sum_{l=1}^{\infty} \sqrt{4\pi(2l+1)}\, T_{23}^{l} \boldsymbol{A}_{2{\rm e}0l}(\theta) \tag{3.50}$$

For an incident P wave the scattering problem is rotationally symmetric and there is thus no dependence on φ.

The scattered energy and the scattering cross sections can also be calculated. The radial traction of the scattered wave in the far field becomes

$$\boldsymbol{t}^{(r)}(\boldsymbol{u}^{\rm sc}(\boldsymbol{r})) = {\rm i}\mu \left(\frac{{\rm e}^{{\rm i}k_{\rm s}r}}{r} \boldsymbol{F}_{\rm s}(\theta,\varphi) + \hat{\boldsymbol{r}} \left(\frac{k_{\rm s}}{k_{\rm p}}\right)^{2} \frac{{\rm e}^{{\rm i}k_{\rm p}r}}{r} F_{\rm p}(\theta,\varphi) \right) \tag{3.51}$$

The mean in time of the total radial energy flow in the far field then becomes (essentially repeating the derivation in Section 2.1.4)

$$\langle J_r \rangle = \frac{\omega\mu}{2k_{\rm s}} \sum_{\tau,\sigma,m,l} |f_{\tau\sigma ml}|^2$$

It is noted that this implies that each out-going wavefunction carries the same amount of energy.

For an incident plane wave this total radial energy flow is normalized with the mean energy flow density of the incident plane wave. For an incident S wave the energy flow density is

$$\langle j_{{\rm s}z} \rangle = \frac{1}{2}\omega\mu k_{\rm s}$$

and for an incident P wave it is

$$\langle j_{{\rm p}z} \rangle = \frac{1}{2}\omega(\lambda + 2\mu) k_{\rm p}$$

This gives the total scattering cross section for an incident S wave

$$\gamma_{\rm s} = \frac{\langle J_r \rangle}{\langle j_{{\rm s}z} \rangle} = \frac{1}{k_{\rm s}^2} \sum_{\tau,\sigma,m,l} |f_{\tau\sigma ml}|^2 \tag{3.52}$$

and for an incident P wave

$$\gamma_{\rm p} = \frac{\langle J_r \rangle}{\langle j_{{\rm p}z} \rangle} = \frac{k_{\rm p}}{k_{\rm s}^3} \sum_{\tau,\sigma,m,l} |f_{\tau\sigma ml}|^2 \tag{3.53}$$

The total scattering cross sections can be divided into two parts due to the scattered S waves ($\tau = 1, 2$) and P waves ($\tau = 3$). The scattering cross sections can be expressed in the T-matrix elements and the expansion coefficients of the incident wave.

3.2.4 Low frequency T-matrix

As in the 2D vector problem for coupled P-SV waves the leading order T-matrix elements for low frequencies has monopole, dipole, and quadropole contributions, corresponding to $l = 0, 1$, and 2, respectively. As noted above the T-matrix is diagonal in the σ-, m-, and l-indices and is independent of the m- and σ-indices, so the elements can be given as $T^l_{\tau\tau'}$. To leading order the elements with $\tau = 1$ (the SH waves) do not contribute at all for low frequencies so only elements with $\tau = 2, 3$ are given.

For $l = 0$ (monopole) only one element exists as there is no SV wave for $l = 0$. This element is

$$T^0_{33} = \mathrm{i}\,(k_\mathrm{p}a)^3\,\frac{3\,(\lambda - \lambda_1) + 2\,(\mu - \mu_1)}{3\,(4\mu + 3\lambda_1 + 2\mu_1)} \tag{3.54}$$

This element does not depend on the density of the sphere, only on its stiffness properties. This is natural as there is no net translation or rotation of the sphere for $l = 0$; this argument also holds for $l = 2$ below. The $l = 0$ motion is spherically symmetric so it is natural that this element mainly depends on the difference in compressibility $K = (3\lambda + 2\mu)/3$ between the obstacle and the surrounding.

For $l = 1$ (dipole) the T-matrix elements are

$$\begin{aligned}
T^1_{22} &= \mathrm{i}\,(k_\mathrm{s}a)^3\,\frac{2}{9}\left(\frac{\rho_1}{\rho} - 1\right)\\
T^1_{23} &= T^1_{32} = \mathrm{i}\sqrt{2k_\mathrm{s}^3k_\mathrm{p}^3}\,\frac{a^3}{9}\left(\frac{\rho_1}{\rho} - 1\right)\\
T^1_{33} &= \mathrm{i}\,(k_\mathrm{p}a)^3\,\frac{1}{9}\left(\frac{\rho_1}{\rho} - 1\right)
\end{aligned} \tag{3.55}$$

As $l = 1$ corresponds to a rigid body translation for low frequencies (where the direction depends on the m-value) these elements only depend on the density of the sphere, more precisely on the mass difference $4\pi a^3(\rho_1 - \rho)/3$, not on the stiffness properties of the sphere.

For $l = 2$ (quadropole) the T-matrix elements are

$$
\begin{aligned}
T^2_{22} &= \mathrm{i}(k_\mathrm{s}a)^3 \frac{2k_\mathrm{s}^2}{k_\mathrm{p}^2} C_2 \\
T^2_{23} &= T^2_{32} = \mathrm{i}\sqrt{k_\mathrm{s}^5 k_\mathrm{p}} a^3 \sqrt{\frac{8}{3}} C_2 \\
T^2_{33} &= \mathrm{i}(k_\mathrm{p}a)^3 \frac{4}{3} C_2
\end{aligned}
\tag{3.56}
$$

where

$$
C_2 = \frac{\mu - \mu_1}{14\mu + 16\mu_1 + 3\lambda(3 + 2\mu_1/\mu)}
$$

These elements depend almost linearly on the difference in shear modulus, but are independent of λ_1 (and ρ_1).

From the listed T-matrix elements the corresponding elements when the spherical obstacle is a fluid (put $\mu_1 = 0$) or a cavity (put $\mu_1 = \lambda_1 = 0$ and $\rho_1 = 0$) are obtained. The rigid spherical obstacle must be treated separately and if it is immovable (which is rather unphysical) it scatters more strongly (going as ka), whereas a movable rigid obstacle scatters as an elastic one, see Olsson [107].

The low frequency T-matrix elements are given by Boström [13]. For plane wave incidence the far field scattering amplitude is given by many authors, for instance Norris [103] and Dassios and Kleinman [39], where further references are given. Dassios and Kleinman [39] also give corresponding results for a general ellipsoid.

3.2.5 Other boundary conditions

More general boundary conditions are sometimes of interest. These can be used to model various situations such as a thin interface layer, a damaged interface, and surface effects due to nanosized obstacles. A thin interface layer can be replaced by an effective boundary condition at the midsurface of the layer, and a common approach is to use a spring boundary condition where the traction is continuous whereas the displacement has a jump proportional to the traction. As shown by Bövik [31] and Boström *et al.* [22] this is not adequate in general; to be "exact" to linear order in layer thickness there must also be a jump in the traction, see below. A damaged interface can also be modelled by spring boundary conditions, see, for example,

[11, 29, 2, 85, 86, 52]. Surface effects for nanosized obstacles are investigated by Wang et al. [137] and Wang [136] using a theory for the equilibrium of the surface due to Gurtin *et al.* [60].

The effective boundary conditions derived by Bövik [31] to model a thin interface layer "exactly" are evaluated in the special case of a spherical obstacle by Boström *et al.* [22]. These general boundary conditions, which have the spring boundary conditions as a special case, are now described. Consider a spherical core of radius $a - h/2$ with density ρ_1 and Lamé constants λ_1 and μ_1 surrounded by a thin layer of thickness h and corresponding material parameters ρ_0, λ_0, and μ_0. Outside the layer for $r > a+h/2$ the material parameters are ρ, λ, and μ as usual. Corresponding wavenumbers are denoted in the obvious way. Replacing the thin layer by equivalent boundary conditions on $r = a$ leads to a jump in both displacement and traction. The jump in displacement is

$$\begin{aligned} \boldsymbol{u}_+ - \boldsymbol{u}_- = \frac{h}{2}\big[&\boldsymbol{B}(\lambda_0, \mu_0, \boldsymbol{u}_+ + \boldsymbol{u}_-, \boldsymbol{t}_+ + \boldsymbol{t}_-) \\ &-\boldsymbol{B}(\lambda,\mu,\boldsymbol{u}_+, \boldsymbol{t}_+) - \boldsymbol{B}(\lambda_1, \mu_1, \boldsymbol{u}_-, \boldsymbol{t}_-)\big] \end{aligned} \tag{3.57}$$

where the vector-valued operator $\boldsymbol{B}$ is

$$\boldsymbol{B}(\lambda, \mu, \boldsymbol{u}, \boldsymbol{t}) = \frac{1}{\mu}\,\boldsymbol{t}_{tan} + \frac{\hat{\boldsymbol{r}}}{\lambda + 2\mu}\,(\hat{\boldsymbol{r}} \cdot \boldsymbol{t} - \lambda \nabla_s \cdot \boldsymbol{u})$$

The notation $\boldsymbol{t}_{tan}$ means the tangential component of $\boldsymbol{t}$. The corresponding jump in traction is

$$\begin{aligned} \boldsymbol{t}_+ - \boldsymbol{t}_- = \frac{h}{2}\big[&\boldsymbol{D}(\rho_0, \lambda_0, \mu_0, \boldsymbol{u}_+ + \boldsymbol{u}_-, \boldsymbol{t}_+ + \boldsymbol{t}_-) \\ &-\boldsymbol{D}(\rho, \lambda,\mu,\boldsymbol{u}_+, \boldsymbol{t}_+) - \boldsymbol{D}(\rho_1, \lambda_1, \mu_1, \boldsymbol{u}_-, \boldsymbol{t}_-)\big] \end{aligned} \tag{3.58}$$

where the operator is

$$\begin{aligned} \boldsymbol{D}(\rho, \lambda, \mu, \boldsymbol{u}, \boldsymbol{t}) &= -\frac{4\mu}{a(\lambda + 2\mu)}\,\hat{\boldsymbol{r}}\,\hat{\boldsymbol{r}} \cdot \boldsymbol{t} - \frac{\lambda}{\lambda + 2\mu}\,\nabla_s(\hat{\boldsymbol{r}} \cdot \boldsymbol{t}) \\ -\mu\nabla_s \times (\nabla_s \times \boldsymbol{u})_{tan} &- 2\mu\nabla_s^2\boldsymbol{u} + \frac{2\mu\lambda}{\lambda + 2\mu}\left(\frac{2}{a}\,\hat{\boldsymbol{r}}\,\nabla_s \cdot \boldsymbol{u} - \nabla_s(\nabla_s \cdot \boldsymbol{u})\right) \\ &+\frac{4\mu}{a}\,\hat{\boldsymbol{r}} \times (\nabla_s \times \boldsymbol{u}) - \mu\nabla_s \times (\nabla_s \times \boldsymbol{u}) - \rho\,\omega^2\boldsymbol{u} \end{aligned}$$

In spherical coordinates the surface gradient ∇_s, etc, are

$$
\begin{aligned}
\nabla_s u &= \frac{1}{a}\left(\hat{\boldsymbol{\theta}}\,\frac{\partial u}{\partial \theta} + \frac{\hat{\boldsymbol{\varphi}}}{\sin\theta}\frac{\partial u}{\partial \varphi}\right) \\
\nabla_s \cdot \boldsymbol{v} &= \frac{1}{a\sin\theta}\left(\frac{\partial}{\partial\theta}(\sin\theta\, v_\theta) + \frac{\partial v_\varphi}{\partial\varphi}\right) \\
\nabla_s \times \boldsymbol{v} &= \frac{\hat{\boldsymbol{r}}}{a\sin\theta}\left(\frac{\partial}{\partial\theta}(\sin\theta\, v_\varphi) - \frac{\partial v_\theta}{\partial\varphi}\right) + \frac{\hat{\boldsymbol{\theta}}}{a\sin\theta}\frac{\partial v_r}{\partial\varphi} - \frac{\hat{\boldsymbol{\varphi}}}{a}\frac{\partial v_r}{\partial\theta} \\
\nabla_a^2 u &= \frac{1}{a^2\sin\theta}\left(\frac{\partial}{\partial\theta}\left(\sin\theta\frac{\partial u}{\partial\theta}\right) + \frac{1}{\sin\theta}\frac{\partial^2 u}{\partial\varphi^2}\right) \\
\nabla_s^2 \boldsymbol{v} &= \hat{\boldsymbol{r}}\left(\nabla_s^2 v_r - \frac{2}{a^2}\frac{\partial v_\theta}{\partial\theta} - \frac{2\cot\theta}{a^2}v_\theta - \frac{2}{a^2\sin\theta}\frac{\partial v_\varphi}{\partial\varphi}\right) \\
&\quad + \hat{\boldsymbol{\theta}}\left(\nabla_s^2 u_\theta + \frac{2}{a^2}\frac{\partial v_r}{\partial\theta} - \frac{1}{a^2\sin^2\theta}v_\theta - \frac{2\cos\theta}{a^2\sin^2\theta}\frac{\partial v_\varphi}{\partial\varphi}\right) \\
&\quad + \hat{\boldsymbol{\varphi}}\left(\nabla_s^2 v_\varphi + \frac{2}{a^2\sin\theta}\frac{\partial v_r}{\partial\varphi} - \frac{1}{a^2\sin^2\theta}v_\varphi + \frac{2\cos\theta}{a^2\sin^2\theta}\frac{\partial v_\theta}{\partial\varphi}\right)
\end{aligned}
$$

It is then straightforward to calculate the effect of operation with the surface operators on the spherical vector wavefunctions (which actually only concerns their effects on the vector spherical harmonics as only tangential derivatives are involved). The results are rather lengthy and are not given [22]. The boundary conditions are then set up in the usual way which essentially solves the problem.

In the case of a spherical obstacle it is actually easier to solve the problem for a layered sphere exactly; this is straightforward and is done in the same manner as for a layered circle in Section 2.1.7.

4 Green functions with applications

In the previous two chapters the scattering by a circle and a sphere is investigated using the classical method of separation of variables. This method is not useful for other obstacle shapes in 3D as the elastodynamic equation of motion does not separate in other coordinate systems in 3D (with the exception of spheroidal coordinates for axisymmetric problems [99, 106]). For 2D problems (both scalar and vector) more shapes can be treated [112], but these problems are of limited interest and are not further investigated.

To proceed with analytical approaches other methods are needed. Green functions and integral representations are very useful in this context, and these topics are thus taken up in this chapter. As an application of volume integral representations the Born approximation is briefly described.

4.1 Green functions

A Green function (also called a fundamental solution [75]) to a differential equation is a solution to the differential equation with a source term that is concentrated to a single point, i.e. it is a delta function source. The differential equation is assumed to be valid in some domain (which may consist of several parts), and the source term for the Green function should then be located inside the domain. To be unique the Green function must satisfy some boundary condition on the boundary of the domain. The most used Green function is the free space Green function, and the domain is then the whole space and the boundary condition is the radiation condition.

In the following free space Green functions are derived in 2D and 3D, first for the scalar Helmholtz equation, then for the elastic vector cases in 2D and 3D. The expansions of the Green function in polar and spherical wavefunctions are also given. For more about Green functions in elastodynamics

Ström [125], de Hoop [41], Aki and Richards [5], and Kausel [75] can be consulted.

4.1.1 Scalar Green functions

For the scalar Helmholtz equation in 2D and 3D the Green functions have, in fact, been derived in Section 2.1.3 and 3.2.2, respectively. In 2D the Helmholtz equation with a delta function source is

$$\nabla^2 u + k^2 u = -\delta(x)\delta(y)$$

where it is assumed that the strength of the source is unity and the source is located at the origin. The solution that is valid in whole space and satisfies the radiation condition is given in Eq. (2.15)

$$u(r) = -\frac{\mathrm{i}}{4}\mathrm{H}_0(kr)$$

If the source is located at a position $\boldsymbol{r}'$ instead of the origin the Helmholtz equation for the Green function is obtained

$$\nabla^2 g(\boldsymbol{r};\boldsymbol{r}') + k^2 g(\boldsymbol{r};\boldsymbol{r}') = -\delta(\boldsymbol{r}-\boldsymbol{r}') \tag{4.1}$$

where the delta function now is the 2D delta function. It should be stressed that the Laplacian operates on the unprimed coordinates; the position of the source $\boldsymbol{r}'$ acts as a parameter in the equation. The scalar free space Green function in 2D is thus

$$g(\boldsymbol{r};\boldsymbol{r}') = -\frac{\mathrm{i}}{4}\mathrm{H}_0(k|\boldsymbol{r}-\boldsymbol{r}'|) \tag{4.2}$$

As this function only depends on the distance between the two points $\boldsymbol{r}$ and $\boldsymbol{r}'$, the following symmetry property is obvious

$$g(\boldsymbol{r}';\boldsymbol{r}) = g(\boldsymbol{r};\boldsymbol{r}')$$

To expand the scalar Green function in the scalar polar wavefunctions, the solution to Helmholtz equation with a point source on the negative x-axis [Eq. (2.16)] is first repeated

$$u(r,\varphi) = \frac{\mathrm{i}}{4}\sum_{m=0}^{\infty}(-1)^m \epsilon_m \mathrm{J}_m(kr_<)\mathrm{H}_m(kr_>)\cos m\varphi$$

For arbitrary location of the source, φ must be replaced by $\varphi - \varphi' + \pi$. Using that $\cos m(\varphi - \varphi' + \pi) = (-1)^m(\cos m\varphi \cos m\varphi' + \sin m\varphi \sin m\varphi')$ and remembering the definition of the scalar wavefunctions in Eq. (2.9) the expansion of the Green function is obtained

$$g(\boldsymbol{r};\boldsymbol{r}') = \mathrm{i}\sum_{\sigma,m} \chi^0_{\sigma m}(\boldsymbol{r}_<)\chi^+_{\sigma m}(\boldsymbol{r}_>) \tag{4.3}$$

where $\boldsymbol{r}_< = \boldsymbol{r}$ and $\boldsymbol{r}_> = \boldsymbol{r}'$ if $r < r'$ and vice versa. This expansion also explains the normalization of the wavefunctions.

Turning to 3D, the differential equation for the Green function is

$$\nabla^2 g(\boldsymbol{r};\boldsymbol{r}') + k^2 g(\boldsymbol{r};\boldsymbol{r}') = -\delta(\boldsymbol{r} - \boldsymbol{r}') \tag{4.4}$$

where the delta function now is the 3D one. The solution to this equation is given in Eq. (3.38) and is

$$g(\boldsymbol{r};\boldsymbol{r}') = \frac{\exp(\mathrm{i}k|\boldsymbol{r} - \boldsymbol{r}'|)}{4\pi|\boldsymbol{r} - \boldsymbol{r}'|} \tag{4.5}$$

With the present normalizations the Green functions in 2D and 3D have different dimension, the 2D one is dimensionless whereas the 3D one has dimension inverse length. In the vector cases to follow all Green functions are chosen dimensionless. Also in 3D the Green function has the symmetry property

$$g(\boldsymbol{r}';\boldsymbol{r}) = g(\boldsymbol{r};\boldsymbol{r}')$$

To expand the 3D Green function in the scalar spherical wavefunctions the solution to a point source on the negative z-axis [Eq. (3.39)] is repeated

$$u(r,\theta) = \mathrm{i}k\sum_{l=0}^{\infty}(-1)^l \frac{2l+1}{4\pi}\mathrm{j}_l(kr_<)\mathrm{h}_l(kr_>)\mathrm{P}_l(\cos\theta)$$

Here the source is assumed to be located on the negative z-axis, i.e. for $\theta' = \pi$. Using the addition theorem for Legendre functions Eq. (3.13), where γ now is the angle between the two directions specified by (θ,φ) and (θ',φ') [and the fact that $\mathrm{P}_l(-1) = (-1)^l$], this gives

$$g(\boldsymbol{r};\boldsymbol{r}') = \mathrm{i}k\sum_{\sigma,m,l} \mathrm{j}_l(kr_<)\mathrm{h}_l(kr_>)\mathrm{Y}_{\sigma ml}(\theta,\varphi)Y_{\sigma ml}(\theta',\varphi') \tag{4.6}$$

where $r_< = \min(r,r')$ and $r_> = \max(r,r')$.

4.1.2 2D dyadic Green function

In tensor notation the elastodynamic equation in 2D with a point force in the x_1-direction can be written

$$\left(k_{\mathrm{p}}^{-2} - k_{\mathrm{s}}^{-2}\right) \partial_i \partial_j u_j + k_{\mathrm{s}}^{-2} \partial_j \partial_j u_i + u_i = -F \delta_{i1} \delta(\boldsymbol{r} - \boldsymbol{r}')$$

where all indices now take on only the values 1 and 2. The Green function should be the displacement response to a point force in any direction so it must be a tensor (of rank two). It is denoted $G_{ij}(\boldsymbol{r};\boldsymbol{r}')$ or in dyadic notation $\boldsymbol{G}(\boldsymbol{r};\boldsymbol{r}')$ and satisfies the equation

$$\left(k_{\mathrm{p}}^{-2} - k_{\mathrm{s}}^{-2}\right) \partial_i \partial_m G_{mj}(\boldsymbol{r};\boldsymbol{r}') + k_{\mathrm{s}}^{-2} \partial_m \partial_m G_{ij}(\boldsymbol{r};\boldsymbol{r}') + G_{ij}(\boldsymbol{r};\boldsymbol{r}') = -k_{\mathrm{s}}^{-2} \delta_{ij} \delta(\boldsymbol{r} - \boldsymbol{r}') \tag{4.7}$$

The factor k_{s}^{-2} on the right-hand side is inserted to make the Green tensor dimensionless. This equation can be solved by Fourier transform techniques [125]. Introduce the spatial Fourier transform

$$\tilde{u}(k) = \int_{-\infty}^{\infty} u(x) \mathrm{e}^{-\mathrm{i}kx}\, \mathrm{d}x \tag{4.8}$$

with inverse

$$u(x) = \frac{1}{2\pi} \int_{-\infty}^{\infty} \tilde{u}(k) \mathrm{e}^{\mathrm{i}kx}\, \mathrm{d}k \tag{4.9}$$

Note that there is a sign change in the exponentials as compared to the temporal Fourier transform [Eqs. (1.6) and (1.7)]. Taking a double Fourier transform with respect to both x_1 and x_2 with corresponding Fourier transform variables k_1 and k_2 gives

$$\left(1 - (k_{\mathrm{s}}/k_{\mathrm{p}})^2\right) k_i k_m \tilde{G}_{mj} + \left(k_{\mathrm{s}}^2 - k^2\right) \tilde{G}_{ij} = -\delta_{ij}$$

where the tilde on G_{ij} denotes the double Fourier transform. Here $k^2 = k_1^2 + k_2^2$ and $\boldsymbol{r}' = 0$ for simplicity ($\boldsymbol{r}' \neq 0$ is inserted at the end). After some algebra the solution can be written

$$\tilde{G}_{ij} = \frac{1}{k_{\mathrm{s}}^2} \left[\frac{k_{\mathrm{s}}^2 \delta_{ij}}{k^2 - k_{\mathrm{s}}^2} - k_i k_j \left(\frac{1}{k^2 - k_{\mathrm{s}}^2} - \frac{1}{k^2 - k_{\mathrm{p}}^2} \right) \right]$$

The inverse Fourier transform can now be performed by inspection. First it is noted that the double Fourier transform of the scalar Green function in Eq. (4.1) is

$$\tilde{g}_n = \frac{1}{k^2 - k_n^2} \qquad n = \mathrm{s}, \mathrm{p}$$

As a multiplication by $\mathrm{i}k_j$ in Fourier space corresponds to a derivative ∂_j in real space, it is seen that the Green tensor becomes

$$G_{ij}(\boldsymbol{r};\boldsymbol{r}') = \delta_{ij} g_{\mathrm{s}}(\boldsymbol{r};\boldsymbol{r}') + \frac{1}{k_{\mathrm{s}}^2}\partial_i \partial_j \left[g_{\mathrm{s}}(\boldsymbol{r};\boldsymbol{r}') - g_{\mathrm{p}}(\boldsymbol{r};\boldsymbol{r}')\right] \tag{4.10}$$

where the second argument $\boldsymbol{r}'$ is again inserted and the scalar Green function is

$$g_{\mathrm{n}}(\boldsymbol{r};\boldsymbol{r}') = -\frac{\mathrm{i}}{4}\mathrm{H}_0(k_{\mathrm{n}}|\boldsymbol{r} - \boldsymbol{r}'|)$$

In dyadic notation the Green tensor is written

$$\boldsymbol{G}(\boldsymbol{r};\boldsymbol{r}') = \boldsymbol{I} g_{\mathrm{s}}(\boldsymbol{r};\boldsymbol{r}') + \frac{1}{k_{\mathrm{s}}^2}\nabla\nabla \left[g_{\mathrm{s}}(\boldsymbol{r};\boldsymbol{r}') - g_{\mathrm{p}}(\boldsymbol{r};\boldsymbol{r}')\right] \tag{4.11}$$

where $\boldsymbol{I}$ is the unit dyadic and ∇ operates on $\boldsymbol{r}$. It is clear that the Green tensor is symmetric, so that

$$G_{ij}(\boldsymbol{r};\boldsymbol{r}') = G_{ji}(\boldsymbol{r};\boldsymbol{r}')$$

and as the scalar Green function is symmetric in the arguments, this is also true for the Green tensor

$$G_{ij}(\boldsymbol{r};\boldsymbol{r}') = G_{ij}(\boldsymbol{r}';\boldsymbol{r})$$

To obtain the expansion of the Green dyadic in terms of the polar vector wavefunctions in Eq. (2.26) the expansion of the scalar Green function in Eq. (4.3) is used. When operating on the Green function it is noticed that $\nabla' = -\nabla$, and if this is used for the P term in Eq. (4.11) it is directly seen that this term becomes

$$-\frac{1}{k_{\mathrm{s}}^2}\nabla\nabla g_{\mathrm{p}}(\boldsymbol{r};\boldsymbol{r}') = \mathrm{i}\sum_{\sigma,m} \boldsymbol{\chi}_{2\sigma m}^{0}(\boldsymbol{r}_<)\boldsymbol{\chi}_{2\sigma m}^{+}(\boldsymbol{r}_>)$$

For the S part an auxiliary Green dyadic $\boldsymbol{G}_{\mathrm{s}}$ is introduced

$$\boldsymbol{G}_{\mathrm{s}}(\boldsymbol{r};\boldsymbol{r}') = \boldsymbol{I}g_{\mathrm{s}}(\boldsymbol{r};\boldsymbol{r}')$$

Working in Cartesian coordinates it is directly seen that this Green dyadic satisfies

$$\nabla^2 \boldsymbol{G}_{\mathrm{s}} + k_{\mathrm{s}}^2 \boldsymbol{G}_{\mathrm{s}} = -\boldsymbol{I}\delta(\boldsymbol{r}-\boldsymbol{r}')$$

This is rewritten

$$\nabla\nabla\cdot\boldsymbol{G}_{\mathrm{s}} - \nabla\times(\nabla\times\boldsymbol{G}_{\mathrm{s}}) + k_{\mathrm{s}}^2\boldsymbol{G}_{\mathrm{s}} = -\boldsymbol{I}\delta(\boldsymbol{r}-r')$$

Remembering that this is a 2D equation it is realized that the second term must be interpreted in a 3D sense, i.e. the curl operator and the unit dyadic (in the definition of $\boldsymbol{G}_{\mathrm{s}}$) must be the 3D ones. Avoiding the singular point $\boldsymbol{r} = \boldsymbol{r}'$ this gives that the two S terms in the Green dyadic in Eq. (4.11) become

$$\boldsymbol{I}g_{\mathrm{s}} + \frac{1}{k_{\mathrm{s}}^2}\nabla\nabla g_{\mathrm{s}} = \boldsymbol{G}_{\mathrm{s}} + \frac{1}{k_{\mathrm{s}}^2}\nabla\nabla\cdot\boldsymbol{G}_{\mathrm{s}} = \frac{1}{k_{\mathrm{s}}^2}\nabla\times\nabla\times\boldsymbol{G}_{\mathrm{s}}$$

Using again that $\nabla' = -\nabla$ this can be written

$$\boldsymbol{I}g_{\mathrm{s}} + \frac{1}{k_{\mathrm{s}}^2}\nabla\nabla g_{\mathrm{s}} = -\frac{1}{k_{\mathrm{s}}^2}\nabla\times\nabla'\times(\hat{\boldsymbol{x}}\hat{\boldsymbol{x}}+\hat{\boldsymbol{y}}\hat{\boldsymbol{y}}+\hat{\boldsymbol{z}}\hat{\boldsymbol{z}})\,g_{\mathrm{s}}$$

where the unit dyadic is written out fully. Only the term $\hat{\boldsymbol{z}}\hat{\boldsymbol{z}}$ contributes and inserting the expansion of the scalar Green function Eq. (4.3) this gives

$$\boldsymbol{I}g_{\mathrm{s}} + \frac{1}{k_{\mathrm{s}}^2}\nabla\nabla g_{\mathrm{s}} = \mathrm{i}\sum_{\sigma,m}\boldsymbol{\chi}^0_{1\sigma m}(\boldsymbol{r}_<)\boldsymbol{\chi}^+_{1\sigma m}(\boldsymbol{r}_>)$$

Summing the two terms the expansion of the 2D Green dyadic is finally obtained

$$\boldsymbol{G}(\boldsymbol{r};\boldsymbol{r}') = \mathrm{i}\sum_{\tau,\sigma,m}\boldsymbol{\chi}^0_{\tau\sigma m}(\boldsymbol{r}_<)\boldsymbol{\chi}^+_{\tau\sigma m}(\boldsymbol{r}_>) \tag{4.12}$$

Due to the symmetries the order of the wavefunctions can be changed. Alternatively, this expansion can be obtained in the same way as in the 3D case below.

4.1.3 3D dyadic Green function

The explicit form of the 3D Green dyadic is obtained in the same way as the 2D counterpart. The differential equation for the 3D Green tensor is

$$\left(k_{\mathrm{p}}^{-2} - k_{\mathrm{s}}^{-2}\right) \partial_i \partial_m G_{mj}(\boldsymbol{r};\boldsymbol{r}') + k_{\mathrm{s}}^{-2} \partial_m \partial_m G_{ij}(\boldsymbol{r};\boldsymbol{r}') \\ + G_{ij}(\boldsymbol{r};\boldsymbol{r}') = -k_{\mathrm{s}}^{-3} \delta_{ij} \delta(\boldsymbol{r} - \boldsymbol{r}') \tag{4.13}$$

The only difference is that the indices now take on the values 1, 2, and 3 and that the normalization is a little different (k_{s}^{-3} on the right-hand side instead of k_{s}^{-2}), although this still makes the Green tensor dimensionless. Taking a triple Fourier transform in all three space coordinates (with corresponding transform variables k_1, k_2, and k_3) the transformed Green tensor becomes

$$\tilde{G}_{ij} = \frac{1}{k_{\mathrm{s}}^3} \left[\frac{k_{\mathrm{s}}^2 \delta_{ij}}{k^2 - k_{\mathrm{s}}^2} - k_i k_j \left(\frac{1}{k^2 - k_{\mathrm{s}}^2} - \frac{1}{k^2 - k_{\mathrm{p}}^2} \right) \right]$$

where now $k^2 = k_1^2 + k_2^2 + k_3^2$. The inversion to the space coordinates then gives

$$G_{ij}(\boldsymbol{r};\boldsymbol{r}') = \frac{1}{k_{\mathrm{s}}} \delta_{ij} g_{\mathrm{s}}(\boldsymbol{r};\boldsymbol{r}') + \frac{1}{k_{\mathrm{s}}^3} \partial_i \partial_j \left[g_{\mathrm{s}}(\boldsymbol{r};\boldsymbol{r}') - g_{\mathrm{p}}(\boldsymbol{r};\boldsymbol{r}') \right] \tag{4.14}$$

In 3D the scalar Green function is

$$g_n(\boldsymbol{r};\boldsymbol{r}') = \frac{\exp\left(\mathrm{i} k_n |\boldsymbol{r} - \boldsymbol{r}'|\right)}{4\pi |\boldsymbol{r} - \boldsymbol{r}'|}, \qquad n = \mathrm{s}, \mathrm{p}$$

In dyadic notation the Green tensor is written

$$\boldsymbol{G}(\boldsymbol{r};\boldsymbol{r}') = \frac{1}{k_{\mathrm{s}}} \boldsymbol{I} g_{\mathrm{s}}(\boldsymbol{r};\boldsymbol{r}') + \frac{1}{k_{\mathrm{s}}^3} \nabla\nabla \left[g_{\mathrm{s}}(\boldsymbol{r};\boldsymbol{r}') - g_{\mathrm{p}}(\boldsymbol{r};\boldsymbol{r}') \right] \tag{4.15}$$

To obtain the expansion of the Green dyadic in spherical vector wavefunctions, the Green dyadic is scalar multiplied by a constant vector $\boldsymbol{a}$ to get the auxiliary vector field $\boldsymbol{g}_a(\boldsymbol{r};\boldsymbol{r}') = \boldsymbol{a} \cdot \boldsymbol{G}(\boldsymbol{r};\boldsymbol{r}')$. This vector field satisfies

$$k_{\mathrm{p}}^{-2} \nabla\nabla \cdot \boldsymbol{g}_a - k_{\mathrm{s}}^{-2} \nabla \times (\nabla \times \boldsymbol{g}_a) + \boldsymbol{g}_a = -k_{\mathrm{s}}^{-3} \boldsymbol{a}\, \delta(\boldsymbol{r} - \boldsymbol{r}')$$

For $r > r'$ the right-hand side vanishes and as the solution must satisfy radiation conditions the field must admit an expansion in terms of the outgoing spherical vector wavefunctions

$$\boldsymbol{g}_a(\boldsymbol{r};\boldsymbol{r}') = \sum_{\tau,\sigma,m,l} c_{\tau\sigma ml} \boldsymbol{\psi}_{\tau\sigma ml}^{+}(\boldsymbol{r})$$

The expansion coefficients $c_{\tau\sigma ml}$ depend on $\boldsymbol{r}'$ and the vector $\boldsymbol{a}$ and to determine these coefficients the reciprocity relation is employed [Eq. (1.9)]

$$\int_S \hat{\boldsymbol{n}} \cdot \left(\boldsymbol{\sigma}^{\mathrm{A}} \cdot \boldsymbol{u}^{\mathrm{B}} - \boldsymbol{\sigma}^{\mathrm{B}} \cdot \boldsymbol{u}^{\mathrm{A}}\right) \mathrm{d}S = \int_V \rho \left(\boldsymbol{u}^{\mathrm{A}} \cdot \boldsymbol{f}^{\mathrm{B}} - \boldsymbol{u}^{\mathrm{B}} \cdot \boldsymbol{f}^{\mathrm{A}}\right) \mathrm{d}V$$

State A is here chosen as the auxiliary field $\boldsymbol{g}_a(\boldsymbol{r};\boldsymbol{r}')$ and state B as a regular spherical vector wavefunction $\boldsymbol{\psi}^0_{\tau\sigma ml}(\boldsymbol{r})$. The surface S is any closed surface enclosing the sphere $r = r'$ and V is its interior. The forces are then $\rho \boldsymbol{f}^{\mathrm{A}} = (\mu/k_{\mathrm{s}})\boldsymbol{a}\,\delta(\boldsymbol{r}-\boldsymbol{r}')$ and $\rho\boldsymbol{f}^{\mathrm{B}} = 0$. This gives

$$\sum_{\tau',\sigma',m',l'} c_{\tau'\sigma'm'l'} \int_S \left[\boldsymbol{t}^{(n)}\left(\boldsymbol{\psi}^+_{\tau'\sigma'm'l'}(\boldsymbol{r})\right) \cdot \boldsymbol{\psi}^0_{\tau\sigma ml}(\boldsymbol{r}) \right.$$

$$\left. - \boldsymbol{t}^{(n)}\left(\boldsymbol{\psi}^0_{\tau\sigma ml}(\boldsymbol{r})\right) \cdot \boldsymbol{\psi}^+_{\tau'\sigma'm'l'}(\boldsymbol{r})\right] \mathrm{d}S = -\frac{\mu}{k_{\mathrm{s}}}\, \boldsymbol{a} \cdot \boldsymbol{\psi}^0_{\tau\sigma ml}(\boldsymbol{r}')$$

The orthogonality of the spherical vector wavefunctions [the last of Eqs. (3.32)] can then be used to obtain

$$c_{\tau\sigma ml} = \mathrm{i}\boldsymbol{a} \cdot \boldsymbol{\psi}^0_{\tau\sigma ml}(\boldsymbol{r}')$$

As the vector $\boldsymbol{a}$ is arbitrary this gives for $r > r'$

$$\boldsymbol{G}(\boldsymbol{r};\boldsymbol{r}') = \mathrm{i} \sum_{\tau,\sigma,m,l} \boldsymbol{\psi}^0_{\tau\sigma ml}(\boldsymbol{r}')\boldsymbol{\psi}^+_{\tau\sigma ml}(\boldsymbol{r})$$

Due to the symmetries this finally gives the expansion of the Green dyadic in 3D

$$\boldsymbol{G}(\boldsymbol{r};\boldsymbol{r}') = \mathrm{i} \sum_{\tau,\sigma,m,l} \boldsymbol{\psi}^0_{\tau\sigma ml}(\boldsymbol{r}_<)\boldsymbol{\psi}^+_{\tau\sigma ml}(\boldsymbol{r}_>) \tag{4.16}$$

Due to the symmetries the order of the wavefunctions can be changed.

4.2 Integral representations

An integral representation is a way to express a field as an integral over a domain (a volume or a surface) where the integrand contains the field, some of its derivatives, and a Green function. Integral representations are useful when formulating a scattering problem as an integral equation, but are also useful in other contexts. Here integral representations are derived in 2D (both scalar and vector cases) and 3D. The development here closely follows Pao and Varatharajulu [113] and Ström [125], see also de Hoop [41].

4.2.1 Scalar integral representation in 2D

As usual the simplest case is the 2D SH waves and this is treated first. The starting point is the differential equation for the displacement

$$\nabla^2 u(\boldsymbol{r}) + k^2 u(\boldsymbol{r}) = 0$$

where the displacement is called u instead of u_z and the wavenumber k instead of k_{s}. For the free space Green function the corresponding equation is

$$\nabla^2 g(\boldsymbol{r};\boldsymbol{r}') + k^2 g(\boldsymbol{r};\boldsymbol{r}') = -\delta(\boldsymbol{r}-\boldsymbol{r}')$$

Multiply the first equation with $g(\boldsymbol{r};\boldsymbol{r}')$ and the second with $u(\boldsymbol{r})$ and take the difference between the two equations

$$\left(\nabla^2 u(\boldsymbol{r})\right) g(\boldsymbol{r};\boldsymbol{r}') - u(\boldsymbol{r})\nabla^2 g(\boldsymbol{r};\boldsymbol{r}') = \delta(\boldsymbol{r}-\boldsymbol{r}')u(\boldsymbol{r})$$

This can be rewritten as

$$\nabla\cdot\left[(\nabla u(\boldsymbol{r}))g(\boldsymbol{r};\boldsymbol{r}') - u(\boldsymbol{r})\nabla g(\boldsymbol{r};\boldsymbol{r}')\right] = \delta(\boldsymbol{r}-\boldsymbol{r}')u(\boldsymbol{r}) \tag{4.17}$$

Consider now a typical scattering situation where the whole space is homogeneous except for two domains, see Figure 4.1. One of these domains is the obstacle, which has a bounding surface (curve) S with outward-pointing normal $\hat{\boldsymbol{n}}$, but need not be further specified. The other domain is where the incident wave is generated, its outer boundary is S_i with outward-pointing normal $\hat{\boldsymbol{n}}_i$. A third, fictitious, surface is also introduced and this is a large circle of radius R, boundary S_R, and outward-pointing normal $\hat{\boldsymbol{n}}_R$. Integrate Eq. (4.17) over the volume (area) V_R between the three surfaces. On the left-hand side Gauss' law in 2D can be used to obtain

$$\int_{S_R-S-S_i}\left[(\nabla u(\boldsymbol{r}))g(\boldsymbol{r};\boldsymbol{r}') - u(\boldsymbol{r})\nabla g(\boldsymbol{r};\boldsymbol{r}')\right]\cdot\hat{\boldsymbol{n}}\,\mathrm{d}S = \begin{cases} u(\boldsymbol{r}'),\ \boldsymbol{r}' \text{ in } V_R \\ 0,\ \boldsymbol{r}' \text{ not in } V_R \end{cases}$$

where the normal $\hat{\boldsymbol{n}}$ is the appropriate one. The difference in sign between S_R and the other two surfaces stems from the difference in the direction of the normals. The integral over $-S_i$ gives the incident wave u^{in} and any influence of the scattered wave on the source is thus neglected in this step. In the limit $R\to\infty$ the integral over the large circle S_R vanishes when both the

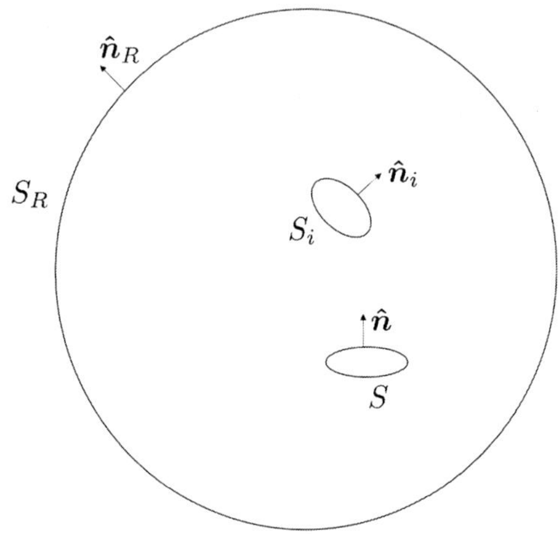

Figure 4.1: The scattering geometry with the obstacle with surface S, the source region bounded by the surface S_i, and a large circle S_R.

displacement u and the Green function g satisfy the radiation condition (to leading order they both behave as $\exp{(ikr)}/\sqrt{kr}$ for large r, so the leading order terms cancel). Exchanging $\boldsymbol{r}$ and $\boldsymbol{r}'$ the final result is

$$u^{\mathrm{in}}(\boldsymbol{r}) + \int_S \left[u(\boldsymbol{r}')\nabla' g(\boldsymbol{r}';\boldsymbol{r}) - (\nabla' u(\boldsymbol{r}'))g(\boldsymbol{r}';\boldsymbol{r})\right] \cdot \hat{\boldsymbol{n}}' \, \mathrm{d}S'$$

$$= \begin{cases} u(\boldsymbol{r}), & \boldsymbol{r} \text{ outside } S \\ 0, & \boldsymbol{r} \text{ inside } S \end{cases} \tag{4.18}$$

This is the desired integral representation. With the field point outside the obstacle it gives the scattered wave $u^{\mathrm{sc}} = u - u^{\mathrm{in}}$ as an integral over the obstacle surface S, where the integrand contains both the field u and its normal derivative $\partial u/\partial n$ (and the known Green function). One of u and $\partial u/\partial n$ may be determined by boundary conditions, but both are not known until the scattering problem is (at least partially) solved. If the obstacle is rigid or a cavity the boundary condition is that u or $\partial u/\partial n$ vanishes and an integral equation for $\partial u/\partial n$ or u, respectively, is obtained if the field point is allowed to approach S. Due to the singularity of the Green function this must be done with care [125]. It should be noted that the surface S need not be the surface of the obstacle; it can be any (fictitious) surface enclosing the obstacle. Furthermore, the surface S need not be a single surface, thus the integral representation is equally valid if there are two (or more) obstacles.

In the surface integral representation Eq. (4.18) the material inside the obstacle can be quite arbitrary. If the material inside the obstacle is anisotropic the most general constitutive relations for a 2D SH problem to exist are (in the reduced notation for the stiffness tensor)

$$\sigma^1_{yz} = C^1_{44}\frac{\partial u^1}{\partial y} + C^1_{45}\frac{\partial u^1}{\partial x}$$

$$\sigma^1_{zx} = C^1_{45}\frac{\partial u^1}{\partial y} + C^1_{55}\frac{\partial u^1}{\partial x}$$

where the superscript "1" on all quantities refers to the obstacle. In tensor notation this is

$$\sigma^1_{i3} = c^1_{i3k3}\partial_k u^1_3$$

where the indices run through $i, k = 1, 2$ and $u^1 = u^1_3$ for clarity. Starting from Eq. (4.18) with the field point outside the obstacle and using continuity of displacement and traction (welded boundary), the scattered wave becomes

$$u_3^{\mathrm{sc}}(\boldsymbol{r}) = \int_S \left[u^1_3(\boldsymbol{r}')\partial'_i g(\boldsymbol{r}';\boldsymbol{r}) - \frac{1}{\mu}\sigma^1_{i3}(\boldsymbol{r}')g(\boldsymbol{r}';\boldsymbol{r})\right] n'_i \, \mathrm{d}S'$$

$$= \int_V \partial'_i \left[u^1_3(\boldsymbol{r}')\partial'_i g(\boldsymbol{r}';\boldsymbol{r}) - \frac{1}{\mu}\sigma^1_{i3}(\boldsymbol{r}')g(\boldsymbol{r}';\boldsymbol{r})\right] \mathrm{d}V'$$

where V is the volume of the obstacle (the volume enclosed by S) and Gauss' law is used in the last step. Using the constitutive relations and the equations of motion for the displacement and the Green function this can be transformed into

$$u_3^{\mathrm{sc}}(\boldsymbol{r}) = \int_V \left[\left(\delta_{ij} - \frac{1}{\mu}c^1_{i3j3}(\boldsymbol{r}')\right)\partial'_j u^1_3(\boldsymbol{r}')\partial'_i g(\boldsymbol{r}';\boldsymbol{r}) - k^2\left(1 - \frac{1}{\rho}\rho_1(\boldsymbol{r}')\right)u^1_3(\boldsymbol{r}')g(\boldsymbol{r}';\boldsymbol{r})\right]\mathrm{d}V' \qquad (4.19)$$

It is explicitly spelled out that the material properties of the obstacle may depend on position (an inhomogeneous obstacle). For the application of Gauss' law, however, the material properties must be differentiable within the obstacle. Equation (4.19) is a volume integral representation which, for instance, is useful when the Born approximation is derived.

4.2.2 Vector integral representation in 2D and 3D

The derivation of the vector integral representations in 2D and 3D are almost identical (Pao and Varatharajulu [113]), so here only the 3D case is considered and the result is just stated in 2D. The derivation is performed "from scratch", but another way is to use the reciprocity relation in Eq. (1.9); the derivation here is essentially the same as the derivation of the reciprocity relation.

At this point it is convenient to introduce the stresses corresponding to the Green tensor. Thus define the Green stress triadic (a tensor of rank three)

$$\Sigma_{ijk}(\boldsymbol{r};\boldsymbol{r}') = \lambda\delta_{ij}\partial_l G_{lk}(\boldsymbol{r};\boldsymbol{r}') + \mu\left(\partial_i G_{jk}(\boldsymbol{r};\boldsymbol{r}') + \partial_j G_{ik}(\boldsymbol{r};\boldsymbol{r}')\right) \quad (4.20)$$

Here and in the whole derivation of the integral representation tensor notation is followed; it is possible to use dyadic notation, but then some nonstandard forms are necessary (Ström[125]). The Green stress triadic is obviously symmetric in the first two indices

$$\Sigma_{ijk}(\boldsymbol{r};\boldsymbol{r}') = \Sigma_{jik}(\boldsymbol{r};\boldsymbol{r}')$$

It follows that the differential equation for the Green tensor [Eq. (4.13)] can be written

$$\frac{1}{\mu k_{\mathrm{s}}^2}\,\partial_k\Sigma_{kij}(\boldsymbol{r};\boldsymbol{r}') + G_{ij}(\boldsymbol{r};\boldsymbol{r}') = -k_{\mathrm{s}}^{-3}\delta_{ij}\delta(\boldsymbol{r}-\boldsymbol{r}')$$

Multiplying this equation with $-u_i(\boldsymbol{r})$ and adding it to the equation of motion for the displacement

$$\frac{1}{\mu k_{\mathrm{s}}^2}\,\partial_k\sigma_{ki}(\boldsymbol{r}) + u_i(\boldsymbol{r}) = 0$$

multiplied with $G_{ij}(\boldsymbol{r};\boldsymbol{r}')$ (and of course summing over i) give

$$\frac{1}{\mu k_{\mathrm{s}}^2}\left[\partial_k\sigma_{ki}(\boldsymbol{r})G_{ij}(\boldsymbol{r};\boldsymbol{r}') - u_i(\boldsymbol{r})\partial_k\Sigma_{kij}(\boldsymbol{r};\boldsymbol{r}')\right] = k_{\mathrm{s}}^{-3}\delta(\boldsymbol{r}-\boldsymbol{r}')u_j(\boldsymbol{r})$$

A few steps show that the left-hand side can be written as a divergence (cf. the derivation of the reciprocity relation in Section 1.1.5)

$$\frac{k_{\rm s}}{\mu}\,\partial_k\left[\sigma_{ki}(\boldsymbol{r})G_{ij}(\boldsymbol{r};\boldsymbol{r}') - u_i(\boldsymbol{r})\Sigma_{ikj}(\boldsymbol{r};\boldsymbol{r}')\right] = \delta(\boldsymbol{r}-\boldsymbol{r}')u_j(\boldsymbol{r})$$

Now the steps from the previous subsection can be repeated and this gives the integral representation

$$u_j^{\rm in}(\boldsymbol{r}) + \frac{k_{\rm s}}{\mu}\int_S n_k\left[u_i(\boldsymbol{r}')\Sigma_{kij}(\boldsymbol{r}';\boldsymbol{r}) - \sigma_{ki}(\boldsymbol{r}')G_{ij}(\boldsymbol{r}';\boldsymbol{r})\right]\mathrm{d}S' = \begin{cases} u_j(\boldsymbol{r}), & \boldsymbol{r} \text{ outside } S \\ 0, & \boldsymbol{r} \text{ inside } S \end{cases} \tag{4.21}$$

where n_k are the components of the normal $\hat{\boldsymbol{n}}$ to the surface S. In vector (and dyadic) notation this equation is

$$\boldsymbol{u}^{\rm in}(\boldsymbol{r}) + \frac{k_{\rm s}}{\mu}\int_S \left[\boldsymbol{u}(\boldsymbol{r}')\cdot(\hat{\boldsymbol{n}}\cdot\boldsymbol{\Sigma}(\boldsymbol{r}';\boldsymbol{r})) - \boldsymbol{t}^{(n)}(\boldsymbol{r}')\cdot\boldsymbol{G}(\boldsymbol{r}';\boldsymbol{r})\right]\mathrm{d}S' = \begin{cases} \boldsymbol{u}(\boldsymbol{r}), & \boldsymbol{r} \text{ outside } S \\ \boldsymbol{0}, & \boldsymbol{r} \text{ inside } S \end{cases} \tag{4.22}$$

where $\boldsymbol{t}^{(n)} = \hat{\boldsymbol{n}}\cdot\boldsymbol{\sigma}$ is the traction vector on S.

In 2D the only difference is that the normalization is slightly different in that the right-hand side in the defining differential equation for the Green tensor has a factor $k_{\rm s}^{-2}$ in 2D instead of $k_{\rm s}^{-3}$ in 3D. The final integral representation in 2D thus looks the same as in 3D except that the factor k_s in front of the integral is removed (and the Green tensor is the 2D one instead of the 3D one).

To get a volume integral representation start from Eq. (4.21), apply the boundary conditions on the obstacle, use Gauss' law to convert the surface integral into a volume integral, and use the general anisotropic constitutive law for the displacement Eq. (1.1) to obtain

$$u_j^{\mathrm{sc}}(\boldsymbol{r}) = \frac{k_{\mathrm{s}}}{\mu}\int_V \left[\left(c_{kilm} - c^1_{kilm}(\boldsymbol{r}')\right)\partial_k u^1_i(\boldsymbol{r}')\partial_l G_{mj}(\boldsymbol{r}';\boldsymbol{r})\right.$$
$$\left. -\omega^2\left(\rho - \rho_1(\boldsymbol{r}')\right)u^1_i(\boldsymbol{r}')G_{ij}(\boldsymbol{r}';\boldsymbol{r})\right]\mathrm{d}V' \tag{4.23}$$

Here the stiffness tensor in the surrounding isotropic medium

$$c_{kilm} = \lambda\delta_{ki}\delta_{lm} + \mu\left(\delta_{kl}\delta_{im} + \delta_{km}\delta_{il}\right)$$

is used for convenience. [Equation (4.23) is in fact valid also when the surrounding is anisotropic provided the Green tensor for the anisotropic material is used.] As spelled out the obstacle may be inhomogeneous. In 2D the same formula is valid if the k_{s} in front is dropped and the indices only run over $1, 2$.

4.3 Born approximation

As an example of the use of volume integral representations the Born approximation is considered. This is an approximate method that can be used when the scattering is small and the contrast in material properties between the obstacle and the surrounding is also small. The Born approximation in elastodynamics is described by, for example, Gubernatis *et al.* [59] and Schmerr [119, Chapter 10]. The Born approximation is widely used in the determination of attenuation and phase velocity in polycrystalline media, for early work see Stanke and Kino [123] and Weaver [141]. More recent work include Thompson *et al.* [128] and Huang *et al.* [65] (where many more references are to be found).

To start as simple as possible 2D SH waves are first considered, followed by 3D waves.

4.3.1 2D SH waves

Recall the volume integral representation of the scattered wave [Eq. (4.19)]

$$u^{\mathrm{sc}}(\boldsymbol{r}) = \int_V \left[\left(\delta_{ij} - \frac{1}{\mu}\,c^1_{i3j3}(\boldsymbol{r}')\right)\partial'_j u^1(\boldsymbol{r}')\partial'_i g(\boldsymbol{r}';\boldsymbol{r})\right.$$
$$\left. - k^2\left(1 - \frac{1}{\rho}\,\rho_1(\boldsymbol{r}')\right)u^1(\boldsymbol{r}')g(\boldsymbol{r}';\boldsymbol{r})\right]\mathrm{d}V'$$

Here the implicit summations are over $i, j = 1, 2$ and $c^1_{i3j3}(\boldsymbol{r}')$ is the stiffness tensor of the obstacle which may be both anisotropic and inhomogeneous. The total wave within the obstacle u^1 is present so the integral representation only gives the scattered wave when u^1 is determined. To obtain u^1 the solution to the whole scattering problem is in general needed, but if the scattering is weak a simple approximation is to put the total wave within the obstacle equal to the incident wave, thus

$$u^{\rm sc}(\boldsymbol{r}) \simeq \int_V \left[\left(\delta_{ij} - \frac{1}{\mu}\, c^1_{i3j3}(\boldsymbol{r}') \right) \partial'_j u^{\rm in}(\boldsymbol{r}') \partial'_i g(\boldsymbol{r}';\boldsymbol{r}) - k^2 \left(1 - \frac{1}{\rho}\, \rho_1(\boldsymbol{r}') \right) u^{\rm in}(\boldsymbol{r}') g(\boldsymbol{r}';\boldsymbol{r}) \right] {\rm d}V' \quad (4.24)$$

This gives the scattered wave explicitly as an integral over the obstacle where the integrand contains the incident wave and the Green function [Eq. (4.2)]

$$g(\boldsymbol{r};\boldsymbol{r}') = -\frac{\rm i}{4}\, {\rm H}_0(k|\boldsymbol{r} - \boldsymbol{r}'|)$$

Because of the Hankel function this integral is seldom possible to perform analytically.

If the scattered wave in the far field is of interest the asymptotic form of the Hankel function can be used [Eq.(2.6)]

$${\rm H}_0(z) \simeq \sqrt{\frac{2}{\pi z}}\, {\rm e}^{{\rm i}(z-\pi/4)}$$

The distance between the obstacle and the field point may in addition be approximated as

$$|\boldsymbol{r} - \boldsymbol{r}'| \simeq r - \hat{\boldsymbol{r}} \cdot \boldsymbol{r}'$$

where the second term can be neglected in the denominator. When taking the gradient of the Green function only the exponential needs to be differentiated and this gives the Born approximation in the far field

$$u^{\rm sc}(\boldsymbol{r}) \simeq \frac{{\rm e}^{{\rm i}(kr+\pi/4)}}{2\sqrt{2\pi kr}} \int_V \left[{\rm i}k_i \left(\delta_{ij} - \frac{1}{\mu}\, c^1_{i3j3}(\boldsymbol{r}') \right) \partial'_j u^{\rm in}(\boldsymbol{r}') + k^2 \left(1 - \frac{1}{\rho}\, \rho_1(\boldsymbol{r}') \right) u^{\rm in}(\boldsymbol{r}') \right] \exp\left(-{\rm i}\boldsymbol{k} \cdot \boldsymbol{r}'\right) {\rm d}V' \quad (4.25)$$

where k_i is the Cartesian component of the wavevector $\boldsymbol{k} = k\hat{\boldsymbol{r}}$ of the scattered wave. If the incident wave is a plane wave with unit amplitude

$$u^{\mathrm{in}}(\boldsymbol{r}) = \exp\left(\mathrm{i}\boldsymbol{k}_0 \cdot \boldsymbol{r}\right)$$

this can be further simplified to (assuming also a homogeneous and isotropic obstacle)

$$u^{\mathrm{sc}}(\boldsymbol{r}) \simeq \frac{\mathrm{e}^{\mathrm{i}(kr+\pi/4)}}{2\sqrt{2\pi kr}} \left[k^2\left(1-\frac{\rho_1}{\rho}\right) - \boldsymbol{k}_0\cdot\boldsymbol{k}\left(1-\frac{\mu_1}{\mu}\right)\right] S(\boldsymbol{\Delta k}) \quad (4.26)$$

Here the form factor is

$$S(\boldsymbol{\Delta k}) = \int_V \exp\left(-\mathrm{i}\boldsymbol{\Delta k}\cdot\boldsymbol{r}\right)\mathrm{d}V$$

where $\boldsymbol{\Delta k} = \boldsymbol{k} - \boldsymbol{k}_0$. In the simple result in Eq. (4.26) the scattering due to the material properties and the shape of the obstacle are completely decoupled, so that two factors independently determine the scattered wave. For low frequencies the form factor becomes the volume (area) of the obstacle, so the scattering becomes proportional to this volume.

If the T-matrix of the obstacle is of interest it is straightforward to calculate it in the Born approximation. Insert the expansion of the Green function Eq. (4.3) in the Born approximation Eq. (4.24) and assume that the field point is outside the circumscribed circle to the obstacle

$$u^{\mathrm{sc}}(\boldsymbol{r}) \simeq \mathrm{i}\sum_{\sigma,m} \chi^{+}_{\sigma m}(\boldsymbol{r}) \int_V \left[\left(\delta_{ij} - \frac{1}{\mu}c^{1}_{i3j3}(\boldsymbol{r}')\right)\partial'_j u^{\mathrm{in}}(\boldsymbol{r}')\partial'_i\chi^{0}_{\sigma m}(\boldsymbol{r}') \right.$$
$$\left. - k^2\left(1-\frac{1}{\rho}\rho_1(\boldsymbol{r}')\right)u^{\mathrm{in}}(\boldsymbol{r}')\chi^{0}_{\sigma m}(\boldsymbol{r}')\right]\mathrm{d}V'$$

Choosing the incident wave as a regular polar wavefunction the T-matrix elements are directly obtained as

$$T_{\sigma m,\sigma' m'} = \mathrm{i}\int_V \left[\left(\delta_{ij} - \frac{1}{\mu}c^{1}_{i3j3}(\boldsymbol{r})\right)\partial_j\chi^{0}_{\sigma' m'}(\boldsymbol{r})\partial_i\chi^{0}_{\sigma m}(\boldsymbol{r}) \right.$$
$$\left. - k^2\left(1-\frac{1}{\rho}\rho_1(\boldsymbol{r})\right)\chi^{0}_{\sigma' m'}(\boldsymbol{r})\chi^{0}_{\sigma m}(\boldsymbol{r})\right]\mathrm{d}V \quad (4.27)$$

The symmetry of the T-matrix is obvious.

In the Born approximation the scattering depends linearly on the difference in material parameters compared to the surrounding. This is certainly not true of the exact solution and it suggest that the Born approximation demands that this difference is small (in addition to the requirement of small scattering), and this is further illuminated in the examples below.

As a first simple example the obstacle is chosen as an isotropic and homogeneous circle with radius a, shear modulus μ_1, and density ρ_1. Using the orthogonality of the trigonometric system the T-matrix elements become

$$T_{\sigma m,\sigma' m'} = \mathrm{i}\delta_{mm'}\delta_{\sigma\sigma'}\frac{\pi}{2}\int_0^a \left[\left(1-\frac{\mu_1}{\mu}\right)\left[k^2\left(\mathrm{J}_m'(kr)\right)^2 + \frac{m^2}{r^2}\left(\mathrm{J}_m(kr)\right)^2\right] + k^2\left(1-\frac{\rho_1}{\rho}\right)\left(\mathrm{J}_m(kr)\right)^2\right] r\,\mathrm{d}r$$

The integrals can be calculated, but concentrating on low frequencies the Bessel functions can be approximated with the first term in their series expansion and the leading order T-matrix elements then become

$$T_{\mathrm{e}0,\mathrm{e}0} = \mathrm{i}(ka)^2\frac{\pi}{4}\left(\frac{\rho_1}{\rho}-1\right)$$

$$T_{\sigma 1,\sigma 1} = \mathrm{i}(ka)^2\frac{\pi}{8}\left(1-\frac{\mu_1}{\mu}\right)$$

for $\sigma = \mathrm{e},\mathrm{o}$. The exact solution for the circle for low frequencies is given in Eqs. (2.23) and (2.24). For $m = 0$ the two T-matrix elements are identical, but for $m = 1$ there is a difference in the denominator which is 8μ in the Born approximation as compared to $4(\mu+\mu_1)$ in the exact solution. The approximation is thus only good when the shear modulus of the obstacle is close to the one of the surrounding.

Consider an isotropic and homogeneous obstacle, but let the shape of the obstacle be arbitrary. For low frequencies the leading order is coming from $m = 0$ (because $\mathrm{J}_0(kr) = 1$ to lowest order) for the term depending on the densities and from $m = 1$ (because $\mathrm{J}_1'(kr) = 1/2$ and $\mathrm{J}_1(kr)/r = k/2$ to lowest order) for the term depending on the shear moduli. For $m = 0$ the result is

$$T_{\mathrm{e}0,\mathrm{e}0} = \mathrm{i}\,\frac{k^2 A}{4}\left(\frac{\rho_1}{\rho}-1\right)$$

where A is the volume of the obstacle. It is only the (relative) mass of the obstacle that enters and this is natural as $m = 0$ corresponds to rigid body motion of the obstacle. For $m = 1$ the two elements become equal

$$T_{\sigma 1,\sigma 1} = \mathrm{i}\,\frac{k^2 A}{8}\left(1 - \frac{\mu_1}{\mu}\right)$$

for $\sigma = \mathrm{e}, \mathrm{o}$. For low frequencies in the Born approximation the scattering by the obstacle depends on the volume of the obstacle, but not on its shape.

As a third example take the circular obstacle that has circular anisotropy so that the constitutive relations are [Eqs. (1.27) and (1.11)]

$$\sigma_{\varphi z} = C_{44}\frac{1}{r}\frac{\partial u}{\partial \varphi}$$

$$\sigma_{zr} = C_{55}\frac{\partial u}{\partial r}$$

The T-matrix elements then become

$$T_{\sigma m,\sigma' m'} = \mathrm{i}\delta_{mm'}\delta_{\sigma\sigma'}\frac{\pi}{2}\int_0^a \Bigg[\left(1 - \frac{C_{44}}{\mu}\right) k^2 \left(\mathrm{J}'_m(kr)\right)^2 + \left(1 - \frac{C_{55}}{\mu}\right)\frac{m^2}{r^2}\left(\mathrm{J}_m(kr)\right)^2 + k^2\left(1 - \frac{\rho_1}{\rho}\right)\left(\mathrm{J}_m(kr)\right)^2\Bigg] r\,\mathrm{d}r$$

For low frequencies the T-matrix element for $m = 0$ becomes the same as in the isotropic case and the $m = 1$ elements become

$$T_{\sigma 1,\sigma 1} = \mathrm{i}(ka)^2\frac{\pi}{8}\left(1 - \frac{C_{44} + C_{55}}{2\mu}\right)$$

for $\sigma = \mathrm{e}, \mathrm{o}$. The corresponding exact element is [Eq. (6.3)]

$$T_1 = \mathrm{i}(ka)^2\frac{\pi(\mu - \sqrt{C_{44}C_{55}})}{4(\mu + \sqrt{C_{44}C_{55}})}$$

That these two elements are equal to lowest order is most easily seen by writing $C_{44} = \mu(1 + \epsilon_4)$ and $C_{55} = \mu(1 + \epsilon_5)$, with ϵ_4 and ϵ_5 small, and expanding the root in the last equation.

4.3.2 3D waves

As in the 2D case the Born approximation is starting from the volume integral representation [Eq. (4.23)]

$$u_j^{\mathrm{sc}}(\boldsymbol{r}) = \frac{k_{\mathrm{s}}}{\mu}\int_V \left[\left(c_{kilm} - c_{kilm}^1(\boldsymbol{r}')\right)\partial_k u_i^1(\boldsymbol{r}')\partial_l G_{mj}(\boldsymbol{r}';\boldsymbol{r})\right.$$
$$\left. -\omega^2\left(\rho - \rho_1(\boldsymbol{r}')\right)u_i^1(\boldsymbol{r}')G_{ij}(\boldsymbol{r}';\boldsymbol{r})\right]\mathrm{d}V'$$

where the stiffness tensor in the surrounding isotropic medium is

$$c_{kilm} = \lambda\delta_{ki}\delta_{lm} + \mu\left(\delta_{kl}\delta_{im} + \delta_{km}\delta_{il}\right)$$

The approximation consists in replacing the total wave u_i^1 inside the obstacle with the incident wave

$$u_j^{\mathrm{sc}}(\boldsymbol{r}) \simeq \frac{k_{\mathrm{s}}}{\mu}\int_V \left[\left(c_{kilm} - c_{kilm}^1(\boldsymbol{r}')\,\partial_k u_i^{\mathrm{in}}(\boldsymbol{r}')\partial_l G_{mj}(\boldsymbol{r}';\boldsymbol{r})\right.\right.$$
$$\left. -\omega^2\left(\rho - \rho_1(\boldsymbol{r}')\right)u_i^{\mathrm{in}}(\boldsymbol{r}')G_{ij}(\boldsymbol{r}';\boldsymbol{r})\right]\mathrm{d}V' \tag{4.28}$$

The Green tensor in this equation is [Eq. (4.14)]

$$G_{ij}(\boldsymbol{r};\boldsymbol{r}') = \frac{1}{k_{\mathrm{s}}}\delta_{ij}g_{\mathrm{s}}(\boldsymbol{r};\boldsymbol{r}') + \frac{1}{k_{\mathrm{s}}^3}\partial_i\partial_j\left[g_{\mathrm{s}}(\boldsymbol{r};\boldsymbol{r}') - g_{\mathrm{p}}(\boldsymbol{r};\boldsymbol{r}')\right]$$

with the 3D scalar Green function

$$g_n(\boldsymbol{r};\boldsymbol{r}') = \frac{\exp\left(\mathrm{i}k_n|\boldsymbol{r}-\boldsymbol{r}'|\right)}{4\pi|\boldsymbol{r}-\boldsymbol{r}'|}, \qquad n = \mathrm{s},\mathrm{p}$$

In the far field the scalar Green tensor can be approximated as

$$g_n(\boldsymbol{r};\boldsymbol{r}') = \frac{\exp\left(\mathrm{i}k_n(r - \hat{\boldsymbol{r}}\cdot\boldsymbol{r}')\right)}{4\pi r}, \qquad n = \mathrm{s},\mathrm{p}$$

The incident wave is assumed to be a plane wave

$$u_j(\boldsymbol{r}) = A_j\exp\left(\mathrm{i}\boldsymbol{k}_{0\mathrm{n}}\cdot\boldsymbol{r}\right)$$

where $\boldsymbol{k}_{0n}$, $n = \mathrm{s}$, p is the wavevector of the incident wave, and A_j is its amplitude which is orthogonal to the wavevector for an incident S wave and parallel to the wavevector for an incident P wave. Assuming constant material parameters inside the obstacle this eventually leads to the scattered wave in the far field

$$u_j^{\mathrm{sc}}(\boldsymbol{r}) \simeq \frac{1}{4\pi r \rho\,\omega^2}\left[\mathrm{e}^{\mathrm{i}k_{\mathrm{p}}r} A_i k_{\mathrm{p}j}\left[\left(c^1_{kilm} - c_{kiml}\right) k_{\mathrm{n}k} k_{\mathrm{p}l} k_{\mathrm{p}m}\right.\right.$$
$$\left. +\omega^2(\rho_1 - \rho) k_{\mathrm{p}i}\right] S(\boldsymbol{\Delta k}_{0\mathrm{np}})$$
$$+\mathrm{e}^{\mathrm{i}k_{\mathrm{s}}r} A_i \left[\left(c^1_{kilm} - c_{kiml}\right)\left(\delta_{mj}k_{\mathrm{s}}^2 - k_{\mathrm{s}m}k_{\mathrm{s}j}\right) k_{\mathrm{n}k}\right.$$
$$\left.\left. +\omega^2(\rho_1 - \rho)\left(\delta_{ij}k_{\mathrm{s}}^2 - k_{\mathrm{s}i}k_{\mathrm{s}j}\right)\right] S(\boldsymbol{\Delta k}_{0\mathrm{ns}})\right] \qquad (4.29)$$

where $\boldsymbol{\Delta k}_{0\mathrm{np}} = \boldsymbol{k}_{0\mathrm{n}} - \boldsymbol{k}_{\mathrm{p}}$ and $\boldsymbol{\Delta k}_{0\mathrm{ns}} = \boldsymbol{k}_{0\mathrm{n}} - \boldsymbol{k}_{\mathrm{s}}$, and where the form factor is

$$S(\boldsymbol{\Delta k}) = \int_V \exp\left(\mathrm{i}\boldsymbol{\Delta k}\cdot\boldsymbol{r}\right)\mathrm{d}V$$

As in 2D the material properties and the shape of the obstacle are decoupled and for low frequencies it follows that the scattering is proportional to the volume.

To obtain the T-matrix elements the expansion of the Green tensor in the spherical vector wavefunctions [Eq. (4.16)] is employed and the incident wave is taken as a regular spherical vector wavefunction

$$T_{\tau\sigma ml,\tau'\sigma'm'l'} \simeq \frac{\mathrm{i}k_{\mathrm{s}}}{\mu}\int_V \left[\left(c_{ijkn} - c^1_{ijkn}(\boldsymbol{r}\right)\partial_i\psi^0_{j,\tau\sigma ml}(\boldsymbol{r})\partial_k\psi^0_{n,\tau'\sigma'm'l'}(\boldsymbol{r})\right.$$
$$\left. -\omega^2\left(\rho - \rho_1(\boldsymbol{r})\right)\psi^0_{i,\tau\sigma ml}(\boldsymbol{r})\psi^0_{i,\tau'\sigma'm'l'}(\boldsymbol{r})\right]\mathrm{d}V \qquad (4.30)$$

Here $\psi^0_{j,\tau\sigma ml}(\boldsymbol{r})$ is the j:th Cartesian component of the regular spherical wavefunction $\psi^0_{\tau\sigma ml}(\boldsymbol{r})$.

As in 2D there are of course rather strict restrictions on the validity of the Born approximation. Both the scattering and the difference in material parameters must be small. See Gubernatis *et al.* [59] and Schmerr [119] for 3D examples and further comments.

5 Incident and scattered waves

Incident waves, primarily plane waves, are given in previous chapters, but only for particular directions of incidence as this is sufficient for the rotationally symmetric obstacles investigated so far. For other obstacles arbitrary directions of incidence are of interest, so here the expansions of incident plane waves from any direction are given. Also point forces and general volume sources are considered.

Scattered waves are also further investigated, in particular in the far field. The consequences of energy conservation and reciprocity are investigated.

The transition (T) matrix is defined and calculated in previous chapters. This matrix has some general properties, due to reciprocity and energy conservation, which are explored in the last section.

5.1 Incident waves

In this section incident waves are considered and the division is then between 2D and 3D cases, and the 2D cases are further divided into the SH case and the P-SV case. The Green functions are then found to be of great value. Incident waves from a point source and plane waves are considered. By superposition general volume sources are then easily treated. However, in some applications sources are better (or more easily) modelled as a boundary condition and such sources are not considered.

5.1.1 2D SH waves

In Section 2.1.3 incident waves from a point source and incident plane waves were considered by deriving the expansion coefficients for them in polar wavefunctions. Due to rotational symmetry of the obstacle it was enough to do this for the source (which is at infinity for a plane wave) on the x-axis. Here this is extended to arbitrary location of the source. Also rather arbitrary volume sources are taken up.

Consider first a point force with arbitrary location specified by the position $\boldsymbol{d}$ with polar coordinates d and φ_0. With this source located in an infinite homogeneous medium (wavenumber k) without any obstacle, the wave u from this source is the solution of

$$\nabla^2 u + k^2 u = -A\delta(\boldsymbol{r} - \boldsymbol{d})$$

where A gives the strength of the force. This is the same differential equation as for the 2D free space scalar Green function, Eq. (4.1). If the source point is located further from the origin than the field point the solution is [Eq. (4.3)]

$$u(\boldsymbol{r}) = \mathrm{i}A \sum_{\sigma,m} \chi^{+}_{\sigma m}(\boldsymbol{d})\chi^{0}_{\sigma m}(\boldsymbol{r})$$

where it is thus assumed that $d > r$. This is an expansion in the regular polar wavefunctions with expansion coefficients

$$b_{\sigma m} = \mathrm{i}A\chi^{+}_{\sigma m}(\boldsymbol{d})$$

or more explicitly

$$\begin{aligned} b_{em} &= \frac{\mathrm{i}\sqrt{\epsilon_m}A}{2}\,\mathrm{H}_m(kd)\cos m\varphi_0, \qquad m = 0, 1, 2, \ldots \\ b_{om} &= \frac{\mathrm{i}A}{\sqrt{2}}\,\mathrm{H}_m(kd)\sin m\varphi_0, \qquad m = 1, 2, \ldots \end{aligned} \tag{5.1}$$

For a plane wave it easiest to use the plane wave expansion [Eq. (2.17)]

$$\mathrm{e}^{\mathrm{i}kr\cos\varphi} = \sum_{m=0}^{\infty} \mathrm{i}^m \epsilon_m \mathrm{J}_m(kr)\cos m\varphi$$

where $x = r\cos\varphi$ is inserted. Exchanging φ to $\varphi - \varphi_0$ it is realized that the expansion coefficients for a plane wave with unit amplitude propagating in the direction given by φ_0 are

$$b_{em} = 2\sqrt{\epsilon_m}\,\mathrm{i}^m\cos m\varphi_0, \qquad m = 0, 1, 2, \ldots$$

$$b_{\mathrm{o}m} = 2\sqrt{2}\,\mathrm{i}^m \sin m\varphi_0, \qquad m = 1, 2, \ldots \tag{5.2}$$

A general volume (which is an area in this 2D case) force is also straightforward to consider. The Helmholtz equation with a volume force can be written

$$\nabla^2 u(\boldsymbol{r}) + k^2 u(\boldsymbol{r}) = -\frac{\rho}{\mu} f(\boldsymbol{r}) = -\frac{\rho}{\mu} \int_{V_i} f(\boldsymbol{r}')\delta(\boldsymbol{r} - \boldsymbol{r}')\,\mathrm{d}V'$$

where the volume force $f(\boldsymbol{r})$ is assumed nonzero in a domain (area) V_i. The second way of writing the force shows that it can be viewed as a superposition of point forces. As the solution to a point force is the Green function, the wave from the volume force is

$$u(\boldsymbol{r}) = \frac{\rho}{\mu} \int_{V_i} f(\boldsymbol{r}') G(\boldsymbol{r}; \boldsymbol{r}')\,\mathrm{d}V' \tag{5.3}$$

If the source volume V_i lies completely outside a circle circumscribing the obstacle it is possible to use the expansion of the Green function Eq. (4.3) with the origin at the centre of this circumscribed circle to get the expansion of the incident wave

$$u^{\mathrm{in}}(\boldsymbol{r}) = \sum_{\sigma,m} b_{\sigma m} \chi^0_{\sigma m}(\boldsymbol{r})$$

with expansion coefficients

$$b_{\sigma m} = \frac{\mathrm{i}\rho}{\mu} \int_{V_i} f(\boldsymbol{r}') \chi^+_{\sigma m}(\boldsymbol{r}')\,\mathrm{d}V' \tag{5.4}$$

5.1.2 2D P-SV waves

For the 2D P-SV waves the approach is the same as for the scalar case. For a point force $\boldsymbol{F}$ at the position $\boldsymbol{d}$ the expansion coefficients (in the polar vector wavefunctions) are

$$b_{\tau\sigma m} = \mathrm{i}\boldsymbol{F} \cdot \boldsymbol{\chi}^+_{\tau\sigma m}(\boldsymbol{d}) \tag{5.5}$$

for $\tau = 1, 2$, $\sigma = \mathrm{e}, \mathrm{o}$, and $m = 0, 1, 2, \ldots$. As for the scalar case this can of course be written more explicitly. It is noted that the waves from a point force consist of both SV and P waves.

The plane wave expansion coefficients can be obtained from the corresponding scalar ones exactly as in Section 2.2.3. For an incident plane SV wave propagating in the φ_0-direction they become

$$b_{1\mathrm{e}m} = 2\sqrt{\epsilon_m}\,\mathrm{i}^{m-1}\cos m\varphi_0, \qquad m = 0, 1, 2, \ldots$$

$$b_{1\mathrm{o}m} = 2\sqrt{2}\,\mathrm{i}^{m-1}\sin m\varphi_0, \qquad m = 1, 2, \ldots \tag{5.6}$$

For an incident P wave they become

$$b_{2\mathrm{e}m} = \frac{2k_\mathrm{s}}{k_\mathrm{p}}\sqrt{\epsilon_m}\,\mathrm{i}^{m-1}\cos m\varphi_0, \qquad m = 0, 1, 2, \ldots$$

$$b_{2\mathrm{o}m} = \frac{2k_\mathrm{s}}{k_\mathrm{p}}\sqrt{2}\,\mathrm{i}^{m-1}\sin m\varphi_0, \qquad m = 1, 2, \ldots \tag{5.7}$$

As in the SH case a general force is also possible to consider and gives the wave

$$\boldsymbol{u}(\boldsymbol{r}) = \frac{\rho}{\mu}\int_{V_i} \boldsymbol{f}(\boldsymbol{r}')\cdot\boldsymbol{G}(\boldsymbol{r};\boldsymbol{r}')\,\mathrm{d}V' \tag{5.8}$$

Inserting the expansion of the Green dyadic gives an expansion in terms of the regular polar vector wavefunctions.

5.1.3 3D waves

Exactly as in 2D the expansion coefficients (in the spherical vector wavefunctions) for a point force $\boldsymbol{F}$ at the position $\boldsymbol{d}$ are

$$b_{\tau\sigma ml} = \mathrm{i}\boldsymbol{F}\cdot\boldsymbol{\psi}^{+}_{\tau\sigma ml}(\boldsymbol{d}) \tag{5.9}$$

for $\tau = 1, 2, 3$, $\sigma = \mathrm{e}, \mathrm{o}$, $m = 0, 1, 2, \ldots$, and $l = m, m+1, \ldots$ (with $l = 0$ omitted for $\tau = 1, 2$). It is noted that the waves from a point force consist of both S and P waves.

To derive the expansion coefficients for a plane wave it is convenient to start from the Green dyadic expanded in the spherical vector wavefunctions [Eq. (4.16)]

$$\boldsymbol{G}(\boldsymbol{r};\boldsymbol{r}') = \mathrm{i}\sum_{\tau,\sigma,m,l}\boldsymbol{\psi}^{0}_{\tau\sigma ml}(\boldsymbol{r}_<)\boldsymbol{\psi}^{+}_{\tau\sigma ml}(\boldsymbol{r}_>)$$

The closed form of the Green dyadic is [Eq. (4.15)]

$$\boldsymbol{G}(\boldsymbol{r};\boldsymbol{r}') = \frac{1}{k_{\rm s}}\boldsymbol{I} g_{\rm s}(\boldsymbol{r};\boldsymbol{r}') + \frac{1}{k_{\rm s}^3}\nabla\nabla\left[g_{\rm s}(\boldsymbol{r};\boldsymbol{r}') - g_{\rm p}(\boldsymbol{r};\boldsymbol{r}')\right]$$

where the scalar Green function is [Eq. (4.5)]

$$g_n(\boldsymbol{r};\boldsymbol{r}') = \frac{\exp\left(\mathrm{i}k_n|\boldsymbol{r}-\boldsymbol{r}'|\right)}{4\pi|\boldsymbol{r}-\boldsymbol{r}'|}, \qquad n = \mathrm{s},\mathrm{p}$$

To get a plane wave the source position $\boldsymbol{r}' = \boldsymbol{r}_>$ is pushed to infinity. If the plane wave is to propagate in the direction $\hat{\boldsymbol{k}}$ the source is situated in the direction $-\hat{\boldsymbol{k}}$ at the large distance $d = |\boldsymbol{r}'|$. If d is very large compared to the distance to the field point $r = |\boldsymbol{r}|$ the scalar Green function can be approximated as

$$g(\boldsymbol{r};\boldsymbol{r}') = \frac{\exp\left(\mathrm{i}k(d+\hat{\boldsymbol{k}}\cdot\boldsymbol{r})\right)}{4\pi d}$$

The gradient then has the action

$$\nabla \exp\left(\mathrm{i}k(d+\hat{\boldsymbol{k}}\cdot\boldsymbol{r})\right) = \mathrm{i}k\hat{\boldsymbol{k}} \exp\left(\mathrm{i}k(d+\hat{\boldsymbol{k}}\cdot\boldsymbol{r})\right)$$

Consider first an incident P wave. Scalar multiplication of the Green dyadic with $\hat{\boldsymbol{k}}$ gives that the two terms in g_s cancel and there remains

$$\hat{\boldsymbol{k}}\cdot\boldsymbol{G}(\boldsymbol{r};\boldsymbol{r}') = \frac{k_{\rm p}^2\hat{\boldsymbol{k}}}{4\pi k_{\rm s}^3 d}\exp\left(\mathrm{i}k_{\rm p}(d+\hat{\boldsymbol{k}}\cdot\boldsymbol{r})\right)$$

In the expansion of the Green dyadic the out-going wavefunction $\boldsymbol{\psi}^+_{\tau\sigma ml}(\boldsymbol{r}'_>)$ has a large argument so the asymptotic forms Eq. (3.30) can be used. The two S wavefunctions for $\tau = 1,2$ are then seen to be orthogonal to $\hat{\boldsymbol{k}}$ and there remains

$$\hat{\boldsymbol{k}}\cdot\boldsymbol{G}(\boldsymbol{r};\boldsymbol{r}') = \mathrm{i}\sum_{\sigma,m,l}\boldsymbol{\psi}^0_{3\sigma ml}(\boldsymbol{r})\left(\frac{k_{\rm p}}{k_{\rm s}}\right)^{3/2}\frac{1}{k_{\rm p}d}\,\mathrm{e}^{\mathrm{i}(k_{\rm p}d-l\pi/2)}\mathrm{Y}_{\sigma ml}(-\hat{\boldsymbol{k}})$$

where the argument of the spherical harmonic denotes a dependence on the two spherical angles specifying the direction of $-\hat{\boldsymbol{k}}$. From the last two equations the expansion of a plane P wave follows as

$$\hat{\boldsymbol{k}}\exp\left(\mathrm{i}k_{\rm p}\hat{\boldsymbol{k}}\cdot\boldsymbol{r}\right) = \sum_{\sigma,m,l} b_{3\sigma ml}\boldsymbol{\psi}^0_{3\sigma ml}(\boldsymbol{r}) \tag{5.10}$$

where the coefficients are [using that $\mathrm{Y}_{\sigma ml}(-\hat{\boldsymbol{k}}) = (-1)^l \mathrm{Y}_{\sigma ml}(\hat{\boldsymbol{k}})$]

$$b_{3\sigma ml} = 4\pi \mathrm{i}^{l-1} \left(\frac{k_\mathrm{s}}{k_\mathrm{p}}\right)^{3/2} \mathrm{Y}_{\sigma ml}(\hat{\boldsymbol{k}}) \tag{5.11}$$

When $\hat{\boldsymbol{k}} = \hat{\boldsymbol{z}}$ this is seen to reduce to the corresponding coefficients in Section 3.2.2.

An incident S wave has a polarization, given by the unit vector $\hat{\boldsymbol{e}}$, which is perpendicular to the direction of propagation so that $\hat{\boldsymbol{e}} \cdot \hat{\boldsymbol{k}} = 0$. Scalar multiplication of the Green dyadic with $\hat{\boldsymbol{e}}$ gives that the last two terms vanish and there remains

$$\hat{\boldsymbol{e}} \cdot \boldsymbol{G}(\boldsymbol{r};\boldsymbol{r}') = \frac{\hat{\boldsymbol{e}}}{4\pi k_\mathrm{s} d} \exp\left(\mathrm{i}k_\mathrm{s}(d + \hat{\boldsymbol{k}} \cdot \boldsymbol{r})\right)$$

In the expansion of the Green dyadic the out-going wavefunction $\boldsymbol{\psi}^+_{\tau\sigma ml}(\boldsymbol{r}_>)$ has a large argument so the asymptotic forms Eq. (3.30) can be used. It is then seen that the P wavefunction for $\tau = 3$ is orthogonal to $\hat{\boldsymbol{e}}$ so that there remains

$$\hat{\boldsymbol{e}} \cdot \boldsymbol{G}(\boldsymbol{r};\boldsymbol{r}') = \mathrm{i} \sum_{\sigma,m,l} \frac{1}{k_\mathrm{s} r} \mathrm{e}^{\mathrm{i}(k_\mathrm{s} d - l\pi/2)} \, \hat{\boldsymbol{e}} \cdot \Big[-\mathrm{i}\boldsymbol{A}_{1\sigma ml}(-\hat{\boldsymbol{k}}) \boldsymbol{\psi}^0_{1\sigma ml}(\boldsymbol{r}) + \boldsymbol{A}_{2\sigma ml}(-\hat{\boldsymbol{k}}) \boldsymbol{\psi}^0_{2\sigma ml}(\boldsymbol{r}) \Big]$$

From the last two equations the expansion of a plane S wave follows as

$$\hat{\boldsymbol{e}} \exp\left(\mathrm{i}k_\mathrm{s} \hat{\boldsymbol{k}} \cdot \boldsymbol{r}\right) = \sum_{\tau=1}^{2} \sum_{\sigma,m,l} b_{\tau\sigma ml} \boldsymbol{\psi}^0_{\tau\sigma ml}(\boldsymbol{r}) \tag{5.12}$$

where the coefficients are [using that $\boldsymbol{A}_{1\sigma ml}(-\hat{\boldsymbol{k}}) = (-1)^l \boldsymbol{A}_{1\sigma ml}(\hat{\boldsymbol{k}})$ and $\boldsymbol{A}_{2\sigma ml}(-\hat{\boldsymbol{k}}) = (-1)^{l+1} \boldsymbol{A}_{2\sigma ml}(\hat{\boldsymbol{k}})$]

$$\begin{aligned} b_{1\sigma ml} &= 4\pi \mathrm{i}^l \, \hat{\boldsymbol{e}} \cdot \boldsymbol{A}_{1\sigma ml}(\hat{\boldsymbol{k}}) \\ b_{2\sigma ml} &= 4\pi \mathrm{i}^{l-1} \, \hat{\boldsymbol{e}} \cdot \boldsymbol{A}_{2\sigma ml}(\hat{\boldsymbol{k}}) \end{aligned} \tag{5.13}$$

When $\hat{\boldsymbol{k}} = \hat{\boldsymbol{z}}$ and $\hat{\boldsymbol{e}} = \hat{\boldsymbol{x}}$ this is seen to reduce to the corresponding coefficients in Section 3.2.2.

As in the 2D cases a general volume force is also possible to consider and gives the wave

$$\boldsymbol{u}(\boldsymbol{r}) = \frac{\rho k_{\mathrm{s}}}{\mu} \int_{V_i} \boldsymbol{f}(\boldsymbol{r}') \cdot \mathbf{G}(\boldsymbol{r};\boldsymbol{r}')\,\mathrm{d}V' \tag{5.14}$$

Inserting the expansion of the Green dyadic gives an expansion in terms of the regular spherical vector wavefunctions.

5.2 Scattered waves

In this section some general properties of scattered waves are investigated. In particular, this concerns the far field and the forward amplitude theorem and the reciprocity relations. As for incident waves the division is between 2D SH waves, 2D P-SV waves, and 3D waves. The developments here are similar to Varatharajulu [135] and Ström [126]. A different approach is followed by Tan [127], see also Schmerr [119].

5.2.1 Asymptotic evaluation of integrals

In the subsections to follow an asymptotic calculation of certain integrals is needed. When the integral is a simple one, it is of the following type

$$I(\xi) = \int_a^b g(t)\,\mathrm{e}^{\mathrm{i}\xi\psi(t)}\,\mathrm{d}t$$

Here ξ is a large, real, and positive parameter, $g(t)$ is a continuous function (real or complex), and the phase function $\psi(t)$ is real and $\psi'(t)$ is continuous. The functions $g(t)$ and $\psi(t)$ must be independent of ξ. When ξ is large the integrand oscillates rapidly and cancels over most of the interval. However, when $\psi'(t) = 0$ there are less cancellations. A point where $\psi'(t) = 0$ is called "stationary" and is therefore denoted t_s. Assume that there is only one stationary point t_s in the integration interval (which is not located at either end point). For large ξ the dominating contribution from the integral becomes

$$I(\xi) = \left(\frac{2\pi}{\xi|\psi''(t_s)|}\right)^{1/2} g(t_s)\,\mathrm{e}^{\mathrm{i}(\xi\psi(t_s)\pm\pi/4)} \tag{5.15}$$

where the upper or lower sign is taken according to $\psi''(t_s)$ being positive or negative. If there is more than one stationary point in the integration interval their contributions are simply added. This method is usually called the stationary-phase method. For a proof and further details see Olver [110, Chapter 3.11].

In this section the phase function $\psi(\varphi) = \cos\varphi$ is encountered on the interval $[0, 2\pi]$ with a periodic integrand. There are two stationary points at $\varphi_t = 0, \pi$. As the integrand is periodic it does not matter that one of the points appears at an end point (the integral could as well be taken over $[-\pi/2, 3\pi/2]$). At these points $\psi''(0) = -1$ and $\psi''(\pi) = 1$ so asymptotically the integral becomes

$$\int_0^{2\pi} g(\varphi)\mathrm{e}^{\mathrm{i}\xi\psi(\varphi)}\,\mathrm{d}\varphi = g(0)\sqrt{\frac{2\pi}{\xi}}\,\mathrm{e}^{\mathrm{i}(\xi-\pi/4)} + g(\pi)\sqrt{\frac{2\pi}{\xi}}\,\mathrm{e}^{-\mathrm{i}(\xi-\pi/4)} \tag{5.16}$$

For double integrals the stationary-phase method is as follows, see Felsen and Marcuvitz [50]. The integral to estimate is

$$I(\xi) = \iint g(r,t)\,\mathrm{e}^{\mathrm{i}\xi\psi(r,t)}\,\mathrm{d}r\,\mathrm{d}t$$

integrated over some area. The requirements are similar as above. A stationary point fulfils

$$\frac{\partial\psi}{\partial r}(r_s, t_s) = \frac{\partial\psi}{\partial t}(r_s, t_s) = 0$$

For large ξ the dominating contribution from the integral becomes

$$I(\xi) = g(r_s, t_s)\frac{2\pi}{\xi\sqrt{|Q|}}\,\mathrm{e}^{\mathrm{i}(\xi\psi(r_s,t_s)+\sigma\pi/4)} \tag{5.17}$$

Here $Q = \det Q_{ij}$ is the determinant of the following matrix

$$(Q_{ij}) = \begin{pmatrix} \frac{\partial^2\psi}{\partial r^2}(r_s, t_s) & \frac{\partial^2\psi}{\partial r\partial t}(r_s, t_s) \\ \\ \frac{\partial^2\psi}{\partial r\partial t}(r_s, t_s) & \frac{\partial^2\psi}{\partial t^2}(r_s, t_s) \end{pmatrix}$$

and $\sigma = \mathrm{sign}(\lambda_1) + \mathrm{sign}(\lambda_2)$, where λ_1 and λ_2 are the eigenvalues of the matrix Q_{ij}.

In this section the following type of double integral appears

$$I(\xi) = \int_0^{\pi} \int_0^{2\pi} g(\hat{\boldsymbol{r}}) \exp(\mathrm{i}\xi \hat{\boldsymbol{k}} \cdot \hat{\boldsymbol{r}}) \sin\theta \, \mathrm{d}\theta \, \mathrm{d}\varphi$$

where ξ is again the large parameter. The notation $g(\hat{\boldsymbol{r}})$ means that g is a function of the two spherical angles θ and φ. The double integral can be calculated asymptotically by the 2D version of the stationary-phase method. Writing out the phase function in full it is

$$\psi(\theta, \varphi) = \sin\theta \cos\varphi \sin\alpha \cos\beta + \sin\theta \sin\varphi \sin\alpha \sin\beta + \cos\theta \cos\alpha$$

where α and β are the spherical angles specifying the direction of $\hat{\boldsymbol{k}}$. Calculating the derivatives it is seen that there are two stationary points at $\theta_{s1} = \alpha$, $\varphi_{s1} = \beta$ and $\theta_{s2} = \pi - \alpha$, $\varphi_{s2} = \pi + \beta$, which can be written as $\hat{\boldsymbol{r}}_{s1} = \hat{\boldsymbol{k}}$ and $\hat{\boldsymbol{r}}_{s2} = -\hat{\boldsymbol{k}}$, respectively. At the stationary points the matrix Q becomes

$$Q_1 = \begin{pmatrix} -1 & 0 \\ 0 & -\sin^2\alpha \end{pmatrix}$$

$$Q_2 = \begin{pmatrix} 1 & 0 \\ 0 & \sin^2\alpha \end{pmatrix}$$

where the eigenvalues are directly identified. The stationary-phase method thus gives the dominating contribution

$$I(\xi) = -\frac{2\pi\mathrm{i}}{\xi} \left(g(\hat{\boldsymbol{k}}) \mathrm{e}^{\mathrm{i}\xi} - g(-\hat{\boldsymbol{k}}) \mathrm{e}^{-\mathrm{i}\xi} \right) \tag{5.18}$$

where $g(\hat{\boldsymbol{k}})$ means that g is a function of the two spherical angles α and β specifying the unit vector $\hat{\boldsymbol{k}}$.

5.2.2 2D SH waves

In Section 2.1.2 the wave scattered from a circular obstacle is expanded in the out-going polar wavefunctions. It seems reasonable that such an expansion is possible for arbitrary bounded obstacles sufficiently outside the obstacle. This can be shown by using the integral representation with the Green function expanded in polar wavefunctions. Thus, take the upper line in Eq. (4.18) and insert Eq. (4.3) to get

$$u^{\mathrm{sc}}(\boldsymbol{r}) = \sum_{\sigma,m} f_{\sigma m} \chi^{+}_{\sigma m}(\boldsymbol{r}) \tag{5.19}$$

where now

$$f_{\sigma m} = \mathrm{i} \int_S \left[u(\boldsymbol{r}') \nabla \chi^0_{\sigma m}(\boldsymbol{r}') - (\nabla u(\boldsymbol{r}')) \chi^0_{\sigma m}(\boldsymbol{r}') \right] \cdot \hat{\boldsymbol{n}}' \, \mathrm{d}S' \tag{5.20}$$

where it must be assumed that $r = |\boldsymbol{r}| > \max |\boldsymbol{r}'|$. The expansion is therefore valid outside the circumscribed circle to the obstacle. In general it cannot be used in the region between the obstacle and its circumscribed circle (but there are exceptions). The scattered wave is thus determined by the value of the total field and its derivative on the surface S. It is noted that the surface S can be any surface circumscribing the obstacle and the expansion is also valid if S is composed of several nonoverlapping surfaces.

In the far field the asymptotic form of the Hankel functions in the expansion of the scattered wave can be used and it is then seen that the scattered wave can be written in the form

$$u^{\mathrm{sc}}(r, \varphi) = \sqrt{\frac{2}{\pi k r}} \, \mathrm{e}^{\mathrm{i}(kr - \pi/4)} F(\varphi)$$

where $F(\varphi)$ is the far field scattering amplitude. Consider now an incident plane wave which is written as

$$u^{\mathrm{in}}(\boldsymbol{r}) = \exp(\mathrm{i}k\hat{\boldsymbol{k}} \cdot \boldsymbol{r})$$

Without loss of generality the incident wave is chosen with unit amplitude. The unit vector $\hat{\boldsymbol{k}}$ gives the direction of propagation and depends on the angle of propagation but is independent of the wavenumber k. The total wave in the far field is now written as

$$u(\hat{\boldsymbol{k}}; \boldsymbol{r}) = \exp(\mathrm{i}k\hat{\boldsymbol{k}} \cdot \boldsymbol{r}) + \sqrt{\frac{2}{\pi k r}} \, \mathrm{e}^{\mathrm{i}(kr - \pi/4)} F(\hat{\boldsymbol{r}}, \hat{\boldsymbol{k}}) \tag{5.21}$$

where the dependence on the direction of incidence is made explicit by the dependence on $\hat{\boldsymbol{k}}$. This dependence and the far field dependence on the direction φ are denoted by the unit vectors $\hat{\boldsymbol{k}}$ and $\hat{\boldsymbol{r}}$, respectively. In 2D this notation may seem clumsy as this corresponds to a single angle, but in 3D this is a convenient and common notation and is therefore followed also here.

First the implication of energy conservation is explored. If the materials are lossless energy is conserved and this can be stated as

$$\int_S \langle j_n \rangle \mathrm{d}S = 0$$

where S is any closed curve in the plane and $\langle j_n \rangle$ is the (mean) energy flow in the normal direction to the curve. Choose the curve as a large circle with radius R, then

$$\langle J_r \rangle = \int_0^{2\pi} \langle j_r \rangle R \, \mathrm{d}\varphi = 0$$

The radial energy flow in the 2D SH case is

$$\langle j_r \rangle = -\frac{\omega}{2} \operatorname{Im} (\boldsymbol{u} \cdot \boldsymbol{\sigma}^\star \cdot \hat{\boldsymbol{r}}) = -\frac{\omega\mu}{2} \operatorname{Im} \left(u \frac{\partial u^\star}{\partial r} \right)$$

Inserting the far field from Eq. (5.21) this becomes

$$\begin{aligned} \langle j_r \rangle = -\frac{\omega\mu}{2} \operatorname{Im} \Bigg[& -\mathrm{i}k\hat{\boldsymbol{k}} \cdot \hat{\boldsymbol{r}} \\ & + \sqrt{\frac{2}{\pi k R}} \Big(-\mathrm{i}k F^\star(\hat{\boldsymbol{r}}, \hat{\boldsymbol{k}}) \exp\left(\mathrm{i}(k\hat{\boldsymbol{k}} \cdot \boldsymbol{r} - kR + \pi/4)\right) \\ & -\mathrm{i}k\hat{\boldsymbol{k}} \cdot \hat{\boldsymbol{r}} F(\hat{\boldsymbol{r}}, \hat{\boldsymbol{k}}) \exp\left(\mathrm{i}(-k\hat{\boldsymbol{k}} \cdot \boldsymbol{r} + kR - \pi/4)\right) \Big) - \frac{2\mathrm{i}}{\pi R} |F(\hat{\boldsymbol{r}}; \hat{\boldsymbol{k}})|^2 \Bigg] \end{aligned}$$

The first term corresponds to the incident wave and integrates to zero. The fourth term is left as it is. The second and third terms can be estimated by the stationary-phase method according to Eq. (5.16) and this gives

$$\begin{aligned} \langle J_r \rangle = -\frac{\omega\mu}{2} \operatorname{Im} \Bigg[& -\frac{2\mathrm{i}}{\pi} \int_0^{2\pi} |F(\hat{\boldsymbol{r}}, \hat{\boldsymbol{k}})|^2 \, \mathrm{d}\varphi - 2\mathrm{i} \Big(F^\star(\hat{\boldsymbol{k}}, \hat{\boldsymbol{k}}) \\ & + \mathrm{i}F^\star(-\hat{\boldsymbol{k}}, \hat{\boldsymbol{k}}) \, \mathrm{e}^{-2\mathrm{i}kR} + F(\hat{\boldsymbol{k}}, \hat{\boldsymbol{k}}) + \mathrm{i}F(-\hat{\boldsymbol{k}}, \hat{\boldsymbol{k}}) \, \mathrm{e}^{2\mathrm{i}kR} \Big) \Bigg] = 0 \end{aligned}$$

and simplifying a little

$$\begin{aligned} \langle J_r \rangle = -\frac{\omega\mu}{2} \operatorname{Im} \Bigg[& -\frac{2\mathrm{i}}{\pi} \int_0^{2\pi} |F(\hat{\boldsymbol{r}}, \hat{\boldsymbol{k}})|^2 \, \mathrm{d}\varphi \\ & -4\mathrm{i} \Big(\operatorname{Re} \big(F(\hat{\boldsymbol{k}}, \hat{\boldsymbol{k}}) \big) + \mathrm{i} \operatorname{Re} \big(F(-\hat{\boldsymbol{k}}, \hat{\boldsymbol{k}}) \, \mathrm{e}^{2\mathrm{i}kR} \big) \Big) \Bigg] = 0 \end{aligned}$$

The last term is seen to vanish when the imaginary part is taken and there remains

$$\int_0^{2\pi} |F(\hat{\boldsymbol{r}}, \hat{\boldsymbol{k}})|^2 \, \mathrm{d}\varphi = -2\pi \mathrm{Re}\left(F(\hat{\boldsymbol{k}}, \hat{\boldsymbol{k}})\right) \tag{5.22}$$

This equation can also be expressed in terms of the total scattering cross section. From Section 2.1.4 this is defined as the total scattered energy flow divided by the energy flow density of the incident plane wave. From the above it is seen that the total scattering cross section can be written

$$\gamma = \frac{2}{\pi k} \int_0^{2\pi} |F(\hat{\boldsymbol{r}}, \hat{\boldsymbol{k}})|^2 \, \mathrm{d}\varphi$$

and it follows that

$$\gamma = -\frac{4}{k} \,\mathrm{Re}\left(F(\hat{\boldsymbol{k}}, \hat{\boldsymbol{k}})\right) \tag{5.23}$$

The two relations Eqs. (5.22) and (5.23) are usually called the forward amplitude theorem (in electromagnetics it is often referred to as "the optical theorem").

Next the implication of reciprocity is explored. The general reciprocity relation is [Eq. (1.9)]

$$\int_{S_0} \hat{\boldsymbol{n}} \cdot \left(\boldsymbol{\sigma}^{\mathrm{A}} \cdot \boldsymbol{u}^{\mathrm{B}} - \boldsymbol{\sigma}^{\mathrm{B}} \cdot \boldsymbol{u}^{\mathrm{A}}\right) \mathrm{d}S = \int_V \rho \left(\boldsymbol{u}^{\mathrm{A}} \cdot \boldsymbol{f}^{\mathrm{B}} - \boldsymbol{u}^{\mathrm{B}} \cdot \boldsymbol{f}^{\mathrm{A}}\right) \mathrm{d}V$$

where two elastodynamic states $\boldsymbol{u}^{\mathrm{A}}$ and $\boldsymbol{u}^{\mathrm{B}}$ are involved and the volume V is enclosed by the surface S_0 with outward-pointing normal $\hat{\boldsymbol{n}}$. It is possible to choose S_0 as two closed surfaces and V as the volume in between. One of these surfaces is chosen as the boundary S of the obstacle and the other as a large circle with radius R. The two elastodynamic states are chosen as above [Eq. (5.21)] so that in the far field they are

$$u^i(\hat{\boldsymbol{k}}_i; \boldsymbol{r}) = \exp\left(\mathrm{i}k\hat{\boldsymbol{k}}_i \cdot \boldsymbol{r}\right) + \sqrt{\frac{2}{\pi k r}}\, \mathrm{e}^{\mathrm{i}(kr - \pi/4)} F(\hat{\boldsymbol{r}}, \hat{\boldsymbol{k}}_i)$$

for $i = 1, 2$. The integral over S then vanishes. If the obstacle is a cavity or rigid this follows directly from the boundary condition. If the obstacle is elastic and the boundary conditions the welded ones, then using these boundary conditions the integrand contains the interior fields and as these are source-free inside the obstacle an application of Gauss' law gives that the integral vanishes (and this is true even if the material parameters inside S are varying). The two states have no sources in the region V so the volume integral also vanishes. There remains the integral over the large circle and simplifying this gives

$$\int_0^{2\pi} \Big(F(\hat{\boldsymbol{r}}, \hat{\boldsymbol{k}}_2) \exp(\mathrm{i}k\hat{\boldsymbol{k}}_1 \cdot \boldsymbol{r}) + \hat{\boldsymbol{k}}_2 \cdot \hat{\boldsymbol{r}} F(\hat{\boldsymbol{r}}, \hat{\boldsymbol{k}}_1) \exp(\mathrm{i}k\hat{\boldsymbol{k}}_2 \cdot \boldsymbol{r})$$

$$- F(\hat{\boldsymbol{r}}, \hat{\boldsymbol{k}}_1) \exp(\mathrm{i}k\hat{\boldsymbol{k}}_2 \cdot \boldsymbol{r}) - \hat{\boldsymbol{k}}_1 \cdot \hat{\boldsymbol{r}} F(\hat{\boldsymbol{r}}, \hat{\boldsymbol{k}}_2) \exp(\mathrm{i}k\hat{\boldsymbol{k}}_1 \cdot \boldsymbol{r}) \Big) \, \mathrm{d}\varphi = 0$$

These integrals can be estimated by the stationary-phase method and the result is

$$\sqrt{\frac{2\pi}{kr}} \Big(\mathrm{e}^{\mathrm{i}(kR-\pi/4)} F(\hat{\boldsymbol{k}}_1, \hat{\boldsymbol{k}}_2) + \mathrm{e}^{-\mathrm{i}(kR-\pi/4)} F(-\hat{\boldsymbol{k}}_1, \hat{\boldsymbol{k}}_2)$$

$$+ \mathrm{e}^{\mathrm{i}(kR-\pi/4)} F(\hat{\boldsymbol{k}}_2, \hat{\boldsymbol{k}}_1) - \mathrm{e}^{-\mathrm{i}(kR-\pi/4)} F(-k_2, \hat{\boldsymbol{k}}_1)$$

$$- \mathrm{e}^{\mathrm{i}(kR-\pi/4)} F(\hat{\boldsymbol{k}}_2, \hat{\boldsymbol{k}}_1) - \mathrm{e}^{-\mathrm{i}(kR-\pi/4)} F(-\hat{\boldsymbol{k}}_2, \hat{\boldsymbol{k}}_1)$$

$$- \mathrm{e}^{\mathrm{i}(kR-\pi/4)} F(\hat{\boldsymbol{k}}_1, \hat{\boldsymbol{k}}_2) + \mathrm{e}^{-\mathrm{i}(kR-\pi/4)} F(-\hat{\boldsymbol{k}}_1, \hat{\boldsymbol{k}}_2) \Big) = 0$$

Here some of the terms cancel and there remains

$$F(-\hat{\boldsymbol{k}}_1, \hat{\boldsymbol{k}}_2) = F(-\hat{\boldsymbol{k}}_2, \hat{\boldsymbol{k}}_1)$$

or equivalently

$$F(\hat{\boldsymbol{k}}_1, \hat{\boldsymbol{k}}_2) = F(-\hat{\boldsymbol{k}}_2, -\hat{\boldsymbol{k}}_1) \tag{5.24}$$

This reciprocity relation is a manifestation of time reversal, see Section 5.3.2 for the corresponding result for the T-matrix.

5.2.3 2D P-SV waves

For the 2D P-SV waves the development is very similar to the SH case. For any obstacle the scattered wave can be expanded in the out-going polar wavefunctions

$$\boldsymbol{u}^{\mathrm{sc}}(\boldsymbol{r}) = \sum_{\tau,\sigma,m} f_{\tau\sigma m} \boldsymbol{\chi}^{+}_{\tau\sigma m}(\boldsymbol{r}) \tag{5.25}$$

where (from the integral representation)

$$f_{\tau\sigma m} = \frac{1}{\mu}\int_S \left[\boldsymbol{u}(\boldsymbol{r}')\cdot \boldsymbol{t}^{(n)}(\boldsymbol{\chi}^0_{\tau\sigma m}(\boldsymbol{r}')) - \boldsymbol{t}^{(n)}(\boldsymbol{u}(\boldsymbol{r}'))\cdot \boldsymbol{\chi}^0_{\tau\sigma m}(\boldsymbol{r}')\right] \mathrm{d}S' \tag{5.26}$$

where $r = |\boldsymbol{r}| > \max|\boldsymbol{r}'|$. The expansion is thus valid outside the circumscribed circle to the obstacle.

For an incident plane SV wave the total far field becomes

$$\begin{aligned}\boldsymbol{u}(\hat{\boldsymbol{k}};\boldsymbol{r}) = \hat{\boldsymbol{e}}\exp\left(\mathrm{i}k_s\hat{\boldsymbol{k}}\cdot\boldsymbol{r}\right) + \boldsymbol{F}_{\mathrm{ss}}(\hat{\boldsymbol{r}},\hat{\boldsymbol{k}})\sqrt{\frac{2}{\pi k_{\mathrm{s}} r}}\,\mathrm{e}^{\mathrm{i}(k_{\mathrm{s}}r-\pi/4)} \\ +\hat{\boldsymbol{r}}F_{\mathrm{ps}}(\hat{\boldsymbol{r}},\hat{\boldsymbol{k}})\sqrt{\frac{2}{\pi k_{\mathrm{p}} r}}\,\mathrm{e}^{\mathrm{i}(k_{\mathrm{p}}r-\pi/4)}\end{aligned} \tag{5.27}$$

where $\hat{\boldsymbol{e}}$ is the polarization of the incident SV wave and $\hat{\boldsymbol{e}}\cdot\hat{\boldsymbol{k}} = 0$. The right index on the far field amplitude gives the incident wave type and the left index the scattered one. In this 2D case the scattered shear wave is polarized in the φ-direction. It is, however, convenient to take the SV far field amplitude as a vector and this also points to the situation in 3D. The corresponding radial traction in the far field is

$$\begin{aligned}\boldsymbol{t}^{(r)} &= \mathrm{i}\mu k_{\mathrm{s}}\left((\hat{\boldsymbol{r}}\cdot\hat{\boldsymbol{k}})\,\hat{\boldsymbol{e}} + (\hat{\boldsymbol{r}}\cdot\hat{\boldsymbol{e}})\,\hat{\boldsymbol{k}}\right)\exp\left(\mathrm{i}k_s\hat{\boldsymbol{k}}\cdot\boldsymbol{r}\right) \\ &\quad +\mathrm{i}\mu k_{\mathrm{s}}\boldsymbol{F}_{\mathrm{ss}}(\hat{\boldsymbol{r}},\hat{\boldsymbol{k}})\sqrt{\frac{2}{\pi k_{\mathrm{s}} r}}\,\mathrm{e}^{\mathrm{i}(k_{\mathrm{s}}r-\pi/4)} \\ &\quad +\mathrm{i}(\lambda+2\mu)k_{\mathrm{p}}\hat{\boldsymbol{r}}F_{\mathrm{ps}}(\hat{\boldsymbol{r}},\hat{\boldsymbol{k}})\sqrt{\frac{2}{\pi k_{\mathrm{p}} r}}\,\mathrm{e}^{\mathrm{i}(k_{\mathrm{p}}r-\pi/4)}\end{aligned}$$

Calculating the energy flow vector and integrating over a large circle is done in the same way as in the preceding subsection. One point worth mentioning is that in the stationary-phase calculation the unit vector $\hat{\boldsymbol{r}}$ goes into $\hat{\boldsymbol{k}}$ or $-\hat{\boldsymbol{k}}$ as in the scalar case and that this means that $\hat{\boldsymbol{e}}\cdot\hat{\boldsymbol{r}} = 0$. It follows that the scattered P wave does not contribute at all in the stationary-phase method. The final results is

$$\int_0^{2\pi}\left(|\boldsymbol{F}_{\mathrm{ss}}(\hat{\boldsymbol{r}},\hat{\boldsymbol{k}})|^2 + \frac{\lambda+2\mu}{\mu}|F_{\mathrm{ps}}(\hat{\boldsymbol{r}},\hat{\boldsymbol{k}})|^2\right)\mathrm{d}\varphi = -2\pi\hat{\boldsymbol{e}}\cdot\mathrm{Re}\left(\boldsymbol{F}_{\mathrm{ss}}(\hat{\boldsymbol{k}},\hat{\boldsymbol{k}})\right)$$

$$(5.28)$$

On the right-hand side the two vectors in the scalar product are parallel. The factor $(\lambda+2\mu)/\mu$ appears in front of the P wave far field amplitude. This has to do with the different energies in the P and SV waves and could have been avoided by putting a factor $k_\mathrm{p}/k_\mathrm{s}$ in front of the P wave scattering amplitude in Eq. (5.27). Note that this factor was put into the definition of the vector wavefunctions in Section 2.2.1 and that this has the consequence that each out-going wavefunction carries the same amount of energy (and that the T-matrix becomes symmetric and that the Green function expansion becomes simpler).

For an incident P wave the development is the same. The total far field is

$$\begin{aligned} \boldsymbol{u}(\hat{\boldsymbol{k}};\boldsymbol{r}) = \hat{\boldsymbol{k}} \exp\left(\mathrm{i}k_p \hat{\boldsymbol{k}}\cdot\boldsymbol{r}\right) + \boldsymbol{F}_\mathrm{sp}(\hat{\boldsymbol{r}},\hat{\boldsymbol{k}})\sqrt{\frac{2}{\pi k_\mathrm{s} r}}\,\mathrm{e}^{\mathrm{i}(k_\mathrm{s}r-\pi/4)} \\ +\hat{\boldsymbol{r}} F_\mathrm{pp}(\hat{\boldsymbol{r}},\hat{\boldsymbol{k}})\sqrt{\frac{2}{\pi k_\mathrm{p} r}}\,\mathrm{e}^{\mathrm{i}(k_\mathrm{p}r-\pi/4)} \end{aligned} \tag{5.29}$$

Performing the calculations the result is

$$\int_0^{2\pi}\left(|F_\mathrm{pp}(\hat{\boldsymbol{r}},\hat{\boldsymbol{k}})|^2 + \frac{\mu}{\lambda+2\mu}|\boldsymbol{F}_\mathrm{sp}(\hat{\boldsymbol{r}},\hat{\boldsymbol{k}})|^2\right)\mathrm{d}\varphi = -2\pi\mathrm{Re}\left(F_\mathrm{pp}(\hat{\boldsymbol{k}},\hat{\boldsymbol{k}})\right) \tag{5.30}$$

Turning next to reciprocity it is instructive to start with the coupling between the P and SV waves. As for the SH case the radial energy flow over a large circle with radius R vanishes, thus

$$\langle J_r\rangle = \int_0^{2\pi} \hat{\boldsymbol{r}}\cdot\left(\boldsymbol{\sigma}^\mathrm{A}\cdot\boldsymbol{u}^\mathrm{B} - \boldsymbol{\sigma}^\mathrm{B}\cdot\boldsymbol{u}^\mathrm{A}\right) R\,\mathrm{d}\varphi = 0$$

Let state A be the total field with an incident P wave, with the far field given in Eq. (5.29), and state B with an incident SV wave, Eq. (5.27), with index 1 or 2, respectively, on $\hat{\boldsymbol{k}}$ and $\hat{\boldsymbol{e}}$. Due to orthogonality only terms that are a product of a plane wave part and an out-going wave part of the same type (P or SV) survives in the stationary-phase approximation. The stationary points appear exactly as in the SH case and the final result becomes (changing sign on $\hat{\boldsymbol{k}}_1$)

$$F_\mathrm{ps}(\hat{\boldsymbol{k}}_1,\hat{\boldsymbol{k}}_2) = -\frac{\mu}{\lambda+2\mu}\hat{\boldsymbol{e}}_2\cdot\boldsymbol{F}_\mathrm{sp}(-\hat{\boldsymbol{k}}_2,-\hat{\boldsymbol{k}}_1) \tag{5.31}$$

On the right-hand side the two vectors in the scalar product are parallel.

For an incident SV wave the calculations are very similar, so the result is just stated (changing sign on both $\hat{\boldsymbol{k}}_1$ and $\hat{\boldsymbol{e}}_1$)

$$\hat{\boldsymbol{e}}_1 \cdot \boldsymbol{F}_{\text{ss}}(\hat{\boldsymbol{k}}_1, \hat{\boldsymbol{k}}_2) = \hat{\boldsymbol{e}}_2 \cdot \boldsymbol{F}_{\text{ss}}(-\hat{\boldsymbol{k}}_2, -\hat{\boldsymbol{k}}_1) \tag{5.32}$$

For an incident P wave the result becomes

$$F_{\text{pp}}(\hat{\boldsymbol{k}}_1, \hat{\boldsymbol{k}}_2) = F_{\text{pp}}(-\hat{\boldsymbol{k}}_2, -\hat{\boldsymbol{k}}_1) \tag{5.33}$$

5.2.4 3D waves

In the 3D case the scattered wave can be expanded in the out-going spherical vector wavefunctions

$$\boldsymbol{u}^{\text{sc}}(\boldsymbol{r}) = \sum_{\tau,\sigma,m,l} f_{\tau\sigma ml} \boldsymbol{\psi}^{+}_{\tau\sigma ml}(\boldsymbol{r}) \tag{5.34}$$

where (from the integral representation)

$$f_{\tau\sigma ml} = \frac{k_{\text{s}}}{\mu} \int_S \left[\boldsymbol{u}(\boldsymbol{r}') \cdot \boldsymbol{t}^{(n)}(\boldsymbol{\psi}^{0}_{\tau\sigma ml}(\boldsymbol{r}')) - \boldsymbol{t}^{(n)}(\boldsymbol{u}(\boldsymbol{r}')) \cdot \boldsymbol{\psi}^{0}_{\tau\sigma ml}(\boldsymbol{r}')\right] \mathrm{d}S' \tag{5.35}$$

where $r = |\boldsymbol{r}| > \max |\boldsymbol{r}'|$. The expansion is thus valid outside the circumscribed sphere to the obstacle.

In the 3D case with an incident S wave the far field can be written

$$\boldsymbol{u}(\hat{\boldsymbol{k}}; \boldsymbol{r}) = \hat{\boldsymbol{e}} \exp{(\mathrm{i}k_s \hat{\boldsymbol{k}} \cdot \boldsymbol{r})} + \boldsymbol{F}_{\text{ss}}(\hat{\boldsymbol{r}}, \hat{\boldsymbol{k}}) \frac{\mathrm{e}^{\mathrm{i}(k_{\text{s}} r - \pi/4)}}{k_{\text{s}} r} + \hat{\boldsymbol{r}} F_{\text{ps}}(\hat{\boldsymbol{r}}, \hat{\boldsymbol{k}}) \frac{\mathrm{e}^{\mathrm{i}(k_{\text{p}} r - \pi/4)}}{k_{\text{p}} r} \tag{5.36}$$

The energy flow is set up through a large sphere and the 2D version of the stationary-phase method is used in the form given in Eq. (5.18). Orthogonality then means that there is no coupling between the S and P waves. The final result is

$$\int_0^{2\pi}\int_0^{\pi}\left(|\boldsymbol{F}_{\rm ss}(\hat{\boldsymbol{r}},\hat{\boldsymbol{k}})|^2+\left(\frac{\lambda+2\mu}{\mu}\right)^{3/2}|F_{\rm ps}(\hat{\boldsymbol{r}},\hat{\boldsymbol{k}})|^2\right)\sin\theta\,{\rm d}\theta\,{\rm d}\varphi$$

$$=4\pi\hat{\boldsymbol{e}}\cdot{\rm Im}\left(\boldsymbol{F}_{\rm ss}(\hat{\boldsymbol{k}},\hat{\boldsymbol{k}})\right) \tag{5.37}$$

For the P waves the procedure is the same. In the far field the P wave is

$$\boldsymbol{u}(\hat{\boldsymbol{k}};\boldsymbol{r})=\hat{\boldsymbol{k}}\exp\left({\rm i}k_p\hat{\boldsymbol{k}}\cdot\boldsymbol{r}\right)+\boldsymbol{F}_{\rm sp}(\hat{\boldsymbol{r}},\hat{\boldsymbol{k}})\frac{{\rm e}^{{\rm i}(k_{\rm s}r-\pi/4)}}{k_{\rm s}r}+\hat{\boldsymbol{r}}F_{\rm pp}(\hat{\boldsymbol{r}},\hat{\boldsymbol{k}})\frac{{\rm e}^{{\rm i}(k_{\rm p}r-\pi/4)}}{k_{\rm s}r} \tag{5.38}$$

and calculating the total energy flow through a large sphere gives

$$\int_0^{2\pi}\int_0^{\pi}\left(|F_{\rm pp}(\hat{\boldsymbol{r}},\hat{\boldsymbol{k}})|^2+\left(\frac{\mu}{\lambda+2\mu}\right)^{3/2}|\boldsymbol{F}_{\rm sp}(\hat{\boldsymbol{r}},\hat{\boldsymbol{k}})|^2\right)\sin\theta\,{\rm d}\theta\,{\rm d}\varphi$$

$$=4\pi{\rm Im}\left(F_{\rm pp}(\hat{\boldsymbol{k}},\hat{\boldsymbol{k}})\right) \tag{5.39}$$

In these results it is seen that a factor $((\lambda+2\mu)/\mu)^{3/2}$ appears instead of the factor $(\lambda+2\mu)/\mu$ in 2D. This is possible to avoid by another normalization of the P wave far field amplitude. It is the same factor that appears in the P vector spherical wavefunctions defined in Eq. (3.25) (inserted to get a symmetric T-matrix and a simple expansion of the Green function).

Reciprocity is set up as in 2D except that the 2D version of the stationary-phase method is employed. This results in

$$F_{\rm ps}(\hat{\boldsymbol{k}}_1,\hat{\boldsymbol{k}}_2)=-\left(\frac{\mu}{\lambda+2\mu}\right)^{3/2}\hat{\boldsymbol{e}}_2\cdot\boldsymbol{F}_{\rm sp}(-\hat{\boldsymbol{k}}_2,-\hat{\boldsymbol{k}}_1) \tag{5.40}$$

$$\hat{\boldsymbol{e}}_1\cdot\boldsymbol{F}_{\rm ss}(\hat{\boldsymbol{k}}_1,\hat{\boldsymbol{k}}_2)=\hat{\boldsymbol{e}}_2\cdot\boldsymbol{F}_{\rm ss}(-\hat{\boldsymbol{k}}_2,-\hat{\boldsymbol{k}}_1) \tag{5.41}$$

$$F_{\rm pp}(\hat{\boldsymbol{k}}_1,\hat{\boldsymbol{k}}_2)=F_{\rm pp}(-\hat{\boldsymbol{k}}_2,-\hat{\boldsymbol{k}}_1) \tag{5.42}$$

5.3 Properties of the T-matrix

If there is no damping anywhere in the scattering configuration energy is conserved. This has consequences for the T-matrix as shown in this section. When there is no damping (so that there are only second order derivatives with respect to time in the wave equation) it follows that time can be reversed and this also leads to consequences for the T-matrix. The development here only considers the 3D case, but the same results apply in 2D. The developments here follow Waterman [139].

5.3.1 Energy conservation

It follows from the integral representation Eq. (4.22) with the expansion of the Green dyadic Eq. (4.16) inserted that outside the circumscribed sphere (with the chosen origin) to the obstacle(s) the scattered wave can always be expanded in the outgoing spherical wavefunctions

$$\boldsymbol{u}^{\mathrm{sc}} = \sum_{\tau,\sigma,m,l} f_{\tau\sigma ml}\boldsymbol{\psi}^{+}_{\tau\sigma ml}$$

The incident wave can be expanded in the regular spherical wavefunctions

$$\boldsymbol{u}^{\mathrm{in}} = \sum_{\tau,\sigma,m,l} b_{\tau\sigma ml}\boldsymbol{\psi}^{0}_{\tau\sigma ml}$$

This expansion is valid out to the sources of the incident wave. If it is assumed that the sources are some way outside the circumscribed sphere to the obstacle there is a region between two spheres where the total wave can be expanded as

$$\boldsymbol{u} = \sum_{\tau,\sigma,m,l} \left(b_{\tau\sigma ml}\boldsymbol{\psi}^{0}_{\tau\sigma ml} + f_{\tau\sigma ml}\boldsymbol{\psi}^{+}_{\tau\sigma ml}\right)$$

If the incident wave is a plane wave its sources are at infinity and the expansion is valid everywhere outside the circumscribed sphere. Instead of using the regular and out-going wavefunctions it is temporarily advantageous to employ the in-going and out-going wavefunctions. The in-going wavefunctions are denoted $\boldsymbol{\psi}^{-}_{\tau\sigma ml}$ and contain Hankel functions of the second kind instead of Hankel functions of the first kind. When there is no damping they are just the complex conjugate of the out-going ones $\psi^-_{\tau\sigma ml} = \psi^{+\star}_{\tau\sigma ml}$. Noting that $2\psi^0_{\tau\sigma ml} = \psi^+_{\tau\sigma ml} + \psi^-_{\tau\sigma ml}$ it follows that

$$\boldsymbol{u} = \frac{1}{2} \sum_{\tau,\sigma,m,l} \left(b_{\tau\sigma ml} \psi^-_{\tau\sigma ml} + c_{\tau\sigma ml} \psi^+_{\tau\sigma ml} \right)$$

where thus $c_{\tau\sigma ml} = b_{\tau\sigma ml} + 2 f_{\tau\sigma ml}$. Define now the scattering matrix S that relates the out-going waves to the in-going ones

$$c_{\tau\sigma ml} = \sum_{\tau',\sigma',m',l'} S_{\tau\sigma ml,\tau'\sigma'm'l'} b_{\tau'\sigma'm'l'}$$

A comparison then shows that (in matrix notation)

$$S = 1 + 2T$$

When there is no damping (the mean in time of) the total radial energy flow through any surface enclosing the obstacle(s) must vanish. The energy flow in the scattered wave is calculated in Eq. (3.51) and as the energy flow in the out-going and in-going waves are independent this gives the total energy flow

$$\frac{\omega\mu}{4k_{\rm s}} \sum_{\tau,\sigma,m,l} \left(|c_{\tau\sigma ml}|^2 - |b_{\tau\sigma ml}|^2 \right) = 0$$

The minus of course comes from the fact that the energy flow of the in-going wave is directed inwards. Expressing the expansion coefficients of the in-going wave in terms of the scattering matrix and the expansion coefficients of the incident wave it follows that the scattering matrix is unitary (as the expansion coefficients of the incident wave $b_{\tau\sigma ml}$ are quite arbitrary)

$$S^\dagger S = I$$

where the dagger "†" denotes the Hermitian conjugate (transpose and complex conjugation). For the T-matrix this means that

$$T^\dagger T = -\mathrm{Re}\, T$$

or

$$\sum_{\tau'',\sigma'',m'',l''} T^*_{\tau''\sigma''m''l'',\tau\sigma ml} T_{\tau''\sigma''m''l'',\tau'\sigma'm'l'} = -\mathrm{Re}\, T_{\tau\sigma ml,\tau'\sigma'm'l'} \quad (5.43)$$

As an illustration the 2D SH case with a circular cavity is considered. The T-matrix elements in this case are given in Section 2.1.2

$$T_m = -\frac{\mathrm{J}'_m(ka)}{\mathrm{H}'_m(ka)}$$

This gives

$$T^\star_m T_m = \frac{(\mathrm{J}'_m(ka))^2}{|\mathrm{H}'_m(ka)|^2}$$

and

$$-\operatorname{Re} T_m = \operatorname{Re}\left(\frac{\mathrm{J}'_m(ka)\mathrm{H}'_m(ka)^\star}{\mathrm{H}'_m(ka)\mathrm{H}'_m(ka)^\star}\right) = \frac{(\mathrm{J}'_m(ka))^2}{|\mathrm{H}'_m(ka)|^2}$$

so the two sides become equal as they should.

5.3.2 Time reversal

When there is no damping a reversal of time is possible (as only second order time derivatives occur) and gives a meaningful scattering problem. In the frequency domain time reversal corresponds to taking the complex conjugate of the displacement and as the out-going and in-going wavefunctions are the complex conjugate of each other

$$\boldsymbol{u}^\star = \frac{1}{2}\sum_{\tau,\sigma,m,l}\left(b^\star_{\tau\sigma ml}\boldsymbol{\psi}^+_{\tau\sigma ml} + c^\star_{\tau\sigma ml}\boldsymbol{\psi}^-_{\tau\sigma ml}\right)$$

In terms of the scattering matrix this means that

$$b^\star_{\tau\sigma ml} = \sum_{\tau',\sigma',m',l'} S_{\tau\sigma ml,\tau'\sigma'm'l'}c^\star_{\tau'\sigma'm'l'}$$

Take the complex conjugate of this equation

$$b_{\tau\sigma ml} = \sum_{\tau',\sigma',m',l'} S^\star_{\tau\sigma ml,\tau'\sigma'm'l'}c_{\tau'\sigma'm'l'}$$

Using the definition of the scattering matrix and the fact that the expansion coefficients of the incident wave are quite arbitrary it follows that $S^\star S = 1$ and as also $S^\dagger S = 1$ it follows that S is symmetric. It follows that also the$T$-matrix is symmetric

$$T_{\tau'\sigma'm'l',\tau\sigma ml} = T_{\tau\sigma ml,\tau'\sigma'm'l'} \tag{5.44}$$

6 Scattering by anisotropic circles and spheres

In Chapters 2 and 3 the classical subject of scattering by isotropic circles (cylinders) and spheres is studied. It is of interest to investigate also the scattering by anisotropic circles and spheres with the surrounding still isotropic, and this is the subject of the present chapter. Two special types of anisotropy are taken up, the first is when the anisotropy is of circular or spherical type and in this situation the elastodynamic wave equation is still separable in polar or spherical coordinates, respectively. The second type is when the obstacle is orthotropic or transversely isotropic (in Cartesian coordinates), and then the method used is only suitable for relatively low frequencies. In some cases it is possible to obtain explicit expressions for the leading order T-matrix elements for low frequencies.

6.1 Circular anisotropy

Cylindrical orthotropy is defined in Eq. (1.27) and is a plausible model for wood (Martin and Berger [89] give further references and employs this model to investigate the 2D eigenfrequencies for a circular pole). Here only 2D problems are investigated, first the simpler 2D SH case and then the 2D P-SV case.

6.1.1 2D SH waves

Consider the scattering of 2D SH waves by a circularly orthotropic circle of radius a. The surrounding is still assumed isotropic with density ρ and shear stiffness μ. Inside the circle the density is ρ_1 and the only relevant shear stiffnesses are C_{44} and C_{55}. For 2D SH waves the displacement only has an out-of-plane component $u(r,\varphi)$. As there is no z-dependence the only nonzero stress components inside the circle are [Eqs. (1.27) and (1.11)]

$$\sigma_{\varphi z} = C_{44}\frac{1}{r}\frac{\partial u}{\partial \varphi}$$

$$\sigma_{zr} = C_{55}\frac{\partial u}{\partial r}$$

This means that the equation for u becomes

$$\frac{C_{55}}{r}\frac{\partial}{\partial r}\left(r\frac{\partial u}{\partial r}\right) + \frac{C_{44}}{r^2}\frac{\partial^2 u}{\partial \varphi^2} = -\rho_1\omega^2 u$$

Using separation of variables it is realized that the angular dependence must be in the form of trigonometric functions $\cos m\varphi$ and $\sin m\varphi$ with $m = 0, 1, 2, \ldots$, exactly as in the isotropic case. The radial equation then becomes

$$\frac{C_{55}}{r}\frac{\mathrm{d}}{\mathrm{d}r}\left(r\frac{\mathrm{d}R_m}{\mathrm{d}r}\right) - \frac{C_{44}m^2}{r^2}R_m + \rho_1\omega^2 R_m = 0$$

This is Bessel's differential equation, but in general of noninteger order. To get a bounded solution the Bessel function $\mathrm{J}_{m\alpha}(k_1 r)$ must be chosen. Here $k_1 = \omega\sqrt{\rho_1/C_{55}}$ is the radial wavenumber and $\alpha = \sqrt{C_{44}/C_{55}}$ is the square root of the quotient between the two shear stiffnesses.

With the solution inside the circle determined, the calculations in Section 2.1.2 can be repeated and the T-matrix elements are easily obtained

$$T_m = -\frac{C_{55}k_1\mathrm{J}'_{m\alpha}(k_1a)\mathrm{J}_m(ka) - \mu k\mathrm{J}_{m\alpha}(k_1a)\mathrm{J}'_m(ka)}{C_{55}k_1\mathrm{J}'_{m\alpha}(k_1a)\mathrm{H}_m(ka) - \mu k\mathrm{J}_{m\alpha}(k_1a)\mathrm{H}'_m(ka)} \tag{6.1}$$

for $m = 0, 1, \ldots$.

For low frequencies the leading order T-matrix elements are still those for $m = 0$ and $m = 1$. To leading order they become

$$T_0 = \mathrm{i}(ka)^2\frac{\pi}{4}\left(\frac{\rho_1}{\rho} - 1\right) \tag{6.2}$$

$$T_1 = \mathrm{i}(ka)^2\frac{\pi(\mu - \sqrt{C_{44}C_{55}})}{4(\mu + \sqrt{C_{44}C_{55}})} \tag{6.3}$$

That T_0 is the same as for an isotropic circle is natural as T_0 only depends on the density of the circle, not on its stiffness properties. Also T_1 is similar to the one for an isotropic circle, the difference is that the shear modulus inside the circle has been replaced by the square root of the product of the two shear moduli.

It is of course straightforward to extend this to the scattering by a layered circle, see Nayfeh [100].

6.1.2 2D P-SV waves

Consider the scattering of 2D P-SV waves by a circularly orthotropic circle of radius a. For the waves inside the circle the developments below closely follow Martin and Berger [89]. The surrounding is still assumed isotropic with density ρ and Lamé constants λ and μ. Inside the circle the density is ρ_1 and the only relevant stiffnesses are C_{11}, C_{22}, C_{12}, and C_{66}. For the in-plane waves there is no z-dependence and only two displacement components u_r and u_φ. Using this and eliminating the stresses and strains between Eqs. (1.10), (1.11), and (1.27) the following two equations are obtained

$$C_{11}\left(\frac{\partial^2 u_r}{\partial r^2}+\frac{1}{r}\frac{\partial u_r}{\partial r}\right)-\frac{C_{22}u_r}{r^2}+\frac{C_{66}}{r^2}\frac{\partial^2 u_r}{\partial \varphi^2}$$

$$+\frac{C_{12}+C_{66}}{r}\frac{\partial^2 u_\varphi}{\partial r\partial\varphi}-\frac{C_{22}+C_{66}}{r^2}\frac{\partial u_\varphi}{\partial\varphi}=-\rho_1\omega^2 u_r$$

$$\frac{C_{12}+C_{66}}{r}\frac{\partial^2 u_r}{\partial r\partial\varphi}+\frac{C_{22}+C_{66}}{r^2}\frac{\partial u_r}{\partial\varphi}$$

$$+C_{66}\left(\frac{\partial^2 u_\varphi}{\partial r^2}+\frac{1}{r}\frac{\partial u_\varphi}{\partial r}-\frac{u_\varphi}{r^2}\right)+\frac{C_{22}}{r^2}\frac{\partial^2 u_\varphi}{\partial\varphi^2}=-\rho_1\omega^2 u_\varphi$$

For the partial waves the φ-dependence is taken as $u_r(r,\varphi)=u_m(r)\cos m\varphi$ and $u_\varphi(r,\varphi)=v_m(r)\sin m\varphi$ [or $u_r(r,\varphi)=u_m(r)\sin m\varphi$ and $u_\varphi(r,\varphi)=-v_m(r)\cos m\varphi$, cf. the isotropic case]. This gives

$$C_{11}\left(r^2u_m''+ru_m'\right)+\left(\rho_1\omega^2r^2-C_{22}-m^2C_{66}\right)u_m$$

$$+\left(C_{12}+C_{66}\right)mrv_m'-\left(C_{22}+C_{66}\right)mv_m=0$$

$$- \left(C_{12} + C_{66}\right) mru'_m - \left(C_{22} + C_{66}\right) mu_m + C_{66}\left(r^2 v''_m + rv'_m\right)$$
$$+ \left(\rho_1 \omega^2 r^2 - C_{66} - m^2 C_{22}\right) v_m = 0$$

where a prime denotes differentiation with respect to r. Except for $m = 0$ this is two coupled equations for u_m and v_m.

For $m = 0$ the two equations decouple and both reduce to Bessel's differential equation [Eq. (2.1)]. Thus there are two independent solutions which can be taken as

$$u_0 = \mathrm{J}_\gamma(k_1 r)$$
$$v_0 = \mathrm{J}_1(k_2 r)$$

where $\gamma = \sqrt{C_{22}/C_{11}}$ and the two wavenumbers are $k_1 = \omega\sqrt{\rho_1/C_{11}}$ and $k_2 = \omega\sqrt{\rho_1/C_{66}}$. The first solution is a pure compressional wave and the second a pure shear wave. For the compressional wave the solution has a nonanalytical behaviour at the origin, thus reflecting the material behaviour there. The shear wave is of torsional type and is well behaved at the origin.

The corresponding T-matrix elements for $m = 0$ are determined as in the isotropic case. For low frequencies the SV ($\tau = 1$) element is not of leading order (it is the same as in the isotropic case). To leading order for low frequencies the P ($\tau = 2$) element becomes

$$T^0_{22} = \mathrm{i}(k_\mathrm{p}a)^2 \frac{\pi\left[2(\lambda+\mu) - \left(\sqrt{C_{11}C_{22}} + C_{12}\right)\right]}{4\left(2\mu + \sqrt{C_{11}C_{22}} + C_{12}\right)} \tag{6.4}$$

Specializing to the isotropic case ($C_{11} = C_{22} = \lambda_1 + 2\mu_1$, $C_{12} = \lambda_1$), this element is seen to reduce to the corresponding isotropic T-matrix element as given in Eq. (2.37).

For $m \neq 0$ the two equations must be solved simultaneously. In general, it seems that there is no analytical solution. The method of Frobenius can be used, however, and thus the following ansatz is made

$$u_m(r) = \sum_{j=0}^{\infty} b_j \left(k_2 r\right)^{\alpha+2j}$$

$$v_m(r) = \sum_{j=0}^{\infty} c_j \left(k_2 r\right)^{\alpha+2j}$$

where $k_2 = \omega\sqrt{\rho_1/C_{66}}$ as above. The coefficients b_j and c_j and the exponent α depend on m, but this dependence is suppressed. Introducing the stiffnesses $c_{11} = C_{11}/C_{66}$, $c_{12} = C_{12}/C_{66}$, and $c_{22} = C_{22}/C_{66}$, this gives

$$\sum_{j=0}^{\infty} \left[\left(c_{11}(\alpha+2j)^2 - c_{22} - m^2\right) b_j + \left((c_{12}+1)(\alpha+2j)\right.\right.$$
$$\left.\left. - c_{22} - 1\right) m c_j\right] (k_2 r)^{\alpha+2j} + \sum_{j=1}^{\infty} b_{j-1}(k_2 r)^{\alpha+2j} = 0$$

$$\sum_{j=0}^{\infty} \left[- \left((c_{12}+1)(\alpha+2j) + c_{22} + 1\right) m b_j\right.$$
$$\left. + \left((\alpha+2j)^2 - 1 - m^2 c_{22}\right) c_j\right] (k_2 r)^{\alpha+2j} + \sum_{j=1}^{\infty} c_{j-1}(k_2 r)^{\alpha+2j} = 0$$

The $j = 0$ terms give

$$\left(c_{11}\alpha^2 - c_{22} - m^2\right) b_0 + \left((c_{12}+1)\alpha - c_{22} - 1\right) m c_0 = 0$$

$$- \left((c_{12}+1)\alpha + c_{22} + 1\right) m b_0 + \left(\alpha^2 - 1 - m^2 c_{22}\right) c_0 = 0$$

This is a homogeneous system in b_0 and c_0, and the determinant must therefore vanish for a solution to exist. After simplification the determinant becomes

$$\alpha^4 c_{11} - \alpha^2 \left[c_{11} + c_{22} + m^2\left(c_{11}c_{22} - c_{12}^2 - 2c_{12}\right)\right] + c_{22}(m^2-1)^2 = 0$$

With some mild restrictions [89] both solutions in α^2 become positive. To get a finite displacement at $r = 0$ only the positive root is chosen so that two real positive solutions α_{m1} and α_{m2} are obtained. For $m = 1$ it is seen that one of the solutions is $\alpha = 0$ (which is the same as in the case of isotropy).

Having determined $\alpha_{mn}, n = 1, 2$, it is possible to choose one of b_0 and c_0 arbitrarily and the other then follows. All the following b_j and c_j, $j = 1, 2, \ldots$, are then obtained recursively. The solution inside the circle is then complete, and applying the boundary conditions and calculating the T-matrix is then done as for the isotropic circle. In general it does not seem possible to obtain closed-form expressions for the T-matrix elements for low frequencies. For $m = 1$, however, the solution is the same as in the case of isotropy [Eq. (2.39)], only depending on the densities, not the stiffnesses.

6.2 Spherical anisotropy

There is one case for which the spherical obstacle is anisotropic (and the surrounding is still isotropic) which can be treated in a similar way as the isotropic sphere and that is when the anisotropy is spherically orthotropic of the special type with transverse isotropy in the two angular coordinates. This case is taken up in this section. Previous work on this subject include Norris and Shuvalov [104], Mittleman *et al.* [98], and Hasheminejad and Maleki [62] (the latter two only consider the rotationally symmetric case with an incident plane P wave), but here the development partly deviates from previous work.

6.2.1 Governing equations

The solution procedure is facilitated by splitting the displacement and differential operators into a scalar radial part and a 2D vector angular part. The sphere has radius a and the isotropic surrounding has density ρ and Lamé constants λ and μ. Inside the sphere the density is ρ_1 and there are five independent stiffness constants C_{11}, C_{22}, C_{12}, C_{23}, and C_{55}.

The stress-strain relations for a transversely isotropic spherically orthotropic solid are given in Eq. (1.28) and inserting the strains from Eq. (1.13) the constitutive relations become

$$\sigma_{rr} = C_{11}\frac{\partial u_r}{\partial r} + C_{12}\left(\frac{1}{r}\frac{\partial u_\theta}{\partial \theta} + \frac{2u_r}{r} + \frac{1}{r\sin\theta}\frac{\partial u_\varphi}{\partial \varphi} + \cot\theta\,\frac{u_\theta}{r}\right)$$

$$\begin{aligned}\sigma_{\theta\theta} = {} & C_{12}\frac{\partial u_r}{\partial r} + C_{22}\left(\frac{1}{r}\frac{\partial u_\theta}{\partial \theta} + \frac{u_r}{r}\right)\\ & +C_{23}\left(\frac{1}{r\sin\theta}\frac{\partial u_\varphi}{\partial \varphi} + \cot\theta\,\frac{u_\theta}{r} + \frac{u_r}{r}\right)\end{aligned}$$

$$\begin{aligned}\sigma_{\varphi\varphi} = {} & C_{12}\frac{\partial u_r}{\partial r} + C_{23}\left(\frac{1}{r}\frac{\partial u_\theta}{\partial \theta} + \frac{u_r}{r}\right)\\ & +C_{22}\left(\frac{1}{r\sin\theta}\frac{\partial u_\varphi}{\partial \varphi} + \cot\theta\,\frac{u_\theta}{r} + \frac{u_r}{r}\right)\end{aligned}$$

$$\sigma_{\theta\varphi} = C_{44}\left(\frac{1}{r}\frac{\partial u_\varphi}{\partial \theta} - \cot\theta\,\frac{u_\varphi}{r} + \frac{1}{r\sin\theta}\frac{\partial u_\theta}{\partial \varphi}\right)$$

$$\sigma_{\varphi r} = C_{55}\left(\frac{1}{r\sin\theta}\frac{\partial u_r}{\partial\varphi} + \frac{\partial u_\varphi}{\partial r} - \frac{u_\varphi}{r}\right)$$

$$\sigma_{r\theta} = C_{55}\left(\frac{\partial u_\theta}{\partial r} - \frac{u_\theta}{r} + \frac{1}{r}\frac{\partial u_r}{\partial\theta}\right) \tag{6.5}$$

where $2C_{44} = C_{22} - C_{23}$. Inserting these stresses into the equations of motion in spherical coordinates [Eqs. (1.12)], quite lengthy equations are obtained. Collecting the two angular components into a 2D vector

$$\boldsymbol{v} = \hat{\boldsymbol{\theta}} u_\theta + \hat{\boldsymbol{\varphi}} u_\varphi$$

the following much more compact equations are obtained

$$\begin{aligned}
&C_{11}\frac{1}{r^2}\frac{\partial}{\partial r}\left(r^2\frac{\partial u_r}{\partial r}\right) - \left(2\left(C_{23}+C_{44}\right)-C_{12}\right)\frac{2u_r}{r^2}\\
&\quad + C_{55}\frac{1}{r^2}\nabla_a^2 u_r + \left(C_{12}+C_{55}\right)\frac{1}{r}\frac{\partial}{\partial r}\nabla_a\cdot\boldsymbol{v}\\
&\quad - \left(2(C_{23}+C_{44})+C_{55}-C_{12}\right)\frac{1}{r^2}\nabla_a\cdot\boldsymbol{v} = -\rho_1\omega^2 u_r\\
&\left(C_{12}+C_{55}\right)\frac{1}{r}\frac{\partial}{\partial r}\nabla_a u_r + 2\left(C_{23}+C_{44}+C_{55}\right)\frac{1}{r^2}\nabla_a u_r\\
&\quad + C_{55}\frac{1}{r^2}\frac{\partial}{\partial r}\left(r^2\frac{\partial\boldsymbol{v}}{\partial r}\right) + \left(C_{23}+C_{44}\right)\frac{1}{r^2}\nabla_a\nabla_a\cdot\boldsymbol{v}\\
&\quad + C_{44}\frac{1}{r^2}\nabla_a^2\boldsymbol{v} + 2\left(C_{44}-C_{55}\right)\frac{\boldsymbol{v}}{r^2} = -\rho_1\omega^2\boldsymbol{v}
\end{aligned} \tag{6.6}$$

Here the following operators on the unit sphere are introduced

$$\nabla_a u_r = \hat{\boldsymbol{\theta}}\frac{\partial u_r}{\partial\theta} + \frac{\hat{\boldsymbol{\varphi}}}{\sin\theta}\frac{\partial u_r}{\partial\varphi}$$

$$\nabla_a\cdot\boldsymbol{v} = \frac{1}{\sin\theta}\frac{\partial}{\partial\theta}(\sin\theta\, u_\theta) + \frac{1}{\sin\theta}\frac{\partial u_\varphi}{\partial\varphi}$$

$$\begin{aligned}
\nabla_a^2\boldsymbol{v} = \frac{1}{\sin^2\theta}&\left[\hat{\boldsymbol{\theta}}\left(\sin^2\theta\,\nabla_a^2 u_\theta - u_\theta - 2\cos\theta\frac{\partial u_\varphi}{\partial\varphi}\right)\right.\\
&\left.+\hat{\boldsymbol{\varphi}}\left(\sin^2\theta\,\nabla_a^2 u_\varphi - u_\varphi + 2\cos\theta\frac{\partial u_\theta}{\partial\varphi}\right)\right]
\end{aligned}$$

where

$$\nabla_a^2 u_\theta = \nabla_a \cdot (\nabla_a u_\theta) = \frac{1}{\sin\theta}\frac{\partial}{\partial\theta}\left(\sin\theta\frac{\partial u_\theta}{\partial\theta}\right) + \frac{1}{\sin^2\theta}\frac{\partial^2 u_\theta}{\partial\varphi^2}$$

It follows that

$$\nabla_a\nabla_a\cdot\boldsymbol{v} = \frac{1}{\sin^2\theta}\left[\hat{\boldsymbol{\theta}}\left(\sin\theta\frac{\partial}{\partial\theta}\left(\sin\theta\frac{\partial u_\theta}{\partial\theta}\right) - u_\theta + \frac{\partial^2 u_\varphi}{\partial\theta\partial\varphi} - \cos\theta\frac{\partial u_\varphi}{\partial\varphi}\right)\right.$$
$$\left. + \hat{\boldsymbol{\varphi}}\left(\frac{\partial^2 u_\theta}{\partial\theta\partial\varphi} + \cos\theta\frac{\partial u_\theta}{\partial\varphi} + \frac{\partial^2 u_\varphi}{\partial\varphi^2}\right)\right]$$

From these expressions it is clear that the equations of motion Eqs. (6.6) become lengthy if written out explicitly in components.

6.2.2 The T-matrix

To proceed, it is convenient to make expansions in the vector spherical harmonics. Then the action of the surface operator ∇_a on the vector spherical harmonics is needed

$$\nabla_a \cdot \boldsymbol{A}_{1\sigma ml}(\theta,\varphi) = 0$$

$$\nabla_a \cdot \boldsymbol{A}_{2\sigma ml}(\theta,\varphi) = -\sqrt{l(l+1)}Y_{\sigma ml}(\theta,\varphi)$$

$$\nabla_a \cdot \boldsymbol{A}_{3\sigma ml}(\theta,\varphi) = 2Y_{\sigma ml}(\theta,\varphi)$$

$$\nabla_a\left(\nabla_a \cdot \boldsymbol{A}_{2\sigma ml}(\theta,\varphi)\right) = -l(l+1)\boldsymbol{A}_{2\sigma ml}(\theta,\varphi)$$

$$\nabla_a\left(\nabla_a \cdot \boldsymbol{A}_{3\sigma ml}(\theta,\varphi)\right) = 2\sqrt{l(l+1)}\boldsymbol{A}_{2\sigma ml}(\theta,\varphi)$$

$$\nabla_a^2\boldsymbol{A}_{1\sigma ml}(\theta,\varphi) = -l(l+1)\boldsymbol{A}_{1\sigma ml}(\theta,\varphi)$$

$$\nabla_a^2\boldsymbol{A}_{2\sigma ml}(\theta,\varphi) = -l(l+1)\boldsymbol{A}_{2\sigma ml}(\theta,\varphi)$$

$$\nabla_a^2\boldsymbol{A}_{3\sigma ml}(\theta,\varphi) = 0$$

From these expressions and the structure of Eqs. (6.6) it is clear that the first vector spherical harmonic does not couple to the other two. The simplest type of solution is thus

$$\boldsymbol{u} = f_1(r)\boldsymbol{A}_{1\sigma ml}(\theta, \varphi)$$

The first of Eqs. (6.6) is then satisfied identically and the second becomes

$$C_{55}\frac{\mathrm{d}}{\mathrm{d}r}\left(r^2\frac{\mathrm{d}f_1}{\mathrm{d}r}\right) - C_{44}l(l+1)f_1 + 2\left(C_{44} - C_{55}\right)f_1 = -\rho_1\omega^2r^2f_1$$

This is the differential equation for spherical Bessel functions, in general of noninteger order. As the spherical Bessel functions are usually not used for noninteger order the solution that is finite at $r = 0$ is written as

$$f_{1l}(r) = \frac{A_{1l}}{\sqrt{k_1 r}}\mathrm{J}_{n_l+1/2}(k_1 r)$$

where $k_1 = \omega\sqrt{\rho_1/C_{55}}$ and n_l is the positive root of $n_l(n_l+1) = 2 + \gamma(l^2 + l - 2)$, where $\gamma = C_{44}/C_{55}$ and $l = 1, 2, \ldots$. For $l = 1$ it is noticed $n_1 = 1$, so that $f_{11}(r) = A_{11}\mathrm{j}_1(k_1 r)$ as in the isotropic case and this corresponds to the purely torsional motion for low frequencies. From here on the determination of the T-matrix follows the same lines as for the SH part for the isotropic sphere. The dominant T-matrix element appears for $l = n_l = 1$ and it is in fact the same as in the isotropic case, and is thus not of leading order for low frequencies.

The other type of solution correspondingly is a linear combination of the second and third vector spherical harmonics

$$\boldsymbol{u} = \sqrt{l(l+1)}f_2(r)\boldsymbol{A}_{2\sigma ml}(\theta, \varphi) + f_3(r)\boldsymbol{A}_{3\sigma ml}(\theta, \varphi)$$

where the factor $\sqrt{l(l+1)}$ is inserted to somewhat simplify the following equations. The two equations of motion then become

$$\begin{aligned} C_{11}\frac{\mathrm{d}}{\mathrm{d}r}\left(r^2\frac{\mathrm{d}f_3}{\mathrm{d}r}\right) &- 2\left(2(C_{23} + C_{44}) - C_{12}\right)f_3 \\ &- C_{55}l(l+1)f_3 - \left(C_{12} + C_{55}\right)l(l+1)r\frac{\mathrm{d}f_2}{\mathrm{d}r} \\ &+ \left(2(C_{23} + C_{44}) + C_{55} - C_{12}\right)l(l+1)f_2 = -\rho_1\omega^2r^2f_3 \end{aligned}$$

$$(C_{12} + C_{55})\, r \frac{\mathrm{d}f_3}{\mathrm{d}r} + 2\,(C_{23} + C_{44} + C_{55})\, f_3$$

$$+ C_{55} \frac{\mathrm{d}}{\mathrm{d}r}\left(r^2 \frac{\mathrm{d}f_2}{\mathrm{d}r} \right) - (C_{23} + 2C_{44})\, l(l+1) f_2$$

$$+ 2\,(C_{44} - C_{55})\, f_2 = -\rho_1 \omega^2 r^2 f_2$$

In the spherically symmetric case for $l = 0$ the second vector spherical harmonic does not exist and only the first equation survives and becomes

$$C_{11} \frac{\mathrm{d}}{\mathrm{d}r}\left(r^2 \frac{\mathrm{d}f_3}{\mathrm{d}r} \right) - 2\,(2(C_{23} + C_{44}) - C_{12})\, f_3 = -\rho_1 \omega^2 r^2 f_3$$

This is again the differential equation for spherical Bessel functions and the solution that is finite at $r = 0$ is

$$f_{30}(r) = \frac{A_{30}}{\sqrt{k_3 r}} \mathrm{J}_{n_0+1/2}(k_3 r)$$

where n_0 is the positive root of $n_0(n_0+1) = 2(2(C_{23} + C_{44}) - C_{12})/C_{11}$ and $k_3 = \omega\sqrt{\rho_1/C_{11}}$. The T-matrix element for $l = 0$ is straightforward to calculate in the same way as in the case of an isotropic obstacle. For low frequencies the result becomes

$$T^0_{33} = \mathrm{i}\,(k_\mathrm{p} a)^3 \, \frac{3\lambda + 2\mu - (n_0 C_{11} + 2C_{12})}{3\,(4\mu + n_0 C_{11} + 2C_{12})} \tag{6.7}$$

In the isotropic limit $C_{11} = \lambda_1 + 2\mu_1$ and $C_{12} = \lambda_1$ (giving $n_0 = 1$), so this result agrees with Eq. (3.54).

For $l = 1, 2, \ldots$ the two ordinary differential equations for f_2 and f_3 couple and in general there seems to be no analytical solution. The method of Frobenius can be employed, but as the procedure is the same as for a cylindrically orthotropic circle no further details are given [98, 62]. Concerning the low frequency T-matrix elements it is expected that the elements for $l = 1$ become the same as in the case of isotropy given in Eq. (3.55), because these only depend on the density.

6.3 Orthotropic circle: 2D SH waves

Turning to cases where the anisotropy is in Cartesian coordinates, consider first the scattering of SH waves by an orthotropic circle in an isotropic surrounding, see Boström [18]. This is thus a 2D scalar problem for the out-of-plane displacement component u. The surrounding has shear modulus μ, density ρ, and wavenumber k. The circle has radius a and density ρ_1. The constitutive equations for an orthotropic material are given in Eq. (1.23) and in the 2D out-of-plane case the only relevant ones are

$$\sigma_{zx} = 2C_{55}\epsilon_{zx}$$

$$\sigma_{yz} = 2C_{44}\epsilon_{yz}$$

Here C_{44} and C_{55} are two shear stiffnesses, which both are equal to the shear modulus in the case of isotropy. To be able to apply the boundary conditions on the circle it is of course best to use polar coordinates (r, φ). Using the transformations for shear stresses and strains between Cartesian and polar coordinates and the definition of the strains, the stresses in polar coordinates become

$$\sigma_{zr} = \frac{\partial u}{\partial r}\left(C_{55}\cos^2\varphi + C_{44}\sin^2\varphi\right) + \frac{1}{r}\frac{\partial u}{\partial \varphi}(C_{44} - C_{55})\sin\varphi\cos\varphi$$

$$\sigma_{\varphi z} = \frac{\partial u}{\partial r}(C_{44} - C_{55})\sin\varphi\cos\varphi + \frac{1}{r}\frac{\partial u}{\partial \varphi}\left(C_{55}\sin^2\varphi + C_{44}\cos^2\varphi\right)$$

Using some trigonometric identities this can be written

$$\sigma_{zr} = C\left((1+\beta\cos 2\varphi)\frac{\partial u}{\partial r} - \beta\sin 2\varphi\frac{1}{r}\frac{\partial u}{\partial \varphi}\right)$$

$$\sigma_{\varphi z} = C\left(-\beta\sin 2\varphi\frac{\partial u}{\partial r} + (1+\beta\cos 2\varphi)\frac{1}{r}\frac{\partial u}{\partial \varphi}\right)$$

The mean shear stiffness is

$$C = (C_{44} + C_{55})/2$$

and the degree of anisotropy is given by the dimensionless constant

$$\beta = \frac{C_{55} - C_{44}}{C_{55} + C_{44}}$$

Inserting the stresses into the equation of motion [the third of Eqs. (1.10)] gives

$$(1+\beta\cos 2\varphi)\frac{\partial^2 u}{\partial r^2}+(1-\beta\cos 2\varphi)\left(\frac{1}{r^2}\frac{\partial^2 u}{\partial \varphi^2}+\frac{1}{r}\frac{\partial u}{\partial r}\right)$$
$$+2\beta\sin 2\varphi\left(\frac{1}{r^2}\frac{\partial u}{\partial \varphi}-\frac{1}{r}\frac{\partial^2 u}{\partial r\partial \varphi}\right)+k_1^2 u=0 \qquad (6.8)$$

where the mean wavenumber is $k_1=\omega\sqrt{\rho_1/C}$.

The equation of motion has factors with trigonometric functions with argument 2φ and it is thus useful to expand the solution in a trigonometric series exactly as in the case of isotropy, although now different azimuthal orders couple. Before doing this, it is convenient to realize that due to symmetry the solution can be divided into four parts that are even or odd with respect to the x- and y-axes.

For the even-even part the expansion is

$$u^1=f_{00}+f_{02}r^2+\ldots+f_{22}r^2\cos 2\varphi+\ldots$$

To be regular at the origin the radial expansion corresponding to $\cos m\varphi$ must start with r^m and continue with r^{m+2}, r^{m+4}, and so on. The terms written out explicitly are the ones needed to determine the T-matrix to leading order for low frequencies, but it is of course straightforward to include more terms. Insertion into Eq. (6.8) gives for the constant terms (independent of both φ and r)

$$4f_{02}+4\beta f_{22}+k_1^2 f_{00}=0$$

The terms in $\cos 2\varphi$ cancel each other to lowest order in r and no further information is obtained without going to higher order terms in the expansion. To apply the boundary condition the radial shear stress is needed and inserting the expansion it becomes

$$\sigma_{zr}^1=2Cr\left[f_{02}+\beta f_{22}+(\beta f_{02}+f_{22})\cos 2\varphi\right]+\ldots$$

The boundary conditions are as usual continuity of displacement and radial shear stress. The expansions of the incident and scattered waves are given in Eqs. (2.10) and (2.11). The displacement component independent of φ gives

$$\frac{b_0}{2}\mathrm{J}_0(ka)+\frac{f_0}{2}\mathrm{H}_0(ka)=f_{00}+f_{02}a^2+\ldots$$

and the one depending on $\cos 2\varphi$ gives

$$\frac{b_2}{\sqrt{2}}\mathrm{J}_2(ka) + \frac{f_2}{\sqrt{2}}\mathrm{H}_2(ka) = f_{22}a^2 + \dots$$

Likewise, two equations are obtained from the continuity of σ_{zr}

$$\frac{\mu k}{2}\left(b_0\mathrm{J}_0'(ka) + f_0\mathrm{H}_0'(ka)\right) = 2Ca\left(f_{02} + \beta f_{22}\right) + \dots$$

$$\frac{\mu k}{\sqrt{2}}\left(b_2\mathrm{J}_2'(ka) + f_2\mathrm{H}_2'(ka)\right) = 2Ca\left(\beta f_{02} + f_{22}\right) + \dots$$

To obtain the T-matrix to leading order for low frequencies the Bessel and Hankel functions are expanded for small arguments to leading order. Solving for f_0 and f_2 in terms of b_0 and b_2 four T-matrix elements can be determined. The lowest order element becomes

$$T_{\mathrm{e}0,\mathrm{e}0} = \mathrm{i}(ka)^2\frac{\pi}{4}\left(\frac{\rho_1}{\rho} - 1\right) \tag{6.9}$$

This is, in fact, the same as in the case of isotropy in Eq. (2.23). As it only depends on the densities this is to be expected. The higher order T-matrix elements, including coupling elements between different azimuthal orders, behave at least as $(ka)^4$ and are thus negligible to lowest order.

For the even-odd part (with respect to the x- and y-axes, respectively) it is enough to take the first term to obtain the lowest order T-matrix element

$$u^1 = f_{11}r\cos\varphi + \dots$$

To lowest order (terms that behave as r^{-1}) the equation of motion is identically satisfied and no further information can be obtained. The corresponding stress is

$$\sigma_{zr}^1 = Cf_{11}(1+\beta)\cos\varphi + \dots$$

Putting up the continuity of displacement and stress then gives

$$\frac{b_1}{\sqrt{2}}\mathrm{J}_1(ka) + \frac{f_1}{\sqrt{2}}\mathrm{H}_1(ka) = f_{11}a + \dots$$

$$\frac{\mu k}{\sqrt{2}}\left(b_1\mathrm{J}_1'(ka) + f_1\mathrm{H}_1'(ka)\right) = Cf_{11}(1+\beta) + \dots$$

Expanding the Bessel and Hankel functions and solving for f_1 in terms of b_1 give the T-matrix element

$$T_{\mathrm{e1,e1}} = \mathrm{i}(ka)^2 \frac{\pi(\mu - C_{55})}{4(\mu + C_{55})} \tag{6.10}$$

This element is independent of both the ρ_1 and C_{44}. In the special case of isotropy it of course reduces to Eq. (2.24).

The odd-even part is essentially the same as the even-odd part and the T-matrix element is

$$T_{\mathrm{o1,o1}} = \mathrm{i}(ka)^2 \frac{\pi(\mu - C_{44})}{4(\mu + C_{44})} \tag{6.11}$$

The odd-odd part starts with azimuthal order $m = 2$ and the corresponding T-matrix elements are thus negligible for low frequencies.

6.4 Orthotropic circle: 2D P-SV waves

Consider the scattering of in-plane waves from an orthotropic circle in an isotropic surrounding (Boström [19]). The surrounding has as usual Lamé constants λ and μ and density ρ. The wavenumbers are k_{p} and k_{s}. The circle has radius a and density ρ_1. The constitutive equations for an orthotropic material are given in Eqs. (1.23) and in the 2D in-plane case they become

$$\begin{aligned}
\sigma_{xx} &= C_{11}\epsilon_{xx} + C_{12}\epsilon_{yy} \\
\sigma_{yy} &= C_{12}\epsilon_{xx} + C_{22}\epsilon_{yy} \\
\sigma_{xy} &= 2C_{66}\epsilon_{xy}
\end{aligned}$$

Here C_{11}, C_{12}, C_{22}, and C_{66} are the only relevant stiffness constants.

6.4.1 Solution inside the circle

The constitutive equations are expressed in Cartesian coordinates, but to apply the boundary conditions on the circle it is much more convenient to use polar coordinates (r, φ). The transformation of the stress components are

$$\sigma_{rr} = \cos^2\varphi\,\sigma_{xx} + \sin^2\varphi\,\sigma_{yy} + \sin 2\varphi\,\sigma_{xy}$$

$$\sigma_{\varphi\varphi} = \sin^2\varphi\,\sigma_{xx} + \cos^2\varphi\,\sigma_{yy} - \sin 2\varphi\,\sigma_{xy}$$
$$\sigma_{r\varphi} = \frac{1}{2}\sin 2\varphi\,(\sigma_{yy} - \sigma_{xx}) + \frac{1}{2}\cos 2\varphi\,\sigma_{xy}$$

Use of these transformations, the corresponding ones for strains (the other way around), and the constitutive equations gives the constitutive equations expressed in polar coordinates

$$\begin{aligned}
\sigma_{rr} &= (\alpha_1 + 2\alpha_2)\epsilon_{rr} + \alpha_1\epsilon_{\varphi\varphi} + 2\beta_1\,(\cos 2\varphi\,\epsilon_{rr} - \sin 2\varphi\,\epsilon_{r\varphi}) \\
&\qquad +2\beta_2\,(\cos 4\varphi\,(\epsilon_{rr} - \epsilon_{\varphi\varphi}) - 2\sin 4\varphi\,\epsilon_{r\varphi}) \\
\sigma_{\varphi\varphi} &= \alpha_1\epsilon_{rr} + (\alpha_1 + 2\alpha_2)\epsilon_{\varphi\varphi} - 2\beta_1\,(\cos 2\varphi\,\epsilon_{\varphi\varphi} + \sin 2\varphi\,\epsilon_{r\varphi}) \\
&\qquad +2\beta_2\,(\cos 4\varphi\,(\epsilon_{\varphi\varphi} - \epsilon_{rr}) + 2\sin 4\varphi\,\epsilon_{r\varphi}) \\
\sigma_{r\varphi} &= 2\alpha_2\epsilon_{r\varphi} - \beta_1\sin 2\varphi\,(\epsilon_{rr} + \epsilon_{\varphi\varphi}) \\
&\qquad +2\beta_2\,(\sin 4\varphi\,(\epsilon_{\varphi\varphi} - \epsilon_{rr}) - 2\cos 4\varphi\,\epsilon_{r\varphi})
\end{aligned}$$

The following stiffness constants are introduced

$$\begin{aligned}
\alpha_1 &= \frac{1}{8}(C_{11} + C_{22} + 6C_{12} - 4C_{66}) \\
\alpha_2 &= \frac{1}{8}(C_{11} + C_{22} - 2C_{12} + 4C_{66}) \\
\beta_1 &= \frac{1}{4}(C_{11} - C_{22}) \\
\beta_2 &= \frac{1}{8}(C_{11} + C_{22} - 2C_{12} - 4C_{66})
\end{aligned}$$

The two first constants, α_1 and α_2, are two means that in the case of isotropy reduce to the Lamé constants λ_1 and μ_1, respectively. The two constants β_1 and β_2 are measures of the degree of anisotropy which are zero in the case of isotropy. The constant β_1 is zero for a cubic material, for which $C_{11} = C_{22}$, and is thus a measure of how "noncubic" the material is.

The definition of the strains in polar coordinates are given in Eqs. (1.11) and the equations of motion (expressed in the stresses) are given in Eqs. (1.10). From these the equations of motion expressed in the displacement components are obtained. The equations become lengthy and are therefore not written out. It is noted, however, that they contain trigonometric functions with argument 2φ and 4φ exactly as the stresses. This dependence makes it natural to expand the displacement components in trigonometric series, exactly as in the case of isotropy, although now the different azimuthal orders couple.

Due to symmetry the solution inside the circle can be divided into four independent parts that are symmetric or antisymmetric with respect to the x- and y-axes. A displacement field that is symmetric with respect to the x-axis has an even x-component and an odd y-component as a function of y, and vice versa for an antisymmetric displacement field. In polar coordinates this means that a displacement field that is symmetric with respect to the x-axis has an even r-component and an odd φ-component as a function of φ, and vice versa for antisymmetry.

For the part that is doubly symmetric the following ansatz can be made for the displacement components

$$\begin{aligned} u_r(r,\varphi) &= f_{01}r + f_{03}r^3 + \ldots + \cos 2\varphi\,(f_{21}r + f_{23}r^3 + \ldots) \\ &\quad + \cos 4\varphi\,(f_{43}r^3 + \ldots) + \ldots \\ u_\varphi(r,\varphi) &= \sin 2\varphi\,(g_{21}r + g_{23}r^3 + \ldots) + \sin 4\varphi\,(g_{43}r^3 + \ldots) + \ldots \end{aligned}$$

Here the unknown expansion coefficients f_{ij} and g_{ij} are labeled so that the first index refers to the azimuthal order and the second to the power of the radial coordinate. The dependence on φ follows from the symmetry requirements and the radial dependence follows from the regularity requirement at $r = 0$. The corresponding stresses are calculated and can be simplified by rewriting products of trigonometric functions as sums of trigonometric functions. Insertion of the stresses into the equations of motion gives for the radial component

$$\begin{aligned} & 2\beta_1(f_{21} + g_{21}) + \big[8(\alpha_1 + 2\alpha_2)f_{03} + 12\beta_1 f_{23} + 12\beta_2(3f_{43} + g_{43}) \\ & \quad + \rho_1\omega^2 f_{01}\big]r^2 + \cdots + \big[-4\alpha_2(f_{21} + g_{21}) + \big(12\beta_1 f_{03} + 6\beta_1(3f_{43} + g_{43}) \\ & \quad + (8\alpha_1 + 12\alpha_2 + 4\beta_2)f_{23} + 4(\alpha_1 - \beta_2)g_{23} + \rho_1\omega^2 f_{21}\big)\,r^2 + \cdots\big] \cos 2\varphi \\ & \quad + \big[-2\beta_1(f_{21} + g_{21}) + 8\alpha_1(f_{43} + g_{43})r^2 + \cdots\big] \cos 4\varphi + \cdots = 0 \end{aligned}$$

and for the tangential component

$$\left[-4(\alpha_1+2\alpha_2)(f_{21}+g_{21})-(12\beta_1 f_{03}+6\beta_1(f_{43}+3g_{43})\right.$$
$$\left.+(8\alpha_1+12\alpha_2+4\beta_2)f_{23}+4(\alpha_1-\beta_2)g_{23}-\rho_1\omega^2 g_{21})r^2+\cdots\right]\sin 2\varphi$$
$$+\left[6\beta_1(f_{21}+g_{21})-8(2\alpha_1+3\alpha_2)(f_{43}+g_{43})r^2+\cdots\right]\sin 4\varphi+\cdots=0$$

If higher order terms are included more expansion coefficients not included explicitly in the expansions enter. Use of the linear independence of the trigonometric functions and the powers of r gives that $g_{21}=-f_{21}$ and $g_{43}=-f_{43}$ and that (where g_{21} and g_{43} have been eliminated)

$$8(\alpha_1+2\alpha_2)f_{03}+12\beta_1 f_{23}+24\beta_2 f_{43}+\rho_1\omega^2 f_{01}=0$$

$$12\beta_1(f_{03}+f_{43})+(8\alpha_1+12\alpha_2+4\beta_2)f_{23}+4(\alpha_1-\beta_2)g_{23}+\rho_1\omega^2 f_{21}=0$$

These two equations are all that is obtained using only the terms explicitly written in the displacement expansions above and this is all that is needed to obtain the T-matrix for low frequencies. There is of course no particular problem to include more terms in the displacement components and obtain more equations that are needed for higher frequencies.

For the part that is symmetric with respect to the x-axis and antisymmetric with respect to the y-axis the proper ansatz is

$$u_r(r,\varphi)=\cos\varphi\,(f_{10}+f_{12}r^2+\ldots)+\cos 3\varphi\,(f_{32}r^2+\ldots)+\ldots$$

$$u_\varphi(r,\varphi)=\sin\varphi\,(g_{10}+g_{12}r^2+\ldots)+\sin 3\varphi\,(g_{32}r^2+\ldots)+\ldots$$

In the same way as for the doubly symmetric case this gives $g_{10}=-f_{10}$ and $g_{32}=-f_{32}$ and (with g_{10} and g_{32} eliminated)

$$(3\alpha_1+5\alpha_2+4\beta_1)f_{12}+(\alpha_1-\alpha_2)g_{12}+4(\beta_1+2\beta_2)f_{32}+\rho_1\omega^2 f_{10}=0$$

The part that is antisymmetric with respect to the x-axis and symmetric with respect to the y-axis is essentially the same with the displacement expansions

$$u_r(r,\varphi)=\sin\varphi\,(d_{10}+d_{12}r^2+\ldots)+\sin 3\varphi\,(d_{32}r^2+\ldots)+\ldots$$

$$u_\varphi(r,\varphi)=\cos\varphi\,(e_{10}+e_{12}r^2+\ldots)+\cos 3\varphi\,(e_{32}r^2+\ldots)+\ldots$$

This gives $d_{10} = -e_{10}$ and $d_{32} = -e_{32}$ and (with d_{10} and d_{32} eliminated)

$$(3\alpha_1 + 5\alpha_2 - 4\beta_1)e_{12} + (\alpha_1 - \alpha_2)d_{12} + 4(-\beta_1 + 2\beta_2)e_{32} + \rho_1\omega^2 e_{10} = 0$$

The change in sign on β_1 comes from an interchange of c_1 and c_2.

For the part that is doubly antisymmetric the proper ansatz is

$$\begin{aligned} u_r(r,\varphi) &= \sin 2\varphi\,(d_{21}r + d_{23}r^3 + \ldots) + \sin 4\varphi\,(d_{43}r^3 + \ldots) + \ldots \\ u_\varphi(r,\varphi) &= e_{01}r + e_{03}r^3 + \ldots + \cos 2\varphi\,(e_{21}r + e_{23}r^3 + \ldots) \\ &\quad + \cos 4\varphi\,(e_{43}r^3 + \ldots) + \ldots \end{aligned}$$

This gives $e_{21} = d_{21}$ and $e_{43} = d_{43}$ and (with e_{21} and e_{43} eliminated)

$$8\alpha_2 e_{03} + 4\beta_1(e_{23} - 2d_{23}) - 24\beta_2 d_{43} + \rho_1\omega^2 e_{01} = 0$$

$$4\beta_1(e_{03} - 3d_{43}) + 4(2\alpha_1 + 3\alpha_2 - \beta_2)d_{23} - 4(\alpha_1 + \beta_2)e_{23} + \rho_1\omega^2 d_{21} = 0$$

6.4.2 Low frequency T-matrix

To determine the T-matrix continuity of displacement and traction at $r = a$ is set up and the resulting systems are solved for the expansion coefficients of the scattered wave. As in the previous subsection the symmetries lead to a decoupling into parts that are symmetric or antisymmetric with respect to the coordinate axes, thus leading to a decomposition of the T-matrix into four parts.

The doubly symmetric case has only even azimuthal orders m. Thus continuity of radial displacement gives one equation for each azimuthal order $m = 0, 2, \ldots$. The two first are

$$af_{01} + a^3 f_{03} = -\left(\frac{k_p a}{2} - \frac{k_p^3 a^3}{16}\right) b_{2\mathrm{e}0} + \frac{2\mathrm{i}}{\pi k_p a} f_{2\mathrm{e}0}$$

$$\begin{aligned} af_{21} + a^3 f_{23} &= \frac{k_s a}{4} b_{1\mathrm{o}2} + \frac{k_p a}{4} b_{2\mathrm{e}2} - \left(\frac{8\mathrm{i}}{\pi k_s^3 a^3} + \frac{2\mathrm{i}}{\pi k_s a}\right) f_{1\mathrm{o}2} \\ &\quad + \frac{8\mathrm{i}}{\pi k_p^3 a^3} f_{2\mathrm{e}2} \end{aligned}$$

where the expansion coefficients for the incident and scattered waves from Eqs. (2.30) and (2.31) enter. As only the leading order terms for low frequencies are included the spherical Bessel and Hankel functions are expanded for small arguments. For the azimuthal displacement component there is no equation for $m = 0$ and the one for $m = 2$ is

$$-af_{21} + a^3 g_{23} = -\frac{k_s a}{4} b_{1o2} - \frac{k_p a}{4} b_{2e2} - \frac{8\mathrm{i}}{\pi k_s^3 a^3} f_{1o2} + \left(\frac{8\mathrm{i}}{\pi k_p^3 a^3} + \frac{2\mathrm{i}}{\pi k_p a} \right) f_{2e2}$$

Turning to continuity of traction it is noted that the traction in the anisotropic circle must be simplified with the help of product formulas for trigonometric functions. As for the displacement only the three first equations are given

$$2(\alpha_1 + \alpha_2) f_{01} + a^2(4\alpha_1 + 6\alpha_2) f_{03} + 2\beta_1 f_{21} + \beta_1 a^2 (4f_{23} - g_{23}) = \mu \left(\left(k_p - \frac{k_s^2}{k_p} + \frac{k_p a^2 (2k_s^2 - k_p^2)}{8} \right) b_{2e0} - \frac{4\mathrm{i}}{\pi k_p a^2} f_{2e0} \right)$$

$$2\beta_1 f_{01} + 6\beta_1 a^2 f_{03} + 2(\alpha_2 + \beta_2) f_{21} + 2a^2 (2\alpha_1 + 3\alpha_2 + \beta_2) f_{23} + 2a^2 (\alpha_1 - \beta_2) g_{23} = \mu \left(\frac{k_s}{2} b_{1o2} + \frac{k_p}{2} b_{2e2} + \left(\frac{48\mathrm{i}}{\pi k_s^3 a^4} + \frac{4\mathrm{i}}{\pi k_s a^2} \right) f_{1o2} - \left(\frac{48\mathrm{i}}{\pi k_p^3 a^4} + \frac{4\mathrm{i}(2k_p^2 - k_s^2)}{\pi k_p^3 a^2} \right) f_{2e2} \right)$$

$$-2\beta_1 f_{01} - 4a^2 \beta_1 f_{03} - 2(\alpha_2 + \beta_2) f_{21} - 2a^2 (\alpha_2 + \beta_2)(f_{23} - g_{23}) = \mu \left(-\frac{k_s}{2} b_{1o2} - \frac{k_p}{2} b_{2e2} + \left(\frac{48\mathrm{i}}{\pi k_s^3 a^4} + \frac{4\mathrm{i}}{\pi k_s a^2} \right) f_{1o2} - \left(\frac{48\mathrm{i}}{\pi k_p^3 a^4} + \frac{4\mathrm{i}}{\pi k_p a^2} \right) f_{2e2} \right)$$

Altogether the boundary conditions give six equations to lowest order. The unknowns are the scattered wave coefficients f_{2e0}, f_{1o2}, and f_{2e2} and the expansion coefficients inside the circle f_{01}, f_{03}, f_{21}, f_{23}, f_{43}, and g_{23}. From the previous subsection two equations relate the coefficients inside the circle. So there are eight equations and nine unknowns. However, going one step further in the continuity equations it is seen that f_{43} must vanish to leading order if there is no incident partial wave of azimuthal order $m = 4$ (which can be assumed if only the T-matrix elements of lower order are of interest). Using this it is possible to solve for the scattered field coefficients in terms of those of the incident wave, and thus the T-matrix elements. The result is finally expanded in powers of ka with only the first term kept. After lengthy calculations the monopole element for $m = m' = 0$ becomes

$$T_{2e0,2e0} = i(k_p a)^2 \frac{\pi}{4} \left(\frac{(\lambda + 2\mu)M}{N} - 1 \right) \tag{6.12}$$

where

$$M = \lambda + \mu + (3 + \lambda/\mu)(\alpha_2 + \beta_2)$$

$$N = (\mu + \alpha_1 + \alpha_2)M - \beta_1^2(3 + \lambda/\mu)$$

In the limit of an isotropic circle this element becomes equal to the corresponding element in Eq. (2.37). The quadrupole elements for $m = m' = 2$ are

$$T_{1o2,1o2} = i \left(\frac{k_s^2 a}{k_p} \right)^2 L_2$$

$$T_{1o2,2e2} = T_{2e2,1o2} = i(k_s a)^2 L_2$$

$$T_{2e2,2e2} = i(k_p a)^2 L_2 \tag{6.13}$$

where

$$L_2 = \frac{\pi}{4N}[\beta_1^2 + (\mu + \alpha_1 + \alpha_2)(\mu - \alpha_2 - \beta_2)]$$

These elements become equal to the ones for the isotropic circle Eqs. (2.40) in the isotropic limit. The elements for $m = 0$ and $m' = 2$ are equal to those for $m = 2$ and $m' = 0$ and are coupling elements between monopole and quadrupole waves and become

$$T_{2e0,2e2} = T_{2e2,2e0} = i(k_p a)^2 L_0$$

$$T_{2e0,1o2} = T_{1o2,2e0} = i(k_s a)^2 L_0 \tag{6.14}$$

where

$$L_0 = \frac{\pi\beta_1(\lambda + 2\mu)}{2\sqrt{2}\,N}$$

These elements of course vanish when the circle becomes isotropic because $\beta_1 = 0$ then. It is noted that none of the doubly symmetric T-matrix elements depends on the shear modulus C_{66} of the circle.

For the other symmetries the developments are very similar so only the final results are given. For the parts that are symmetric with respect to one coordinate axis and antisymmetric with respect to the other only $m = 1$ enters. For the part that is symmetric with respect to the x-axis and antisymmetric with respect to the y-axis the T-matrix elements become

$$\begin{aligned} T_{1\mathrm{o}1,1\mathrm{o}1} &= \mathrm{i}(k_\mathrm{s}a)^2 C_1 \\ T_{1\mathrm{o}1,2\mathrm{e}1} &= T_{2\mathrm{e}1,1\mathrm{o}1} = \mathrm{i}k_\mathrm{p}k_\mathrm{s}a^2 C_1 \\ T_{2\mathrm{e}1,2\mathrm{e}1} &= \mathrm{i}(k_\mathrm{p}a)^2 C_1 \end{aligned} \tag{6.15}$$

where

$$C_1 = \frac{\pi}{8}\left(\frac{\rho_1}{\rho} - 1\right)$$

These are, in fact, the same as for the isotropic circle as given in Eqs. (2.39). This is to be expected as these T-matrix elements are independent of the stiffness parameters of the circle and correspond to rigid body motion of the circle. The part that is antisymmetric with respect to the x-axis and symmetric with respect to the y-axis gives the same T-matrix elements

$$\begin{aligned} T_{1\mathrm{e}1,1\mathrm{e}1} &= \mathrm{i}(k_\mathrm{s}a)^2 C_1 \\ T_{1\mathrm{e}1,2\mathrm{o}1} &= T^1_{2\mathrm{o}1,1\mathrm{e}1} = \mathrm{i}k_\mathrm{p}k_\mathrm{s}a^2 C_1 \\ T_{2\mathrm{o}1,2\mathrm{o}1} &= \mathrm{i}(k_\mathrm{p}a)^2 C_1 \end{aligned} \tag{6.16}$$

The doubly antisymmetric part has only even m, exactly as the doubly symmetric part. The element $T_{1\mathrm{o}0,1\mathrm{o}0}$ is the same as for the isotropic case as given in Eq. (2.37) and is thus very small going as $(k_s a)^4$ and only depends on the density of the circle. Also the elements coupling $m = 0$ and $m' = 2$, which are the same as for $m = 2$ and $m' = 0$, are very small, going as $(k_s a)^4$, and are therefore not given. The quadrupole elements for $m = m' = 2$ become

$$T_{1e2,1e2} = \mathrm{i}\left(\frac{k_s^2 a}{k_p}\right)^2 K_2$$

$$T_{1e2,2o2} = T_{2o2,1e2} = \mathrm{i}(k_s a)^2 K_2$$

$$T_{2o2,2o2} = \mathrm{i}(k_p a)^2 K_2 \tag{6.17}$$

where

$$K_2 = \frac{\pi(\mu - \alpha_2 + \beta_2)}{4M}$$

where M is defined above. These elements are very similar to the corresponding ones in the isotropic case as given in Eqs. (2.40), to which they reduce in the limit of an isotropic circle. It is noted that $\alpha_2 - \beta_2 = C_{66}$ so the scattering depends almost linearly on the difference in shear moduli between the circle and the surrounding (with only a weak dependence on the other stiffness constants of the circle), exactly as in the isotropic case.

6.5 Transversely isotropic sphere

Consider the scattering by a transversely isotropic (in Cartesian coordinates) sphere in an isotropic surrounding, see Jafarzadeh *et al.* [70]. Due to rotational symmetry of the material parameters this is the simplest case of scattering by an anisotropic sphere. The surrounding has Lamé constants λ and μ and density ρ and the corresponding wavenumbers are k_p and k_s.

The incident and scattered waves are expanded in the spherical vector wavefunctions as in the case of the scattering by an isotropic sphere. These expansions are given in Eqs. (3.33) and (3.34) and the expansion coefficients $b_{\tau\sigma ml}$ of the incident wave are regarded as known whereas the expansion coefficients $f_{\tau\sigma ml}$ of the scattered wave are to be determined. As usual the solution is given in terms of the T-matrix.

Due to the rotational symmetry the T-matrix becomes diagonal in the azimuthal m-index. From the trigonometric functions in the spherical vector wavefunctions it follows that the T-matrix, furthermore, decouples into two parts, where $\tau\sigma = 1\text{o}, 2\text{e}, 3\text{e}$ for the symmetric (with respect to the xz-plane) part and $\tau\sigma = 1\text{e}, 2\text{o}, 3\text{o}$ for the antisymmetric part. These two parts have the same T-matrix elements (when all indices except the σ-indices are equal) except for the elements with $\tau, \tau' = 12, 13, 21, 31$ where there is a sign change between the two parts. In the following the part with $\tau\sigma = 1\text{o}, 2\text{e}, 3\text{e}$ is treated and the σ-index is then suppressed and the T-matrix element is thus denoted $T^m_{\tau l,\tau' l'}$. For $\tau = 1$ (SH waves) and $m = 0$ only $\sigma = \text{e}$ exists and then the SH waves decouples from the coupled P-SV waves ($\tau = 2, 3$) so that the parts with $\tau\sigma = 1\text{e}$ (which belongs to the antisymmetric part) and $\tau\sigma = 2\text{e}, 3\text{e}$ are treated separately.

There is also mirror symmetry with respect to the xy-plane and this gives a decoupling into symmetric and antisymmetric parts with respect to this plane. The symmetric part has odd values of l for $\tau = 1$ and even values of l for $\tau = 2, 3$ and vice versa for the antisymmetric part. This decoupling is exploited below when explicit expressions are given for the T-matrix elements for low frequencies.

6.5.1 Solution inside the sphere

The sphere has radius a and density ρ_1 and the stress-strain relations are [Eq.(1.24)]

$$\sigma_{xx} = C_{11}\varepsilon_{xx} + C_{12}\varepsilon_{yy} + C_{13}\varepsilon_{zz}$$

$$\sigma_{yy} = C_{12}\varepsilon_{xx} + C_{11}\varepsilon_{yy} + C_{13}\varepsilon_{zz}$$

$$\sigma_{zz} = C_{13}\varepsilon_{xx} + C_{13}\varepsilon_{yy} + C_{33}\varepsilon_{zz}$$

$$\sigma_{yz} = 2C_{44}\varepsilon_{yz}$$

$$\sigma_{zx} = 2C_{44}\varepsilon_{zx}$$

$$\sigma_{xy} = 2C_{66}\varepsilon_{xy}$$

where $2C_{66} = C_{11} - C_{12}$. There are thus five independent stiffness constants.

It is of course natural to use spherical coordinates (r, θ, φ) and the stress-strain relations are therefore transformed to spherical coordinates using the usual rule for transformation of second order tensors

$$\boldsymbol{S}_s = \boldsymbol{R}^T \boldsymbol{S}_c \boldsymbol{R}$$

where $\boldsymbol{S}_s$ and $\boldsymbol{S}_c$ are stress or strain tensors in spherical and Cartesian coordinates, respectively, and $\boldsymbol{R}$ is the rotation matrix

$$\boldsymbol{R} = \begin{bmatrix} \cos\varphi\sin\theta & \cos\varphi\cos\theta & -\sin\varphi \\ \sin\varphi\sin\theta & \sin\varphi\cos\theta & \cos\varphi \\ \cos\theta & -\sin\theta & 0 \end{bmatrix} \tag{6.18}$$

The stress-strain relations in spherical coordinates become

$$\begin{aligned}
\sigma_{rr} &= (\alpha_1 + 2\alpha_2)\,\epsilon_{rr} + \alpha_1\epsilon_{\theta\theta} + (\alpha_1 - \beta_2 + \beta_3)\,\epsilon_{\varphi\varphi} \\
&\quad -2\beta_1\,(\epsilon_{rr}\cos 2\theta - \epsilon_{r\theta}\sin 2\theta) - \beta_3\,(\epsilon_{\varphi\varphi}\cos 2\theta) \\
&\quad +\beta_2\,[(\epsilon_{rr} - \epsilon_{\theta\theta})\cos 4\theta - 2\epsilon_{r\theta}\sin 4\theta] \\
\sigma_{\theta\theta} &= (\alpha_1 + 2\alpha_2)\,\epsilon_{\theta\theta} + \alpha_1\epsilon_{rr} + (\alpha_1 - \beta_2 + \beta_3)\,\epsilon_{\varphi\varphi} \\
&\quad +2\beta_1\,(\epsilon_{\theta\theta}\cos 2\theta + \epsilon_{r\theta}\sin 2\theta) + \beta_3\,(\epsilon_{\varphi\varphi}\cos 2\theta) \\
&\quad +\beta_2\,((\epsilon_{\theta\theta} - \epsilon_{rr})\cos 4\theta + 2\epsilon_{r\theta}\sin 4\theta) \\
\sigma_{\varphi\varphi} &= (\alpha_1 + 2\alpha_2 + 2\beta_1 + \beta_2)\,\epsilon_{\varphi\varphi} + (\alpha_1 - \beta_2 + \beta_3)\,(\epsilon_{\theta\theta} + \epsilon_{rr}) \\
&\quad +\beta_3\,((\epsilon_{\theta\theta} - \epsilon_{rr})\cos 2\theta + 2\epsilon_{r\theta}\sin 2\theta) \\
\sigma_{\theta\varphi} &= (2\alpha_2 + \beta_1 - \beta_3)\epsilon_{\theta\varphi} + (2\beta_2 + \beta_1 - \beta_3)\,(\epsilon_{\theta\varphi}\cos 2\theta + \epsilon_{\varphi r}\sin 2\theta) \\
\sigma_{\varphi r} &= (2\alpha_2 + \beta_1 - \beta_3)\epsilon_{\varphi r} - (2\beta_2 + \beta_1 - \beta_3)\,(\epsilon_{\varphi r}\cos 2\theta - \epsilon_{\theta\varphi}\sin 2\theta) \\
\sigma_{r\theta} &= 2\alpha_2\epsilon_{r\theta} + \beta_1\,((\epsilon_{\theta\theta} + \epsilon_{rr})\sin 2\theta) + \beta_3\,(\epsilon_{\varphi\varphi}\sin 2\theta) \\
&\quad +\beta_2\,[(\epsilon_{\theta\theta} - \epsilon_{rr})\sin 4\theta - 2\epsilon_{r\theta}\cos 4\theta]
\end{aligned} \tag{6.19}$$

where

$$\begin{aligned}
\alpha_1 &= \frac{1}{8}(C_{11} + 6C_{13} + C_{33} - 4C_{44}) \\
\alpha_2 &= \frac{1}{8}(C_{11} - 2C_{13} + C_{33} + 4C_{44}) \\
\beta_1 &= \frac{1}{4}(C_{11} - C_{33}) \\
\beta_2 &= \frac{1}{8}(C_{11} - 2C_{13} + C_{33} - 4C_{44})
\end{aligned}$$

$$\beta_3 = \frac{1}{2}(C_{12} - C_{13}) \tag{6.20}$$

Here α_1, α_2, β_1, β_2, and β_3 are new stiffness constants, such that in the isotropic limit $\alpha_1 = \lambda_1$, $\alpha_2 = \mu_1$, and $\beta_1 = \beta_2 = \beta_3 = 0$, where λ_1 and μ_1 are the Lamé constants of the obstacle. It is noted that the stress-strains relations contain explicit trigonometric functions in the polar angle θ, but are independent of the azimuthal angle φ due to the rotational symmetry of the transversely isotropic material.

From the stress-strain relations Eq. (6.19), the definition of strain Eq. (1.13), and the equations of motion Eq. (1.12), all in spherical coordinates, the equations of motion in terms of the displacement components can be derived. The equations become very long and are not given. However, they contain trigonometric functions in 2θ and 4θ, exactly as the stress-strain relations.

To solve the equations of motion it is convenient to expand the displacement field in the vector spherical harmonics (with the σ-index suppressed)

$$\boldsymbol{u}_1(r,\theta,\varphi) = \sum_{\tau,m,l} F_{\tau ml}(r)\boldsymbol{A}_{\tau ml}(\theta,\varphi) \tag{6.21}$$

It is noted that the sum in l starts with $l = 1$ for $\tau = 1, 2$ and with $l = 0$ for $\tau = 3$. The coefficients $F_{\tau ml}(r)$ are expanded in power series in r

$$F_{1ml}(r) = \sum_{j=l,l+2,\dots}^{\infty} f_{1ml,j}\, r^j$$

$$F_{2ml}(r) = \sum_{j=l-1,l+1,\dots}^{\infty} f_{2ml,j}\, r^j$$

$$F_{3ml}(r) = \sum_{j=l-1,l+1,\dots}^{\infty} f_{3ml,j}\, r^j \tag{6.22}$$

where $f_{3m0,-1} = 0$. The powers are chosen so that regularity at the origin is fulfilled. When the displacement ansatz is inserted into the equations of motion there appear terms which are a product of a trigonometric function in 2θ or 4θ times a vector spherical harmonic. These can be expanded in the vector spherical harmonics so that the equations of motion have the structure

$$\sum_{\tau,m,l} H_{\tau ml}(r)\boldsymbol{A}_{\tau ml}(\theta,\varphi) = \boldsymbol{0}$$

As the vector spherical harmonics are orthogonal this leads to

$$H_{\tau ml}(r) = 0 \tag{6.23}$$

These equations are power series in r and as the powers of r are linearly independent the coefficient in front of each power must vanish and this leads to recursion relations among the coefficients $f_{\tau ml,j}$. Because of the trigonometric functions in 2θ and 4θ there is coupling between different values of l, so that $H_{\tau ml}$ contains $f_{\tau ml,j}$, $f_{\tau m\pm 2,j}$, and $f_{\tau ml\pm 4}$ (there is also coupling between different values of τ and j, but this happens also in the isotropic case). These recursion relations give all the higher order coefficients $f_{\tau ml,j}$ in terms of these of lowest order $j = l$ for $\tau = 1$ and $j = l - 1$ for $\tau = 2, 3$ ($j = 1$ for $m = 0$ and $\tau = 3$) and this completes the solution inside the sphere.

To determine the T-matrix the boundary conditions on the sphere are set up, and these are taken as continuity of displacement and traction. For the displacement this is straightforward as the expansions inside and outside the sphere are both in terms of the vector spherical harmonics. To apply the boundary condition on the traction, the traction inside the sphere is expanded in the vector spherical harmonics and the factors with trigonometric functions in 2θ and 4θ then give a coupling between l, $l \pm 2$, and $l \pm 4$, exactly as in the equations of motion.

All the equations necessary to solve for the T-matrix are now at hand, but the equations are very long and are not given explicitly. In principle there is no difficulty in solving them numerically for low and intermediate frequencies (higher frequencies lead to troubles because of convergence issues). In the following low frequencies are considered as this can be done analytically (although the calculations are long).

6.5.2 Low frequency T-matrix

For low frequencies there are T-matrix elements of leading order for $l = 0, 1, 2$, exactly as in the case of an isotropic sphere. As opposed to the isotropic case the elements now become different for different values of m and as $l \geq m$ there are fewer elements for higher values of m.

Starting with $m = 0$ there is a decoupling into $\tau = 1$ and $\tau = 2, 3$. However, $\tau = 1$ gives no leading order elements for low frequencies (as is the case also for an isotropic sphere), so only $\tau = 2, 3$ is considered. There is the further decoupling into symmetric and antisymmetric parts (with respect to the xy-plane). For the symmetric part the displacement field is expanded as

$$\boldsymbol{u}_1(r,\theta,\varphi) = F_{202}(r)\boldsymbol{A}_{202}(\theta,\varphi) + F_{204}(r)\boldsymbol{A}_{204}(\theta,\varphi) + \ldots$$
$$+F_{300}(r)\boldsymbol{A}_{300}(\theta,\varphi)+F_{302}(r)\boldsymbol{A}_{302}(\theta,\varphi)+F_{304}(r)\boldsymbol{A}_{304}(\theta,\varphi)+\ldots \tag{6.24}$$

where the r-dependent coefficients are expanded in power series in r according to Eqs. (6.22)

$$\begin{aligned}
F_{202}(r) &= f_{202,1}\, r + f_{202,3}\, r^3 + \ldots \\
F_{204}(r) &= f_{204,3}\, r^3 + \ldots \\
F_{300}(r) &= f_{300,1}\, r + f_{300,3}\, r^3 + \ldots \\
F_{302}(r) &= f_{302,1}\, r + f_{302,3}\, r^3 + \ldots \\
F_{304}(r) &= f_{304,3}\, r^3 + \ldots
\end{aligned} \tag{6.25}$$

Here exactly those terms needed for low frequencies are given. Insertion into the equations of motion gives

$$\begin{aligned}
H_{300} &= \frac{4}{7\sqrt{5}}\left(7\beta_1 + 2\beta_2\right)\left(\sqrt{6}f_{202,1} - 3f_{302,1}\right) r \\
&+\frac{1}{3}\Big(3\rho_1\omega^2 f_{300,1} + f_{300,3}(30\alpha_1 + 60\alpha_2 + 20\beta_1 - 2\beta_2) \\
&+32\sqrt{5}\beta_2 f_{204,3} - 4\sqrt{5}\left(7\beta_1 + 2\beta_2\right) f_{302,3} + 64\beta_2 f_{304,3}\Big) r^3 + \ldots = 0 \\
H_{302} &= \frac{2}{21}\left(\sqrt{6}f_{202,1} - 3f_{302,1}\right)\left(21\alpha_2 + 6\beta_1 - \beta_2\right) r \\
&+\Bigg(\frac{2}{3}f_{302,3}(15\alpha_1 + 21\alpha_2 - 4\beta_1 + 4\beta_2) + 2\sqrt{6}f_{202,3}\left(\beta_2 - \alpha_1\right) \\
&+8\beta_1 f_{204,3} + \frac{272}{33}\beta_2 f_{204,3} - \frac{56\beta_1 f_{300,3}}{3\sqrt{5}} - \frac{16\beta_2 f_{300,3}}{3\sqrt{5}} \\
&-\frac{56\beta_1 f_{304,3}}{\sqrt{5}} + \frac{112\beta_2 f_{304,3}}{33\sqrt{5}} + \rho_1\omega^2 f_{302,1}\Bigg) r^3 + \ldots = 0
\end{aligned}$$

The linear independence of the powers of r then gives a number of equations which can be solved to give

$$f_{302,1} = \sqrt{\frac{2}{3}} f_{202,1}$$

$$f_{304,1} = \frac{2}{\sqrt{5}} f_{204,3}$$

$$f_{302,3} = \frac{1}{20\sqrt{5}(7\beta_1 + 2\beta_2)} \Big(10 f_{300,3}(15\alpha_1 + 30\alpha_2 + 10\beta_1 - \beta_2) + 3\left(96\sqrt{5} f_{204,3}\beta_2 + 5 f_{300,1}\rho_1\omega^2\right)\Big)$$

$$f_{202,3} = \frac{1}{330\sqrt{6}\,(\alpha_1 - \beta_2)} \Big(110 f_{302,3}\,(15\alpha_1 + 21\alpha_2 - 4\beta_1 + 4\beta_2) + 40\,(33\beta_1 + 34\beta_2)\, f_{204,3} - 88\sqrt{5}\,(7\beta_1 + 2\beta_2)\, f_{300,3} + 56\sqrt{5}\,(2\beta_2 - 33\beta_1)\, f_{304,3} + 165\rho_1\omega^2 f_{302,1}\Big)$$

There remain four unknown expansion coefficients $f_{300,1}$, $f_{300,3}$, $f_{202,1}$, and $f_{204,3}$ inside the sphere. To determine these and the unknown expansion coefficients g_{300}, g_{302}, and g_{202} of the scattered wave, the boundary conditions on the sphere are set up. Continuity of displacement and traction gives one equation each for $l = 0, 2$, $\tau = 3$ and $l = 2$, $\tau = 2$, thus six equations in all. To obtain one more relation it is, in fact, necessary to go one step further in the hierarchy and it then follows that $f_{204,3} = 0$ if there is no incident partial wave for $l = 4$ (which can be assumed if only the T-matrix elements for $l = 0, 2$ are of interest).

Using also the series expansion of the spherical Bessel and Hankel functions and solving the seven equations give the leading order T-matrix elements

$$T^0_{30,30} = -\mathrm{i}\,(k_\mathrm{p}a)^3 \left(\frac{1}{3} - \frac{\mu k_\mathrm{s}^2 \left(4k_\mathrm{p}^2(\gamma - 3\mu) + 3k_\mathrm{s}^2(2\gamma + 9\mu)\right)}{k_\mathrm{p}^2(4k_\mathrm{p}^2 M_0 + 3k_\mathrm{s}^2 N_0)} \right)$$

$$T^0_{22,30} = T^0_{30,22} = -\mathrm{i}\,(k_\mathrm{s}a)^3 \sqrt{\frac{10k_\mathrm{p}}{3k_\mathrm{s}}}\, \frac{2\mu k_\mathrm{s}^2\,(2\beta_1 + \beta_3)}{4k_\mathrm{p}^2 M_0 + 3k_\mathrm{s}^2 N_0}$$

$$
\begin{aligned}
T^0_{32,30} &= T^0_{30,32} = -\mathrm{i}\,(k_\mathrm{p}a)^3\,\frac{4\sqrt{5}\mu k_\mathrm{s}^2\,(2\beta_1+\beta_3)}{3(4k_\mathrm{p}^2M_0+3k_\mathrm{s}^2N_0)}\\
T^0_{22,22} &= -\mathrm{i}\,(k_\mathrm{s}a)^3\,\frac{2k_\mathrm{s}^2M_0}{4k_\mathrm{p}^2M_0+3k_\mathrm{s}^2N_0}\\
T^0_{32,22} &= T^0_{22,32} = -\mathrm{i}\,(k_\mathrm{s}a)^3\sqrt{\frac{2k_\mathrm{p}}{3k_\mathrm{s}}}\,\frac{2k_\mathrm{p}^2M_0}{4k_\mathrm{p}^2M_0+3k_\mathrm{s}^2N_0}\\
T^0_{32,32} &= -\mathrm{i}\,(k_\mathrm{p}a)^3\,\frac{4k_\mathrm{p}^2M_0}{3(4k_\mathrm{p}^2M_0+3k_\mathrm{s}^2N_0)}
\end{aligned}
\tag{6.26}
$$

where

$$
\begin{aligned}
M_0 &= 2\alpha_2^2-2\beta_1^2+\beta_1\beta_2-\beta_2^2-2\beta_1\beta_3+\beta_2\beta_3+\alpha_1\,(\gamma-3\mu)\\
&\quad -2\beta_1\mu+5\beta_2\mu-4\mu^2+\alpha_2\,[\beta_2+2\,(\beta_3+\mu)]\\
N_0 &= 2M_0+\frac{5}{3}\mu\,(9\alpha_1+6\alpha_2+2\beta_1-3\beta_2+4\beta_3+12\mu)\\
\gamma &= 3\alpha_2-\beta_1+3\beta_2+\beta_3
\end{aligned}
$$

It is noted that there are leading order elements of the T-matrix that couple $l=0$ and $l=2$, contrary to the situation for an isotropic sphere where the T-matrix is diagonal in l. In the isotropic limit the T-matrix elements of course reduce to those of the isotropic sphere as given in Eqs. (3.54) and (3.56).

For the antisymmetric (with respect to the xy-plane) part for $m=0$ the proper ansatz for the displacement field within the sphere is

$$
\boldsymbol{u}_1(r,\theta,\varphi) = F_{201}(r)\boldsymbol{A}_{201}(\theta,\varphi)+F_{301}(r)\boldsymbol{A}_{301}(\theta,\varphi) \tag{6.27}
$$

where

$$
\begin{aligned}
F_{201}(r) &= f_{201,0}+f_{201,2}\,r^2\\
F_{301}(r) &= f_{301,0}+f_{301,2}\,r^2
\end{aligned}
\tag{6.28}
$$

The equations of motion then give two recursion relations and the boundary conditions give the leading order T-matrix elements

$$
T^m_{21,01} = \mathrm{i}\,(k_\mathrm{s}a)^3\,\frac{2}{9}\left(\frac{\rho_1}{\rho}-1\right)
$$

$$T^m_{31,21} = T^m_{21,31} = \mathrm{i}\sqrt{2k_\mathrm{s}^3 k_\mathrm{p}^3}\,\frac{a^3}{9}\left(\frac{\rho_1}{\rho} - 1\right)$$

$$T^m_{31,31} = \mathrm{i}\,(k_\mathrm{p}a)^3\,\frac{1}{9}\left(\frac{\rho_1}{\rho} - 1\right) \tag{6.29}$$

for $m = 0$. These elements are the same as in the case of isotropy [Eq. (3.55)], which is natural as they do not depend on the stiffness properties of the sphere (and correspond to rigid body motion of the sphere for low frequencies). They are in fact also valid for $m = 1$ (as $l \geq m$ no further m-values can be relevant).

For $m = 1$ the symmetric (with respect to the xy-plane) part remains. The displacement is expanded in the vector spherical harmonics as

$$\boldsymbol{u}_1(r,\theta,\varphi) = F_{111}(r)\boldsymbol{A}_{111}(\theta,\varphi) + F_{212}(r)\boldsymbol{A}_{212}(\theta,\varphi) + F_{312}(r)\boldsymbol{A}_{312}(\theta,\varphi) \tag{6.30}$$

where the coefficients are expanded as

$$F_{111}(r) = f_{111,1}\,r + f_{111,3}\,r^3$$

$$F_{212}(r) = f_{212,1}\,r + f_{212,3}\,r^3$$

$$F_{312}(r) = f_{312,1}\,r + f_{312,3}\,r^3 \tag{6.31}$$

This leads to three recursion relations among the coefficients and after applying the boundary conditions this gives the leading order T-matrix elements

$$T^1_{22,22} = \mathrm{i}\,\frac{2k_\mathrm{s}^5 a^3}{k_\mathrm{p}^2}\,C_1$$

$$T^1_{32,22} = T^1_{22,32} = \mathrm{i}a^3\sqrt{\frac{8k_\mathrm{s}^5 k_\mathrm{p}}{3}}\,C_1$$

$$T^1_{32,32} = \mathrm{i}\,(k_\mathrm{p}a)^3\,\frac{4}{3}\,C_1 \tag{6.32}$$

where

$$C_1 = \frac{\mu - C_{44}}{14\mu + 16C_{44} + 3\lambda(3 + 2C_{44}/\mu)}$$

These expression are almost identical to those of the isotropic case in Eq. (3.56); it is just the shear modulus C_{44} that appears instead of the shear modulus μ_1 for the isotropic case.

For $m = 2$ only T-matrix elements with $l = 2$ are of leading order as $l \geq m$. For the antisymmetric part with respect to the xy-plane l is even for $\tau = 1$ and odd for $\tau = 2, 3$ and these elements are therefore not of leading order as $\tau = 1$ gives no leading order elements. Thus only the symmetric part remains and a proper ansatz in this case is

$$\boldsymbol{u}_1(r,\theta,\varphi) = F_{123}(r)\boldsymbol{A}_{123}(\theta,\varphi)+F_{222}(r)\boldsymbol{A}_{222}(\theta,\varphi)+F_{322}(r)\boldsymbol{A}_{322}(\theta,\varphi) \tag{6.33}$$

where

$$\begin{aligned} F_{123}(r) &= f_{123,3}\, r^3 \\ F_{222}(r) &= f_{222,1}\, r + f_{222,3}\, r^3 \\ F_{322}(r) &= f_{322,1}\, r + f_{322,3}\, r^3 \end{aligned} \tag{6.34}$$

This leads to two recursion relations among the coefficients and after applying the boundary conditions this gives the leading order T-matrix elements

$$\begin{aligned} T^2_{22,22} &= \mathrm{i}\,\frac{2k_\mathrm{s}^5 a^3}{k_\mathrm{p}^2}\, C_2 \\ T^2_{32,22} &= T^1_{22,32} = \mathrm{i}a^3\sqrt{\frac{8k_\mathrm{s}^5 k_\mathrm{p}}{3}}\, C_2 \\ T^2_{32,32} &= \mathrm{i}(k_\mathrm{p}a)^3 \frac{4}{3}\, C_2 \end{aligned} \tag{6.35}$$

where

$$C_2 = \frac{\mu - C_{66}}{14\mu + 16C_{66} + 3\lambda(3 + 2C_{66}/\mu)}$$

These elements are almost identical to those of the isotropic case in Eq. (3.56); it is just the shear modulus C_{66} that appears instead of the shear modulus μ_1 for the isotropic case. The leading order T-matrix elements are thus very similar for $m = 1$ and $m = 2$. The only difference is that C_{44} appears for $m = 1$ whereas C_{66} appears for $m = 2$.

This approach can be generalized to other types of anisotropy. If the spherical obstacle is made of a cubic or orthotropic material this is pursued by Jafarzadeh *et al.* [71, 72].

7 Scattering by two or more obstacles

The scattering problems investigated so far involve only one obstacle in infinite space. In many situations it is of interest to investigate cases with two or more obstacles and this is treated in the present chapter. It is natural to use a coordinate system in each obstacle to describe the scattering by that obstacle, but as the effects of the other obstacles should be taken into account, it becomes necessary to consider the translational properties of the wavefunctions. These and many further aspects of scattering by a number of obstacles are treated by Martin [88] (who also gives numerous references). For more general shapes the scattering by more than one obstacle is considered by Peterson and Ström [114], Boström [12] and Varadan and Varadan [134] (all using the T-matrix method).

7.1 Two or more circles: 2D SH waves

Being the simplest case the 2D SH waves are as usual taken up first. As a start two circles are considered and then this is generalized to more than two. The circles can be of another material or can be rigid or a cavity (or a combination).

7.1.1 Two circles

Consider the scattering by two circles in infinite space, see Figure 7.1 for the geometry. Circle i, $i = 1, 2$, has radius a_i and centre at $\boldsymbol{r} = \boldsymbol{d}_i$ with respect to an arbitrary system. Each circle has its local coordinate system $\boldsymbol{r}_i$, so that $\boldsymbol{r} = \boldsymbol{d}_i + \boldsymbol{r}_i$. Corresponding polar coordinates are denoted (r_i, φ_i) (where it is assumed that the polar angles φ_i are measured from a common direction). The vector $\boldsymbol{d}_{12} = \boldsymbol{d}_1 - \boldsymbol{d}_2$ then goes from circle 2 to circle 1. For simplicity

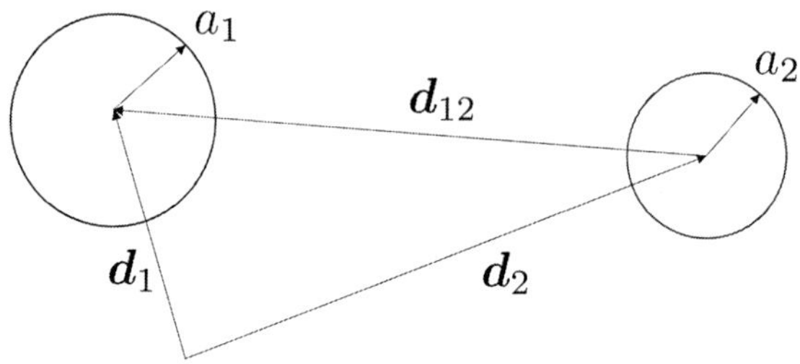

Figure 7.1: The scattering geometry with two obstacles.

the boundary condition on the circles is taken as the simplest possible, i.e. that the displacement vanishes. This is not essential and generalization to other cases is taken up later.

The first point is how the scattered wave should be expanded. As shown in Section 5.2.1 the integral representation gives that the scattered wave can be expanded in terms of out-going polar wavefunctions outside the circumscribed circle to the obstacle. As noted this can be generalized to more than one obstacle and if the Green function is expanded in polar wavefunctions with respect to each circle it follows that the scattered wave can be expanded as a sum of out-going wavefunctions from each circle

$$u^{\text{sc}} = \sum_{\sigma,m} f_{1\sigma m}\chi^{+}_{\sigma m}(\boldsymbol{r}_1) + \sum_{\sigma,m} f_{2\sigma m}\chi^{+}_{\sigma m}(\boldsymbol{r}_2) \tag{7.1}$$

where the coefficients $f_{1\sigma m}$ and $f_{2\sigma m}$ are to be determined by the boundary conditions on the circles. When this is to be done it is clearly of great value if the expansion can be transformed to the local coordinates in question. This can be accomplished by the translational properties of the polar wavefunctions

$$\chi^{+}_{\sigma m}(\boldsymbol{r}_2) = \sum_{\sigma',m'} P_{\sigma m,\sigma' m'}(\boldsymbol{d}_{12})\chi^{0}_{\sigma' m'}(\boldsymbol{r}_1) \tag{7.2}$$

which is valid for $r_1 < d_{12} = |\boldsymbol{d}_{12}|$. This is essentially a version of Graf's addition theorem [88], which is further complicated by the use of real trigonometric functions (instead of exponentials). The translation matrix $P_{\sigma m,\sigma' m'}(\boldsymbol{d})$ can be written

$$P_{\sigma m,\sigma m'}(\boldsymbol{d}) = \sqrt{\frac{\epsilon_m \epsilon_{m'}}{4}} \left[\mathrm{H}_{m-m'}(kd)\cos{(m-m')\psi} + (-1)^{m'+\sigma}\mathrm{H}_{m+m'}(kd)\cos{(m+m')\psi}\right]$$

$$P_{\sigma m,\sigma' m'}(\boldsymbol{d}) = \sqrt{\frac{\epsilon_m \epsilon_{m'}}{4}} \left[(-1)^{\sigma'}\mathrm{H}_{m-m'}(kd)\sin{(m-m')\psi} + (-1)^{m'}\mathrm{H}_{m+m'}(kd)\sin{(m+m')\psi}\right], \qquad \sigma \neq \sigma' \quad (7.3)$$

where (d, ψ) are the polar coordinates of $\boldsymbol{d}$. For two obstacles it is natural to choose coordinate systems so that $\psi = 0$ and it is then seen that $P_{\sigma m,\sigma' m'}(\boldsymbol{d}) = 0$ when $\sigma \neq \sigma'$.

Expanding the incident wave in the regular polar wavefunctions in the usual way (in this case about the origin in obstacle 1) the total wave can be written

$$u(\boldsymbol{r}_1) = \sum_{\sigma,m} \left[b_{1\sigma m}\chi^0_{\sigma m}(\boldsymbol{r}_1) + f_{1\sigma m}\chi^+_{\sigma m}(\boldsymbol{r}_1) + \chi^0_{\sigma m}(\boldsymbol{r}_1)\sum_{\sigma',m'} P_{\sigma' m',\sigma m}(\boldsymbol{d}_{12}) f_{2\sigma' m'}\right] \quad (7.4)$$

where $b_{1\sigma m}$ are the (known) expansion coefficients of the incident wave. Here all terms are expanded in the same trigonometric system so the vanishing displacement at $r_1 = a_1$ gives

$$b_{1\sigma m}\mathrm{J}_m(ka_1) + f_{1\sigma,m}\mathrm{H}_m(ka_1) + \mathrm{J}_m(ka_1)\sum_{\sigma',m'} P_{\sigma' m',\sigma m}(\boldsymbol{d}_{12}) f_{2\sigma' m'} = 0$$

valid for $\sigma = \mathrm{e}, \mathrm{o}$ and $m = 0, 1, 2, \ldots$. Similarly the boundary condition on the other circle gives

$$b_{2\sigma m}\mathrm{J}_m(ka_2) + f_{2\sigma m}\mathrm{H}_m(ka_2) + \mathrm{J}_m(ka_2)\sum_{\sigma',m'} P_{\sigma' m',\sigma m}(-\boldsymbol{d}_{12}) f_{1\sigma' m'} = 0$$

These equations can be divided by $\mathrm{H}_m(ka_1)$ and $\mathrm{H}_m(ka_2)$, respectively, which gives

$$b_{1\sigma m}T^1_m - f_{1\sigma m} + T^1_m \sum_{\sigma',m'} P_{\sigma' m',\sigma,m}(\boldsymbol{d}_{12}) f_{2\sigma' m'} = 0$$

$$b_{2\sigma m}T_m^2 - f_{2\sigma m} + T_m^2 \sum_{\sigma',m'} P_{\sigma'm',\sigma,m}(-\boldsymbol{d}_{12})f_{1\sigma'm'} = 0 \tag{7.5}$$

where the T-matrix elements for a rigid circle

$$T_m^i = -\frac{\mathrm{J}_m(ka_i)}{\mathrm{H}_m(ka_i)}, \qquad i = 1, 2$$

are introduced. Exactly as for one obstacle the influence of each obstacle is fully described by its T-matrix. This is a system of equations which is straightforward to solve for $f_{1\sigma m}$ and $f_{2\sigma m}$.

Another way to look at the scattering by two obstacles is to say that the total wave incident on obstacle 1 is the incident wave plus the wave scattered by obstacle 2. The latter also includes all multiple scattering between the two circles. Thus the total wave incident on obstacle 1 is

$$u^{\mathrm{in},1}(\boldsymbol{r}_1) = \sum_{\sigma,m}\left[b_{1\sigma m}\chi^0_{\sigma m}(\boldsymbol{r}_1) + \chi^0_{\sigma m}(\boldsymbol{r}_1)\sum_{\sigma',m'} P_{\sigma'm',\sigma m}(\boldsymbol{d}_{12})f_{2\sigma'm'}\right]$$

where the translation to the origin in obstacle 1 is already performed. The expansion coefficients for this total incident wave is thus

$$g_{1\sigma m} = b_{1\sigma m} + \sum_{\sigma',m'} P_{\sigma'm',\sigma m}(\boldsymbol{d}_{12})f_{2\sigma'm'}$$

The expansion coefficients for the wave scattered by obstacle 1 are then

$$f_{1\sigma m} = T_m^1 g_{1\sigma m} = T_m^1\left[b_{1\sigma m} + \sum_{\sigma',m'} P_{\sigma m,\sigma'm'}(\boldsymbol{d}_{12})f_{2\sigma'm'}\right] \tag{7.6}$$

This is exactly the first of Eqs. (7.5).

It is possible to express the solution to the Eqs. (7.5) as a T-matrix for the whole system of two obstacles. To be meaningful this should be with respect to a single origin and then it is not natural to use any of the coordinate systems of the obstacles. Instead an arbitrary origin is used and then the incident and scattered waves should be expanded with respect to this origin. To this end further translational properties of the polar wavefunctions are needed

$$\chi^+_{\sigma m}(\boldsymbol{r}_i) = \sum_{\sigma',m'} R_{\sigma m,\sigma'm'}(-\boldsymbol{d}_i)\chi^+_{\sigma'm'}(\boldsymbol{r}), \qquad i = 1, 2 \tag{7.7}$$

which is valid for $r > d_i$, and

$$\chi^0_{\sigma m}(\boldsymbol{r}) = \sum_{\sigma',m'} R_{\sigma m,\sigma' m'}(\boldsymbol{d}_i)\chi^0_{\sigma' m'}(\boldsymbol{r}_i), \qquad i = 1, 2 \tag{7.8}$$

which is valid for all r_i. Here the matrix $R_{\sigma m,\sigma' m'}$ is the same as $P_{\sigma m,\sigma' m'}$ if Bessel functions are inserted instead of Hankel functions. The incident wave can thus be expanded as

$$\begin{aligned} u^{\text{in}}(\boldsymbol{r}) &= \sum_{\sigma,m} b_{\sigma m}\chi^0_{\sigma m}(\boldsymbol{r}) \\ &= \sum_{\sigma,m} b_{\sigma m} \sum_{\sigma',m'} R_{\sigma m,\sigma' m'}(\boldsymbol{d}_i)\chi^0_{\sigma' m'}(\boldsymbol{r}_i), \qquad i = 1, 2 \end{aligned}$$

This means that the expansion coefficients with respect to the origin inside each obstacle is

$$b_{i\sigma m} = \sum_{\sigma',m'} b_{\sigma' m'} R_{\sigma' m',\sigma m}(\boldsymbol{d}_i), \qquad i = 1, 2$$

For the scattered wave the expansion becomes

$$\begin{aligned} u^{\text{sc}}(\boldsymbol{r}) &= \sum_{\sigma,m} f_{1\sigma m}\chi^+_{\sigma m}(\boldsymbol{r}_1) + \sum_{\sigma,m} f_{2\sigma m}\chi^+_{\sigma m}(\boldsymbol{r}_2) \\ &= \sum_{\sigma,m} f_{1\sigma m} \sum_{\sigma',m'} R_{\sigma m,\sigma' m'}(-\boldsymbol{d}_1)\chi^+_{\sigma' m'}(\boldsymbol{r}) \\ &\quad + \sum_{\sigma,m} f_{2\sigma m} \sum_{\sigma',m'} R_{\sigma m,\sigma' m'}(-\boldsymbol{d}_2)\chi^+_{\sigma' m'}(\boldsymbol{r}) \end{aligned}$$

Using this and solving Eqs. (7.5) this results in a total T-matrix for both obstacles that can be written as

$$\begin{aligned} T^{12} = R(\boldsymbol{d_1}) &\Big(T^1 \left[I - P(-\boldsymbol{d}_{12})T^2 P(\boldsymbol{d}_{12})T^1\right]^{-1} \\ &\times \left[I + P(-\boldsymbol{d}_{12})T^2 R(\boldsymbol{d}_{12})\right]\Big) R(-\boldsymbol{d}_1) \\ +R(\boldsymbol{d_2}) &\Big(T^2 \left[I - P(\boldsymbol{d}_{12})T^1 P(-\boldsymbol{d}_{12})T^2\right]^{-1} \\ &\times \left[I + P(\boldsymbol{d}_{12})T^1 R(-\boldsymbol{d}_{12})\right]\Big) R(-\boldsymbol{d}_2) \end{aligned} \tag{7.9}$$

in obvious matrix notation [114]. Here I is the unit matrix. For the circular obstacles considered here the T-matrices are diagonal. However, Eqs. (7.5) and (7.9) have a wider applicability and can be used in all cases where a T-matrix can be calculated and the circumscribed circles to the two obstacles do not overlap (this is a sufficient condition, but in some cases not a necessary condition). This is particularly clear in the derivation leading to Eq. (7.6), see Martin [88].

From a numerical point of view it is of advantage if the matrices can be kept as small as possible, and this means that the maximum value of the azimuthal index m should be as small as possible. This maximum closely depends on the radius of the circumscribed circle to the obstacle. However, for the T-matrix for two obstacles given in Eq. (7.9), it is the radius of the circle enclosing both obstacles that governs the maximum, so it is good to choose the origin so that this radius is as small as possible. Especially for obstacles that are well separated this means a large radius of the enclosed circle and therefore a large maximum m value. In Eqs. (7.5), however, the restriction is only on the individual T matrices, so these equations need a (much) smaller maximum m-value.

7.1.2 Generalizations

In the previous subsection the obstacle was assumed to be rigid, although it was also mentioned that the T-matrix for any obstacle could be used. This is now shown for a circular obstacle of a different material. Consider the scattering configuration as in the previous subsection and let the circles have shear modulus μ_i and wavenumber k_i, $i = 1, 2$. The surrounding has shear modulus μ and wavenumber k.

The expansion Eq. (7.4) for the total wave around origin 1 is still valid. Inside each obstacle an expansion in regular polar wavefunctions is performed

$$u^i(\boldsymbol{r}_i) = \sum_{\sigma,m} c_{i\sigma m} \chi^0_{i,\sigma m}(\boldsymbol{r}_i), \qquad i = 1, 2$$

The boundary conditions on obstacle 1 are that the displacement u and traction $\mu\,\partial u/\partial r$ are continuous for $r_1 = a_1$ and this gives

$$
\begin{aligned}
& b_{1\sigma m}\mathrm{J}_m(ka_1) + f_{1\sigma m}\mathrm{H}_m(ka_1) \\
& \qquad + \mathrm{J}_m(ka_1)\sum_{\sigma',m'} P_{\sigma'm',\sigma m}(\boldsymbol{d}_{12}) f_{2\sigma'm'} = c_{1\sigma m}\mathrm{J}_m(k_1a_1) \\
& \mu k \Bigg[b_{1\sigma m}\mathrm{J}'_m(ka_1) + f_{1\sigma m}\mathrm{H}'_m(ka_1) \\
& \qquad + \mathrm{J}'_m(ka_1)\sum_{\sigma',m'} P_{\sigma'm',\sigma m}(\boldsymbol{d}_{12}) f_{2\sigma'm'} \Bigg] = \mu_1 k_1 c_{1\sigma m}\mathrm{J}'_m(k_1a_1)
\end{aligned}
$$

valid for $\sigma = \mathrm{e}, \mathrm{o}$ and $m = 0, 1, 2, \ldots$. If $c_{1\sigma m}$ is eliminated between these two equations, the first of Eqs. (7.5) is again obtained, but now with the T-matrix element

$$
T^1_m = -\frac{\mu_1 k_1 \mathrm{J}'_m(k_1a_1)\mathrm{J}_m(ka_1) - \mu k \mathrm{J}_m(k_1a_1)\mathrm{J}'_m(ka_1)}{\mu_1 k_1 \mathrm{J}'_m(k_1a_1)\mathrm{H}_m(ka_1) - \mu k \mathrm{J}_m(k_1a_1)\mathrm{H}'_m(ka_1)}
$$

This is exactly the T-matrix element given in Eq. (2.14) for a circular obstacle of a different material and this shows that Eqs. (7.5) can be used for this case. As mentioned above, the validity is much wider and the equations can be used for any obstacle as long as its T-matrix can be computed (and the circumscribed circles of the two obstacles do not overlap).

Another generalization is to more than two obstacles. This is straightforward; thus the generalization of Eqs. (7.5) to a case with N obstacles is

$$
b_{i\sigma m}T^i_m - f_{i\sigma m} + \sum_{\substack{i'=1 \\ i'\neq i}}^{N} T^i_m \sum_{\sigma',m'} P_{\sigma m,\sigma'm'}(\boldsymbol{d}_{ii'}) f_{i'\sigma'm'} = 0
$$

for $i = 1, 2, \ldots, N$. Here the prime on the summation means that the term $i' = i$ is omitted. These equations are thus a system for the unknowns $f_{i\sigma m}$. It is possible to calculate a total T-matrix for the system of N obstacles, but explicit expressions become lengthy and are not given [114].

7.2 Two spheres

For the scattering by two (or more) spheres the development is similar as in the previous section. This is a vector problem and thus the translational properties of the spherical vector wavefunctions are needed.

7.2.1 Translations of spherical vector wavefunctions

The translational properties of the spherical vector wavefunctions are well-known but quite complicated, see Stein [124] and Cruzan [38] and further references in Martin [88]. If the translation is given by $\boldsymbol{r}' = \boldsymbol{d} + \boldsymbol{r}$ the task is to express the spherical vector wavefunctions with argument $\boldsymbol{r}'$ in terms of those with argument $\boldsymbol{r}$. The translation is given by the vector $\boldsymbol{d}$ whose spherical coordinates are denoted (d, η, ψ). The following then applies

$$\boldsymbol{\psi}_n^0(\boldsymbol{r}') = \sum_{n'} R_{nn'}(\boldsymbol{d})\boldsymbol{\psi}_{n'}^0(\boldsymbol{r}) \tag{7.10}$$

$$\boldsymbol{\psi}_n^+(\boldsymbol{r}') = \sum_{n'} R_{nn'}(\boldsymbol{d})\boldsymbol{\psi}_{n'}^+(\boldsymbol{r}), \qquad r > d \tag{7.11}$$

$$\boldsymbol{\psi}_n^+(\boldsymbol{r}') = \sum_{n'} P_{nn'}(\boldsymbol{d})\boldsymbol{\psi}_{n'}^0(\boldsymbol{r}), \qquad r < d \tag{7.12}$$

For simplicity the multi-index $n = \tau\sigma ml$ is introduced, so a summation is performed over all these indices. As there is no coupling between S and P waves in a translation the nonzero elements of the translation matrix $R_{nn'}$ are

$$\begin{aligned}
R_{1\sigma ml,1\sigma m'l'}(\boldsymbol{d}) &= (-1)^{m'} C_{ml,m'l'}(d,\eta)\cos{(m-m')\psi} \\
&\quad +(-1)^{\sigma} C_{ml,-m'l'}(d,\eta)\cos{(m+m')\psi} \\
R_{1\sigma ml,1\sigma' m'l'}(\boldsymbol{d}) &= (-1)^{m'+\sigma'} C_{ml,m'l'}(d,\eta)\sin{(m-m')\psi} \\
&\quad +C_{ml,-m'l'}(d,\eta)\sin{(m+m')\psi} \\
R_{1\sigma ml,2\sigma' m'l'}(\boldsymbol{d}) &= (-1)^{m'+\sigma} D_{ml,m'l'}(d,\eta)\cos{(m-m')\psi} \\
&\quad -D_{ml,-m'l'}(d,\eta)\cos{(m+m')\psi} \\
R_{1\sigma ml,2\sigma m'l'}(\boldsymbol{d}) &= (-1)^{m'} D_{ml,m'l'}(d,\eta)\sin{(m-m')\psi} \\
&\quad +(-1)^{\sigma} D_{ml,-m'l'}(d,\eta)\sin{(m+m')\psi} \\
R_{2\sigma ml,\tau'\sigma' m'l'} &= R_{1\sigma ml,\tau\sigma' m'l'}, \qquad \tau \neq \tau' = 1,2,\ \sigma,\sigma' = \mathrm{e},\mathrm{o} \\
R_{3\sigma ml,3\sigma m'l'}(\boldsymbol{d}) &= (-1)^{m'} B_{ml,m'l'}(d,\eta)\cos{(m-m')\psi} \\
&\quad +(-1)^{\sigma} B_{ml,-m'l'}(d,\eta)\cos{(m+m')\psi}
\end{aligned}$$

$$R_{3\sigma ml,3\sigma' m'l'}(\boldsymbol{d}) = (-1)^{m'+\sigma'} B_{ml,m'l'}(d,\eta)\sin(m-m')\psi$$
$$+B_{ml,-m'l'}(d,\eta)\sin(m+m')\psi$$

where $\sigma' \neq \sigma$ (except on the fifth line for $R_{2\sigma ml,\tau'\sigma' m'l'}$). Here

$$C_{ml,m'l'}(d,\eta) = (-1)^{m+m'}\frac{\sqrt{\epsilon_m\epsilon_{m'}}}{4}\sum_{\lambda=|l-l'|}^{l+l'}(-1)^{(l-l'+\lambda)/2}$$

$$\times(2\lambda+1)\left[\frac{(2l+1)(2l'+1)(\lambda-(m-m')!)}{l(l+1)l'(l'+1)(\lambda+m-m')!}\right]^{1/2}$$

$$\times\begin{pmatrix} l & l' & \lambda \\ 0 & 0 & 0 \end{pmatrix}\begin{pmatrix} l & l' & \lambda \\ m & -m' & m'-m \end{pmatrix}$$

$$\times[l(l+1)+l'(l'+1)-\lambda(\lambda+1)]\mathrm{j}_\lambda(k_{\mathrm{s}}d)\mathrm{P}_\lambda^{m-m'}(\cos\eta)$$

$$D_{ml,m'l'}(d,\eta) = (-1)^{m+m'}\frac{\sqrt{\epsilon_m\epsilon_{m'}}}{4}\sum_{\lambda=|l-l'|+1}^{l+l'}\mathrm{i}^{l-l'+\lambda+1}$$

$$\times(2\lambda+1)\left[\frac{(2l+1)(2l'+1)(\lambda-(m-m'))!}{l(l+1)l'(l'+1)(\lambda+m-m')!}\right]^{1/2}$$

$$\times\begin{pmatrix} l & l' & \lambda-1 \\ 0 & 0 & 0 \end{pmatrix}\begin{pmatrix} l & l' & \lambda \\ m & -m' & m'-m \end{pmatrix}$$

$$\times\left(\lambda^2-(l-l')^2\right)^{1/2}\mathrm{j}_\lambda(k_{\mathrm{s}}d)\mathrm{P}_\lambda^{m-m'}(\cos\eta)$$

$$B_{ml,m'l'}(d,\eta) = (-1)^{m+m'}\frac{\sqrt{\epsilon_m\epsilon_{m'}}}{2}\sum_{\lambda=|l-l'|}^{l+l'}(-1)^{(l-l'+\lambda)/2}$$

$$\times(2\lambda+1)\left[\frac{(2l+1)(2l'+1)(\lambda-(m-m'))!}{(\lambda+m-m')!}\right]^{1/2}$$

$$\times\begin{pmatrix} l & l' & \lambda \\ 0 & 0 & 0 \end{pmatrix}\begin{pmatrix} l & l' & \lambda \\ m & -m' & m'-m \end{pmatrix}$$

$$\times[l(l+1)+l'(l'+1)-\lambda(\lambda+1)]\mathrm{j}_\lambda(k_{\mathrm{p}}d)\mathrm{P}_\lambda^{m-m'}(\cos\eta)$$

Here $\left(\begin{smallmatrix} \cdot & \cdot & \cdot \\ \cdot & \cdot & \cdot \end{smallmatrix}\right)$ is Wigner's 3-j symbol, see Edmonds [46]. The matrix $P_{nn'}(\boldsymbol{d})$ is obtained from $R_{nn'}(\boldsymbol{d})$ if spherical Bessel functions are replaced by spherical Hankel functions. For two obstacles it is natural to use a translation along the z-axis (passing through the centres of the spheres). This corresponds to $\eta = 0$ (and the value of ψ does not matter) and this somewhat simplifies the expressions.

7.2.2 Two spherical obstacles

Consider the scattering by two spheres in infinite space. Sphere i, $i = 1, 2$, has radius a_i and centre at $\boldsymbol{r} = \boldsymbol{d}_i$ with respect to an arbitrary system. Each sphere has its local coordinate system $\boldsymbol{r}_i$, so that $\boldsymbol{r} = \boldsymbol{d}_i + \boldsymbol{r}_i$. Corresponding spherical coordinates are denoted $(r_i, \theta_i, \varphi_i)$ (where it is assumed that the spherical angles are measured from common directions). The vector $\boldsymbol{d}_{12} = \boldsymbol{d}_1 - \boldsymbol{d}_2$ then goes from sphere 2 to sphere 1. For simplicity the spheres are taken as cavities so that the boundary condition is that the traction vanishes. This is not essential and generalization to other cases are mentioned at the end.

In the same way as in the previous section it follows from the integral representation Eq. (4.20) with the Green dyadic expanded in the spherical vector wavefunctions [Eq. (4.16)] around each sphere that the scattered wave can be expanded as a sum of out-going vector wavefunctions from each sphere

$$\boldsymbol{u}^{\mathrm{sc}} = \sum_n f_{1n} \boldsymbol{\psi}_n^+(\boldsymbol{r}_1) + \sum_n f_{2n} \boldsymbol{\psi}_n^+(\boldsymbol{r}_2) \tag{7.13}$$

where the multi-index $n = \tau\sigma ml$ is used as above. Here the coefficients f_{1n} and f_{2n} are to be determined by the boundary conditions on the spheres. To apply the boundary condition on sphere 1 the translation formula Eq. (7.14) is used so that the scattered wave is written

$$\boldsymbol{u}^{\mathrm{sc}}(\boldsymbol{r}_1) = \sum_n \left[f_{1n} \boldsymbol{\psi}_n^+(\boldsymbol{r}_1) + \boldsymbol{\psi}_n^0(\boldsymbol{r}_1) \sum_{n'} P_{n'n}(\boldsymbol{d}_{12}) f_{2n'} \right] \tag{7.14}$$

Adding the expansion (around sphere 1) of the incident wave

$$\boldsymbol{u}^{\mathrm{in}}(\boldsymbol{r}_1) = \sum_n b_{1n} \boldsymbol{\psi}_n^0(\boldsymbol{r}_1)$$

and applying the boundary condition give

$$\sum_{\tau''=1}^{3}\left(S^0_{\tau,\tau''l}(a_1)b_{1\tau''\sigma ml}+S^+_{\tau,\tau''l}(a_1)f_{1\tau''\sigma ml}\right.$$

$$\left.+S^0_{\tau,\tau''l}(a_1)\sum_{\tau'=1}^{3}\sum_{\sigma'=\text{e,o}}\sum_{m'=0}^{\infty}\sum_{l'=m}^{\infty}P_{\tau'\sigma'm'l',\tau\sigma ml}(\boldsymbol{d}_{12})f_{2\tau'\sigma'm'l'}\right)=0$$

for all τ, σ, m, and l. Here all indices and summations are written out for clarity. The vector spherical harmonics components of the tractions $S^0_{\tau,\tau'l}$ are given in Eqs. (3.29). Solving the three equations for $\tau = 1,2,3$ for $f_{1\tau\sigma ml}$, $\tau = 1,2,3$, gives

$$\sum_{\tau''=1}^{3}T^{1l}_{\tau\tau''}b_{1\tau''\sigma ml}-f_{1\tau\sigma ml}$$

$$+\sum_{\tau''=1}^{3}T^{1l}_{\tau\tau''}\sum_{\tau'=1}^{3}\sum_{\sigma'=\text{e,o}}\sum_{m'=0}^{\infty}\sum_{l'=m}^{\infty}P_{\tau'\sigma'm'l',\tau\sigma ml}(\boldsymbol{d}_{12})f_{2\tau'\sigma'm'l'}=0 \tag{7.15}$$

for all τ, σ, m, and l. Here the transition matrix $T^{1l}_{\tau\tau'}$ from Section 3.2.1 is introduced (with an extra superscript "1" added). The boundary condition on sphere 2 gives a similar equation

$$\sum_{\tau''=1}^{3}T^{2l}_{\tau\tau''}b_{2\tau''\sigma ml}-f_{2\tau\sigma ml}$$

$$+\sum_{\tau''=1}^{3}T^{2l}_{\tau\tau''}\sum_{\tau'=1}^{3}\sum_{\sigma'=\text{e,o}}\sum_{m'=0}^{\infty}\sum_{l'=m}^{\infty}P_{\tau'\sigma'm'l',\tau\sigma ml}(-\boldsymbol{d}_{12})f_{1\tau'\sigma'm'l'}=0 \tag{7.16}$$

Equations (7.15) and (7.16) have exactly the same structure as Eqs. (7.5) and can be used to solve for $f_{1\tau\sigma ml}$ and $f_{2\tau\sigma ml}$. For the two cavities a total T-matrix can be calculated and Eq. (7.9) is still valid.

As in the previous section generalizations to other obstacles is straightforward as long as their T matrices are available and their circumscribed spheres do not overlap. More than two obstacles are of course also straightforward to consider.

8 Scattering by obstacles in layered structures

So far one (or more) bounded obstacle in an unbounded space is considered. However, in many situations of interest the obstacle is located in some sort of structure. In simple cases such problems can be investigated with more or less analytical methods. In this chapter geometries with an obstacle in the presence of infinite planar interfaces are considered. As scattering by an obstacle is naturally calculated using spherical (polar in 2D) wavefunctions and scattering by a plane surface is using plane waves, the transformation between spherical (polar) wavefunctions and plane waves becomes an important ingredient.

8.1 Obstacle in layered structures: 2D SH waves

As usual the simplest case with 2D SH waves is taken up first. The 2D acoustic (which is the same as SH) wave scattering by an obstacle (of general shape) in a layered medium is treated by Karlsson [73] with the T-matrix method, but here a more intuitive and direct method is employed. First plane waves are introduced and the transformations between plane and polar waves are given. The scattering by an obstacle in a half-space is considered and thereafter the general case with an obstacle in a layered medium.

8.1.1 Plane waves

Consider a circular obstacle in a half-space with a stress-free surface, so that waves are restricted to the half-space. For the scattering by the obstacle it is natural to use the scalar polar wavefunctions. For the scattering (reflection)

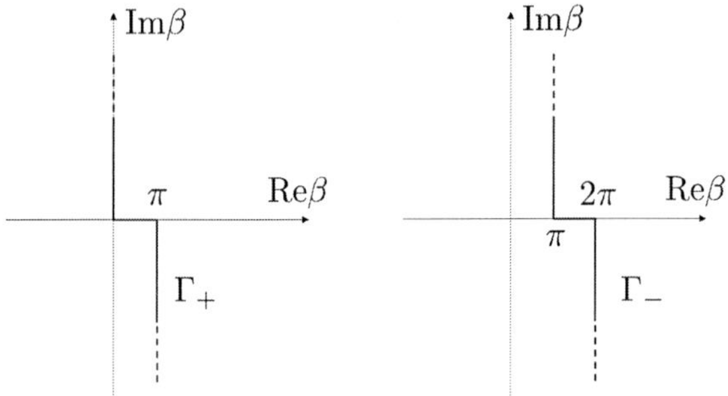

Figure 8.1: The integration contours Γ_+ and Γ_-, where Γ_+ runs from $\mathrm{i}\infty$ to $\pi - \mathrm{i}\infty$ and Γ_- runs from $\pi + \mathrm{i}\infty$ to $2\pi - \mathrm{i}\infty$.

from the plane surface of the half-space it is likewise natural to use plane waves. Plane wavefunctions can be introduced, but in this 2D scalar case these are so simple so a particular notation for them is avoided (but when the vector case is considered such wavefunctions are introduced). An up-going (in the y-coordinate) plane wave can be written

$$\mathrm{e}^{\mathrm{i}(qx+hy)} = \mathrm{e}^{\mathrm{i}k(x\cos\beta+y\sin\beta)} = \mathrm{e}^{\mathrm{i}kr\cos(\varphi-\beta)}$$

Here q and h are wavenumbers in the x- and y-directions, respectively, and $q^2 + h^2 = k^2$. The angle β specifies the direction of propagation of the plane wave and $q = k\cos\beta$ and $h = k\sin\beta$. When using plane waves in a half-space $y > 0$ it is necessary to include also evanescent waves, meaning that $q > k$ so that h becomes purely imaginary (if there are no losses) and the convention here is that $\operatorname{Im} h \geq 0$. For the evanescent waves the angle β becomes complex.

When both plane waves and polar wavefunctions are employed it becomes necessary to transform between the two sets. The transformation from a plane wave to polar wavefunctions is given in Eq. (2.17)

$$\mathrm{e}^{\mathrm{i}kr\cos(\varphi-\beta)} = \sum_{\sigma,m} 2\mathrm{i}^m\sqrt{\epsilon_m}\left\{\begin{array}{c}\cos m\beta\\ \sin m\beta\end{array}\right\}\chi^0_{\sigma m}(r,\varphi) \tag{8.1}$$

where the upper or lower row is used for $\sigma = \mathrm{e}$ or $\sigma = \mathrm{o}$, respectively. Rewriting the left-hand side in Cartesian coordinates this equation can be given in the form

$$\mathrm{e}^{\mathrm{i}(qx \pm hy)} = \sum_{\sigma,m} 2\mathrm{i}^m \sqrt{\epsilon_m} \left\{ \begin{array}{c} \cos m\beta_\pm \\ \sin m\beta_\pm \end{array} \right\} \chi^0_{\sigma m}(r,\varphi) \tag{8.2}$$

where $+$ $(-)$ is for up-going (down-going) waves and $\beta_\pm$ is determined from $q = k\cos\beta_\pm$ and $h = \pm k \sin\beta_\pm$ with $\mathrm{Im}\, h \geq 0$. The transformation from a regular or outgoing polar wavefunction to plane waves is essentially an integral representation of a Bessel or Hankel function, respectively, and they are [27]

$$\chi^0_{\sigma m}(r,\varphi) = \int_0^{2\pi} \mathrm{i}^{-m} \frac{\sqrt{\epsilon_m}}{4\pi} \left\{ \begin{array}{c} \cos m\beta \\ \sin m\beta \end{array} \right\} \mathrm{e}^{\mathrm{i}kr\cos(\varphi-\beta)} \mathrm{d}\beta \tag{8.3}$$

$$\chi^+_{\sigma m}(r,\varphi) = \int_{\Gamma_\pm} \mathrm{i}^{-m} \frac{\sqrt{\epsilon_m}}{2\pi} \left\{ \begin{array}{c} \cos m\beta \\ \sin m\beta \end{array} \right\} \mathrm{e}^{\mathrm{i}kr\cos(\varphi-\beta)} \mathrm{d}\beta \tag{8.4}$$

The contours Γ_+ and Γ_- are shown in Figure 8.1; Γ_+ should be chosen when $y > 0$ and Γ_- when $y < 0$. These contours are such that $q = k\cos\beta$ belongs to $(-\infty, \infty)$ when β belongs to Γ_+ or Γ_-. The transformation can also be written

$$\chi^+_{\sigma m}(r,\varphi) = \int_{-\infty}^{\infty} \mathrm{i}^{-m} \frac{\sqrt{\epsilon_m}}{2\pi} \left\{ \begin{array}{c} \cos m\beta_\pm \\ \sin m\beta_\pm \end{array} \right\} \mathrm{e}^{\mathrm{i}(qx \pm hy)} \frac{\mathrm{d}q}{h} \tag{8.5}$$

where $+$ or $-$ in the exponential is chosen according to $y > 0$ or $y < 0$ and $\beta_\pm$ is determined from $q = k\cos\beta_\pm$ and $h = \pm k\sin\beta_\pm$.

To get a representation of the scattered wave in the half-space with the obstacle the integral representation with the Green function expanded in both polar and plane waves is used. The 2D scalar Green function expanded in polar wavefunctions is given in Eq. (4.3) and is repeated here

$$g(\boldsymbol{r};\boldsymbol{r}') = \mathrm{i} \sum_{\sigma,m} \chi^0_{\sigma m}(\boldsymbol{r}_<) \chi^+_{\sigma m}(\boldsymbol{r}_>)$$

To get the expansion of the Green function in plane waves the differential equation for the Green function is used [Eq. (4.1)]

$$\nabla^2 g(\boldsymbol{r};\boldsymbol{r}') + k^2 g(\boldsymbol{r};\boldsymbol{r}') = -\delta(\boldsymbol{r} - \boldsymbol{r}')$$

Taking the Fourier transform with respect to x the equation becomes

$$\tilde{g}''(q,y;x',y') + h^2\tilde{g}(q,y;x',y') = -\delta(y-y')\mathrm{e}^{-\mathrm{i}qx'}$$

where a prime denotes differentiation with respect to y and $h = \sqrt{k^2-q^2}$, with root chosen so that $\operatorname{Im} h \geq 0$. Outside $y = y'$ the solution must be of exponential type and to fulfil the radiation condition it must be

$$\tilde{g}(q,y;x',y') = \begin{cases} A\,\mathrm{e}^{\mathrm{i}h(y-y')}, & y > y' \\ \\ B\,\mathrm{e}^{-\mathrm{i}h(y-y')}, & y < y' \end{cases}$$

The delta function on the right-hand side means that $\tilde{g}(q,y;x',y')$ must be continuous while its derivative must have a jump $-\exp(-\mathrm{i}qx')$ at $y = y'$. This determines $A = B = (\mathrm{i}/2h)\exp(-\mathrm{i}qx')$ and after the inverse transform the expansion of the Green function becomes

$$g(\boldsymbol{r};\boldsymbol{r}') = \frac{\mathrm{i}}{4\pi}\int_{-\infty}^{\infty}\frac{1}{h}\,\mathrm{e}^{\mathrm{i}(q(x-x')+h|y-y'|)}\,\mathrm{d}q \tag{8.6}$$

8.1.2 Circular obstacle in half-space

Consider the scattering of an incident wave in a half-space with a circular obstacle. The obstacle has radius a and is situated below the surface of the half-space at the depth d to its centre. For simplicity the surface of the half-space and the obstacle are both taken as stress-free, so the normal derivative of the displacement vanishes. An xy-coordinate system is introduced with the origin at the centre of the obstacle and the y-axis normal to the half-space, see Figure 8.2. The incident wave u^{in} is left arbitrary.

The scalar 2D integral representation Eq. (4.18) is given for one bounded obstacle with surface S, but as remarked can be used also when there are more than one obstacle. One of the "obstacles" can be an infinite one; thus the integral representation can be used in the present situation with a circular obstacle and an infinite plane. Expanding the Green function in the integral representation in polar wavefunctions in the integral over the circle and in plane waves in the integral over the plane, it is realized that the scattered wave in the half-space with a circular obstacle can be expanded as a sum of out-going polar waves and down-going plane waves

$$u^{\mathrm{sc}} = \sum_{\sigma,m} f_{\sigma m}\chi^{+}_{\sigma m}(r,\varphi) + \int_{-\infty}^{\infty} g(q)\mathrm{e}^{\mathrm{i}(qx-h(y-d))}\,\mathrm{d}q \tag{8.7}$$

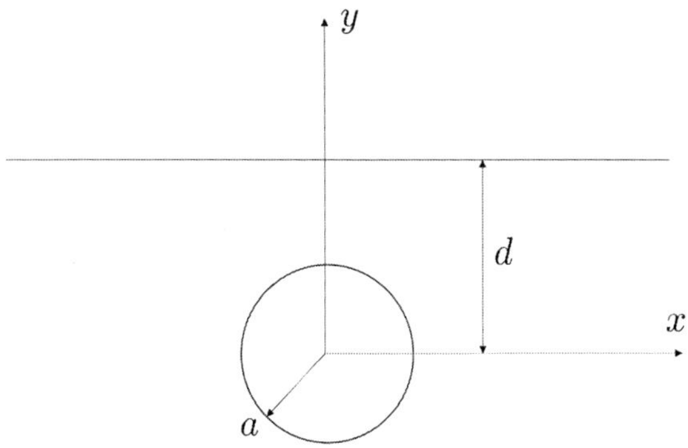

Figure 8.2: The scattering geometry with an obstacles in a half-space.

Here $f_{\sigma m}$ are unknown expansion coefficients and $g(q)$ is an unknown function, both to be determined by boundary conditions. The extra d in the exponential is inserted so that the exponential factor never becomes greater than unity.

When applying the boundary condition on the plane it is convenient if the scattered wave is expressed solely in plane waves. To this end the first term in Eq. (8.7) is transformed to up-going plane waves according to Eq. (8.5) to get the total wave

$$u = u^{\text{in}} + \int_{-\infty}^{\infty} \mathrm{e}^{\mathrm{i}qx} \left[g(q)\mathrm{e}^{-\mathrm{i}h(y-d)} \right.$$
$$\left. + \sum_{\sigma,m} \mathrm{i}^{-m} \frac{\sqrt{\epsilon_m}}{2\pi h} f_{\sigma m} \left\{ \begin{array}{c} \cos m\beta_+ \\ \sin m\beta_+ \end{array} \right\} \mathrm{e}^{\mathrm{i}hy} \right] \mathrm{d}q$$

where the upper row is chosen for $\sigma = \mathrm{e}$ and the lower for $\sigma = \mathrm{o}$. The boundary condition is that the normal derivative is zero on the plane $y = d$ and inverting the Fourier transform this gives

$$g(q) = -\frac{\mathrm{i}}{2\pi h} \int_{-\infty}^{\infty} \frac{\partial u^{\text{in}}(x, d)}{\partial y} \mathrm{e}^{-\mathrm{i}qx} \mathrm{d}x$$
$$+ \sum_{\sigma,m} \mathrm{i}^{-m} \frac{\sqrt{\epsilon_m}}{2\pi h} f_{\sigma m} \left\{ \begin{array}{c} \cos m\beta_+ \\ \sin m\beta_+ \end{array} \right\} \mathrm{e}^{\mathrm{i}hd} \qquad (8.8)$$

When applying the boundary condition on the circle, on the other hand, it is best to express the scattered wave in polar wavefunctions. The plane wave in the second term in the scattered wave in Eq. (8.7) is transformed to polar waves by Eq. (8.2) to give the total wave in the form

$$u = u^{\text{in}} + \sum_{\sigma,m} \Big[f_{\sigma m} \chi^{+}_{\sigma m}(r,\varphi) \\ + 2\mathrm{i}^m \sqrt{\epsilon_m} \int_{-\infty}^{\infty} g(q)\, \mathrm{e}^{\mathrm{i}hd} \left\{ \begin{array}{c} \cos m\beta_- \\ \sin m\beta_- \end{array} \right\} \mathrm{d}q\, \chi^{0}_{\sigma m}(r,\varphi) \Big]$$

It is convenient to expand the incident wave in the polar wavefunctions in the usual way

$$u^{\text{in}} = \sum_{\sigma,m} b_{\sigma m} \chi^{0}_{\sigma m}(r,\varphi)$$

The boundary condition is that the normal (radial) derivative is zero on the circle $r = a$ and using the orthogonality of the polar wavefunctions this gives

$$k\mathrm{J}'_m(ka) b_{\sigma m} + k\mathrm{H}'_m(ka) f_{\sigma m} \\ + 2\mathrm{i}^m \sqrt{\epsilon_m}\, k\mathrm{J}'_m(ka) \int_{-\infty}^{\infty} g(q)\, \mathrm{e}^{\mathrm{i}hd} \left\{ \begin{array}{c} \cos m\beta_- \\ \sin m\beta_- \end{array} \right\} \mathrm{d}q = 0$$

Inserting $g(q)$ from Eq. (8.8) and rearranging a little the result is

$$f_{\sigma m} - T_m \sum_{\sigma',m'} R_{\sigma m,\sigma' m'} f_{\sigma' m'} = T_m \Big[b_{\sigma m} \\ + \sqrt{\epsilon_m} \frac{\mathrm{i}^{m+1}}{\pi} \int_{-\infty}^{\infty} \left\{ \begin{array}{c} \cos m\beta_+ \\ \sin m\beta_+ \end{array} \right\} \int_{-\infty}^{\infty} \frac{\partial u^{\text{in}}(x,d)}{\partial y}\, \mathrm{e}^{-\mathrm{i}qx} \mathrm{d}x \frac{\mathrm{d}q}{h} \Big] \quad (8.9)$$

where

$$R_{\sigma m,\sigma' m'} = \mathrm{i}^{m-m'} \sqrt{\epsilon_m \epsilon_{m'}} \int_{-\infty}^{\infty} \mathrm{e}^{2\mathrm{i}hd} \left\{ \begin{array}{c} \cos m\beta_- \\ \sin m\beta_- \end{array} \right\} \left\{ \begin{array}{c} \cos m'\beta_+ \\ \sin m'\beta_+ \end{array} \right\} \frac{\mathrm{d}q}{h} \quad (8.10)$$

and

$$T_m = -\frac{\mathrm{J}'_m(ka)}{\mathrm{H}'_m(ka)}$$

are the T-matrix elements for the circular cavity from Section 2.1.2. In $R_{\sigma m,\sigma' m'}$ $\sigma = \mathrm{e}$ (o) gives the upper (lower) row in the first factor and similarly for σ' and the second factor. The factor $\exp(2\mathrm{i}hd)$ always gives an exponential decrease of the integral for large $|q|$ so the convergence is fast for $|q| > k$.

In principle the scattering problem is hereby solved; Eq. (8.9) is solved for the unknowns $f_{\sigma m}$ and this is turn gives $g(q)$ from Eq. (8.8) and the scattered wave from Eq. (8.7). It can be of advantage to take the incident wave to satisfy the boundary condition on the plane, for an incident plane wave this simply amounts to the addition of the reflected plane wave. In this way the second term on the right-hand side in Eq. (8.9) vanishes. In this case the scattered wave can be written

$$u^{\mathrm{sc}} = \sum_{\sigma,m} f_{\sigma m}\Bigg[\chi^{+}_{\sigma m}(r,\varphi) - \mathrm{i}^{-m}\frac{\sqrt{\epsilon_m}}{2\pi}\int_{-\infty}^{\infty}\left\{\begin{array}{c}\cos m\beta_+\\ \sin m\beta_+\end{array}\right\}\mathrm{e}^{\mathrm{i}(qx-h(y-d))}\frac{\mathrm{d}q}{h}\Bigg]$$

As was the case for the scattering by two obstacles in the previous chapter, the developments are valid also for other obstacles than a circular hole. An obstacle made of another material can thus be treated as in the previous chapter; the result is that the appropriate T-matrix should be used. Also noncircular obstacles can be treated in this way, it is only to substitute the relevant T-matrix (which is then nondiagonal so that more sums are needed). In this case a sufficient condition is that the circumscribed circle to the obstacle does not touch the surface of the half-space.

8.1.3 Layered half-space

Consider now a generalization, namely a circular obstacle in a layer with outside half-spaces of other materials, see Figure 8.3. Let the xy-system have origin at the centre of the circle and let the planes specifying the layers be $y = -d_1$ and $y = d_2$. The circle has radius a and for simplicity it is still assumed stress-free. In the half-space $y < -d_1$ the stiffness is μ_1 and the wavenumber is k_1, in the upper half-space the corresponding quantities are μ_2 and k_2, and in the layer $-d_1 < y < d_2$ they are μ and k. The boundary conditions between the layer and the surrounding half-spaces are the continuity of displacement and normal stress. It is assumed that the incident

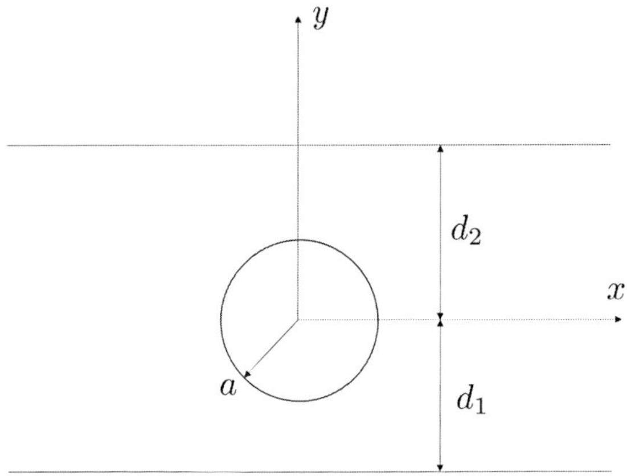

Figure 8.3: The scattering geometry with an obstacles in a layer between two half-spaces.

wave u^{in} satisfies the boundary conditions on the two planes $y = -d_1$ and $y = d_2$ (this means that the problem with the given source in the layered space without the circular obstacle is solved first and gives the solution u^{in}).

It should be clear that in the layer with the obstacle the scattered wave is composed of three parts: an out-going wave from the obstacle, an up-going wave from the lower surface, and a down-going wave from the upper surface

$$u^{\mathrm{sc}} = \sum_{\sigma,m} f_{\sigma m} \chi^{+}_{\sigma m}(r,\varphi) + \int_{-\infty}^{\infty} f_1(q)\, \mathrm{e}^{\mathrm{i}(qx+h(y+d_1))} \mathrm{d}q$$

$$+ \int_{-\infty}^{\infty} f_2(q)\, \mathrm{e}^{\mathrm{i}(qx-h(y-d_2))} \mathrm{d}q \tag{8.11}$$

In the lower and upper half-spaces the scattered wave is expanded in down-going or up-going plane waves, respectively; thus

$$u^{\mathrm{sc}} = \int_{-\infty}^{\infty} g_1(q)\, \mathrm{e}^{\mathrm{i}(qx-h_1(y+d_1))} \mathrm{d}q, \qquad y < -d_1 \tag{8.12}$$

$$u^{\mathrm{sc}} = \int_{-\infty}^{\infty} g_2(q)\, \mathrm{e}^{\mathrm{i}(qx+h_2(y-d_2))} \mathrm{d}q, \qquad y > d_2 \tag{8.13}$$

where $h_i = (k_i^2 - q^2)^{1/2}$ with $\operatorname{Im} h_i \geq 0$, $i = 1, 2$. Transforming the polar waves to plane waves as in the preceding subsection, the continuity of displacement and stress on the two surfaces $y = -d_1$ and $y = d_2$ gives

$$\sum_{\sigma,m} f_{\sigma m}\, \mathrm{e}^{\mathrm{i}hd_1}\, \mathrm{i}^{-m} \frac{\sqrt{\epsilon_m}}{2\pi} \left\{ \begin{array}{c} \cos m\beta_- \\ \sin m\beta_- \end{array} \right\} + f_1 + f_2\, \mathrm{e}^{\mathrm{i}h(d_1+d_2)} = g_1$$

$$\mathrm{i}\mu h \Bigg[-\sum_{\sigma,m} f_{\sigma m}\, \mathrm{e}^{\mathrm{i}hd_1}\, \mathrm{i}^{-m} \frac{\sqrt{\epsilon_m}}{2\pi} \left\{ \begin{array}{c} \cos m\beta_- \\ \sin m\beta_- \end{array} \right\} + f_1 - f_2\, \mathrm{e}^{\mathrm{i}h(d_1+d_2)} \Bigg] = -\mathrm{i}\mu_1 h_1 g_1$$

$$\sum_{\sigma,m} f_{\sigma m}\, \mathrm{e}^{\mathrm{i}hd_2}\, \mathrm{i}^{-m} \frac{\sqrt{\epsilon_m}}{2\pi} \left\{ \begin{array}{c} \cos m\beta_+ \\ \sin m\beta_+ \end{array} \right\} + f_1\, \mathrm{e}^{\mathrm{i}h(d_1+d_2)} + f_2 = g_2$$

$$\mathrm{i}\mu h \Bigg[\sum_{\sigma,m} f_{\sigma m}\, \mathrm{e}^{\mathrm{i}hd_2}\, \mathrm{i}^{-m} \frac{\sqrt{\epsilon_m}}{2\pi} \left\{ \begin{array}{c} \cos m\beta_+ \\ \sin m\beta_+ \end{array} \right\} + f_1\, \mathrm{e}^{\mathrm{i}h(d_1+d_2)} - f_2 \Bigg] = \mathrm{i}\mu_2 h_2 g_2$$

These four equations can be solved for f_1 and f_2 which gives

$$f_1(q) = \frac{R_1\, \mathrm{e}^{\mathrm{i}hd_1}}{(1 - R_1 R_2 \mathrm{e}^{2\mathrm{i}hd})} \sum_{\sigma,m} f_{\sigma m} \mathrm{i}^{-m} \frac{\sqrt{\epsilon_m}}{2\pi} \left(\left\{ \begin{array}{c} \cos m\beta_- \\ \sin m\beta_- \end{array} \right\} - R_2\, \mathrm{e}^{2\mathrm{i}hd_2} \left\{ \begin{array}{c} \cos m\beta_+ \\ \sin m\beta_+ \end{array} \right\} \right)$$

$$f_2(q) = \frac{R_2\, \mathrm{e}^{\mathrm{i}hd_2}}{(1 - R_1 R_2 \mathrm{e}^{2\mathrm{i}hd})} \sum_{\sigma,m} f_{\sigma m} \mathrm{i}^{-m} \frac{\sqrt{\epsilon_m}}{2\pi} \left(\left\{ \begin{array}{c} \cos m\beta_+ \\ \sin m\beta_+ \end{array} \right\} - R_1\, \mathrm{e}^{2\mathrm{i}hd_1} \left\{ \begin{array}{c} \cos m\beta_- \\ \sin m\beta_- \end{array} \right\} \right)$$

Here $d = d_1 + d_2$ is the total thickness of the layer and

$$R_i = \frac{\mu h - \mu_i h_i}{\mu h + \mu_i h_i}, \qquad i = 1, 2$$

is the plane wave reflection coefficient (from the layer side) for the interface between the layer and lower or upper half-space, respectively.

The stress-free boundary condition on the circular obstacle remains to be satisfied. Transforming the plane waves in Eq. (8.11) to polar wavefunctions according to Eq. (8.2) and applying the boundary condition gives

$$\begin{aligned} & k\mathrm{J}'_m(ka)b_{\sigma m} + k\mathrm{H}'_m(ka)f_{\sigma m} \\ & \qquad + 2\mathrm{i}^m\sqrt{\epsilon_m}\,k\mathrm{J}'_m(ka)\int_{-\infty}^{\infty} f_1(q)\,\mathrm{e}^{ihd_1}\left\{\begin{array}{c}\cos m\beta_+ \\ \sin m\beta_+\end{array}\right\}\mathrm{d}q \\ & + 2\mathrm{i}^m\sqrt{\epsilon_m}\,k\mathrm{J}'_m(ka)\int_{-\infty}^{\infty} f_2(q)\,\mathrm{e}^{ihd_2}\left\{\begin{array}{c}\cos m\beta_- \\ \sin m\beta_-\end{array}\right\}\mathrm{d}q = 0 \end{aligned} \quad (8.14)$$

where $b_{\sigma m}$ are the expansion coefficients of the incident wave as in the previous subsection. After insertion of f_1 and f_2, this is an equation that can be solved for $f_{\sigma m}$ and then the plane wave expansion coefficients $f_i(q)$ and $g_i(q)$, $i = 1, 2$, follow. The scattering problem is thus solved as the scattered wave is determined.

It should be apparent that this approach can be extended to an arbitrary number of layers. In each layer without an obstacle a combination of up-going and down-going waves is assumed and the layer with the obstacle is treated exactly as above. Applying all the boundary conditions all the expansion coefficients can be determined essentially as above.

8.2 Spherical obstacle in half-space

When 3D problems with an obstacle in a layered structure are considered it is useful to introduce plane vector wavefunctions. The transformations between the spherical and plane vector wavefunctions are then an important tool. The 3D problem with an obstacle in a half-space is investigated by Boström and Kristensson [26] and the generalization to a layered half-space by Boström and Karlsson [25].

8.2.1 Plane vector wavefunctions

When defining the plane vector wavefunctions the recipe in Eq. (1.18) is used

$$\boldsymbol{u} = \nabla\Phi + \nabla\times(\hat{\boldsymbol{z}}\Psi_1) + \nabla\times\nabla\times(\hat{\boldsymbol{z}}\Psi_2)$$

where the potentials Φ, Ψ_1, and Ψ_2 satisfy scalar Helmholtz equations. Thus the following plane vector wavefunctions are defined

$$\boldsymbol{\phi}_1(\alpha,\beta;\boldsymbol{r}) = \frac{1}{4\pi k_{\rm s}\sin\alpha}\nabla\times(\hat{\boldsymbol{z}}\exp{(\mathrm{i}k_{\rm s}\hat{\boldsymbol{\gamma}}\cdot\boldsymbol{r})}) = -\frac{\mathrm{i}\hat{\boldsymbol{\beta}}}{4\pi}\exp{(\mathrm{i}k_{\rm s}\hat{\boldsymbol{\gamma}}\cdot\boldsymbol{r})}$$

$$\boldsymbol{\phi}_2(\alpha,\beta;\boldsymbol{r}) = \frac{1}{4\pi k_{\rm s}^2\sin\alpha}\nabla\times\nabla\times(\hat{\boldsymbol{z}}\exp{(\mathrm{i}k_{\rm s}\hat{\boldsymbol{\gamma}}\cdot\boldsymbol{r})}) = -\frac{\hat{\boldsymbol{\alpha}}}{4\pi}\exp{(\mathrm{i}k_{\rm s}\hat{\boldsymbol{\gamma}}\cdot\boldsymbol{r})}$$

$$\begin{aligned}\boldsymbol{\phi}_3(\alpha,\beta;\boldsymbol{r}) &= \left(\frac{k_{\rm p}}{k_{\rm s}}\right)^{3/2}\frac{1}{4\pi k_p}\nabla\left(\exp{(\mathrm{i}k_p\hat{\boldsymbol{\gamma}}\cdot\boldsymbol{r})}\right)\\ &= \left(\frac{k_{\rm p}}{k_{\rm s}}\right)^{3/2}\frac{\mathrm{i}\hat{\boldsymbol{\gamma}}}{4\pi}\exp{(\mathrm{i}k_p\hat{\boldsymbol{\gamma}}\cdot\boldsymbol{r})}\end{aligned} \tag{8.15}$$

Here $\hat{\boldsymbol{\gamma}}$, $\hat{\boldsymbol{\alpha}}$, and $\hat{\boldsymbol{\beta}}$ are the spherical unit vectors

$$\hat{\boldsymbol{\gamma}} = \hat{\boldsymbol{x}}\sin\alpha\cos\beta + \hat{\boldsymbol{y}}\sin\alpha\sin\beta + \hat{\boldsymbol{z}}\cos\alpha$$

$$\hat{\boldsymbol{\alpha}} = \hat{\boldsymbol{x}}\cos\alpha\cos\beta + \hat{\boldsymbol{y}}\cos\alpha\sin\beta - \hat{\boldsymbol{z}}\sin\alpha$$

$$\hat{\boldsymbol{\beta}} = -\hat{\boldsymbol{x}}\sin\beta + \hat{\boldsymbol{y}}\cos\beta$$

and α and β are the spherical angles of the direction of propagation of the plane wave. In the present context when up- and down-going waves in the z-direction are of interest β belongs to the interval $[0,2\pi]$ and α belongs to C_+ or C_-. These contours in the complex α-plane are shown in Figure 8.4, and an α belonging to C_+ (C_-) gives up- (down-)going waves. Note that on both C_+ and C_- $\sin\alpha \in [0,\infty)$. The mode index on the wavefunctions is denoted j, thus writing $\boldsymbol{\phi}_j(\alpha,\beta;\boldsymbol{r})$. The modes are called SH, SV, and P for $j = 1,2,3$, respectively. The function $\boldsymbol{\phi}_j^\dagger(\alpha,\beta;\boldsymbol{r})$ has a sign change in the exponential (made before the curl or gradient is applied). The wavefunctions are dimensioness and are normalized to give a simple expansion of the Green function.

The plane wavefunctions above are written as functions of the spherical angles of the direction of propagation α and β and this is, in particular, convenient when the transformations to spherical wavefunctions are stated. In other situations it is more convenient to use the Cartesian components of the wavevector. Thus, $q = k_j\sin\alpha\cos\beta$ and $p = k_j\sin\alpha\sin\beta$, where $k_1 = k_2 = k_{\rm s}$ and $k_3 = k_{\rm p}$, are the x- and y-components, respectively, of

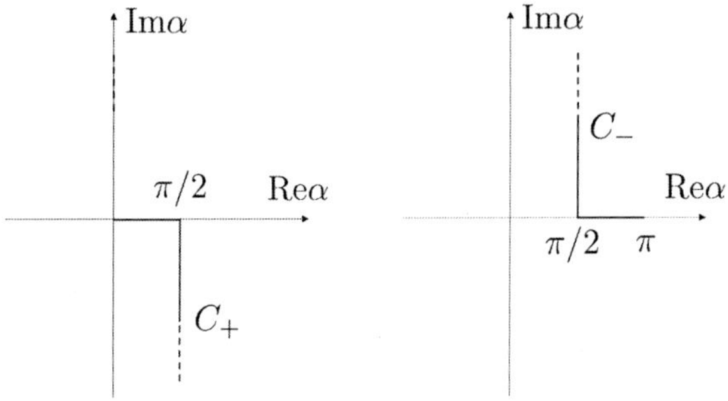

Figure 8.4: The integration contours C_+ and C_-, where C_+ runs from the origin to $\pi/2 - \mathrm{i}\infty$ and C_- runs from $\pi/2 + \mathrm{i}\infty$ to π.

the wavevector and are therefore also the Fourier transform variables corresponding to the x- and y-coordinates. Written in q and p the wavefunctions are

$$\begin{aligned}
\boldsymbol{\phi}_1^+(q,p;\boldsymbol{r}) &= \frac{\mathrm{i}}{4\pi s}\,(\hat{\boldsymbol{x}}p - \hat{\boldsymbol{y}}q)\,\mathrm{e}^{\mathrm{i}(qx+py+h_\mathrm{s}z)} \\
\boldsymbol{\phi}_2^+(q,p;\boldsymbol{r}) &= \frac{1}{4\pi k_\mathrm{s} s}\,\left(-\hat{\boldsymbol{x}}qh_\mathrm{s} - \hat{\boldsymbol{y}}ph_\mathrm{s} + \hat{\boldsymbol{z}}s^2\right)\mathrm{e}^{\mathrm{i}(qx+py+h_\mathrm{s}z)} \\
\boldsymbol{\phi}_3^+(q,p;\boldsymbol{r}) &= \left(\frac{k_\mathrm{p}}{k_\mathrm{s}}\right)^{3/2}\frac{\mathrm{i}}{4\pi k_p}\,(\hat{\boldsymbol{x}}q + \hat{\boldsymbol{y}}p + \hat{\boldsymbol{z}}h_\mathrm{p})\,\mathrm{e}^{\mathrm{i}(qx+py+h_\mathrm{p}z)}
\end{aligned} \tag{8.16}$$

where $s^2 = q^2+p^2$, $h_\mathrm{s} = (k_\mathrm{s}^2-s^2)^{1/2}$ and $h_\mathrm{p} = (k_\mathrm{p}^2-s^2)^{1/2}$ with $\operatorname{Im} h_\mathrm{s} \geq 0$ and $\operatorname{Im} h_\mathrm{p} \geq 0$. The upper index $+$ means that these wavefunctions are up-going; corresponding down-going wavefunctions have an upper index $-$ and have a sign change in h_s or h_p (in all places). The tractions of the wavefunctions are also needed and on a plane $z = d$ and they become

$$\begin{aligned}
\boldsymbol{t}^{(z)}(\boldsymbol{\phi}_1^+) &= \frac{\mu}{4\pi s}\big[-\hat{\boldsymbol{x}}ph_\mathrm{s} + \hat{\boldsymbol{y}}qh_\mathrm{s}\big]\mathrm{e}^{\mathrm{i}(qx+py+h_\mathrm{s}d)} \\
\boldsymbol{t}^{(z)}(\boldsymbol{\phi}_2^+) &= \frac{\mathrm{i}\mu}{4\pi k_\mathrm{s} s}\big[\hat{\boldsymbol{x}}q(2s^2-k_\mathrm{s}^2) + \hat{\boldsymbol{y}}p(2s^2-k_\mathrm{s}^2) \\
&\qquad + \hat{\boldsymbol{z}}\,2s^2h_s\big]\mathrm{e}^{\mathrm{i}(qx+py+h_\mathrm{s}d)}
\end{aligned}$$

$$\boldsymbol{t}^{(z)}(\boldsymbol{\phi}_3^+) = \left(\frac{k_\mathrm{p}}{k_\mathrm{s}}\right)^{3/2} \frac{\mu}{4\pi k_p}\big[-\hat{\boldsymbol{x}}\,2qh_\mathrm{p} - \hat{\boldsymbol{y}}\,2ph_\mathrm{p} + \hat{\boldsymbol{z}}(2s^2 - k_\mathrm{s}^2)\big]\mathrm{e}^{\mathrm{i}(qx+py+h_\mathrm{p}d)} \tag{8.17}$$

The expansion in plane vector wavefunctions of the free space Green function can be obtained in a similar way as the expansion in spherical vector wavefunctions. To this end the orthogonality properties of the plane wavefunctions are first derived

$$\int_S \Big[\boldsymbol{t}^{(n)}\left(\boldsymbol{\phi}_j^+(q,p;\boldsymbol{r})\right)\cdot\boldsymbol{\phi}_{j'}^+(-q',-p';\boldsymbol{r}) - \boldsymbol{t}^{(n)}\left(\boldsymbol{\phi}_{j'}^+(-q',-p';\boldsymbol{r})\right)\cdot\boldsymbol{\phi}_j^+(q,p;\boldsymbol{r})\Big]\,\mathrm{d}S = 0$$

$$\int_S \Big[\boldsymbol{t}^{(n)}\left(\boldsymbol{\phi}_j^-(q,p;\boldsymbol{r})\right)\cdot\boldsymbol{\phi}_{j'}^-(-q',-p';\boldsymbol{r}) - \boldsymbol{t}^{(n)}\left(\boldsymbol{\phi}_{j'}^-(-q',-p';\boldsymbol{r})\right)\cdot\boldsymbol{\phi}_j^-(q,p;\boldsymbol{r})\Big]\,\mathrm{d}S = 0$$

$$\int_S \Big[\boldsymbol{t}^{(n)}\left(\boldsymbol{\phi}_j^+(q,p;\boldsymbol{r})\right)\cdot\boldsymbol{\phi}_{j'}^-(-q',-p';\boldsymbol{r}) - \boldsymbol{t}^{(n)}\left(\boldsymbol{\phi}_{j'}^-(-q',-p';\boldsymbol{r})\right)\cdot\boldsymbol{\phi}_j^+(q,p;\boldsymbol{r})\Big]\,\mathrm{d}S = \frac{\mathrm{i}\mu k_j h_j}{2k_\mathrm{s}}\,\delta_{jj'}\delta(q-q')\delta(p-p') \tag{8.18}$$

where $h_1 = h_2 = h_\mathrm{s}$ and $h_3 = h_\mathrm{p}$. The integration is over any infinite surface S with normal $\hat{\boldsymbol{n}}$ that is situated between two planes parallel to the xy-plane; as a special case it may be an infinite plane $z = d$ for any constant d. When the integration is over an infinite plane the integrals are straightforward to calculate with the expressions for the wavefunctions and their tractions given above and use of the integral

$$\int_{-\infty}^{\infty} \mathrm{e}^{\mathrm{i}xt}\,\mathrm{d}x = 2\pi\delta(t)$$

As for the spherical vector wavefunctions in Section 4.1.3, the Green dyadic is scalar multiplied by an arbitrary constant vector $\boldsymbol{a}$ to get the auxiliary vector field $\boldsymbol{g}_a = \boldsymbol{a}\cdot\boldsymbol{G}(\boldsymbol{r};\boldsymbol{r}')$. This vector field satisfies

$$k_\mathrm{p}^{-2}\nabla\nabla\cdot\boldsymbol{g}_a - k_\mathrm{s}^{-2}\nabla\times(\nabla\times\boldsymbol{g}_a) + \boldsymbol{g}_a = -k_\mathrm{s}^{-3}\boldsymbol{a}\,\delta(\boldsymbol{r}-\boldsymbol{r}')$$

For $z > z'$ the right-hand side vanishes and as the solution must satisfy radiation conditions the field must admit an expansion in terms of the up-going plane vector wavefunctions

$$\boldsymbol{g}_a = \sum_j \int_{-\infty}^{\infty}\int_{-\infty}^{\infty} c_j(q,p)\boldsymbol{\phi}_j^+(q,p;\boldsymbol{r})\,\mathrm{d}q\,\mathrm{d}p$$

The expansion coefficients $c_j(q,p)$ depend on $\boldsymbol{r}'$ and the vector $\boldsymbol{a}$ and to determine these coefficients the reciprocity relation is employed [Eq. (1.9)]

$$\int_S \hat{\boldsymbol{n}}\cdot\left(\boldsymbol{\sigma}^{\mathrm{A}}\cdot\boldsymbol{u}^{\mathrm{B}} - \boldsymbol{\sigma}^{\mathrm{B}}\cdot\boldsymbol{u}^{\mathrm{A}}\right)\mathrm{d}S = \int_V \rho\left(\boldsymbol{u}^{\mathrm{A}}\cdot\boldsymbol{f}^{\mathrm{B}} - \boldsymbol{u}^{\mathrm{B}}\cdot\boldsymbol{f}^{\mathrm{A}}\right)\mathrm{d}V$$

State A is here chosen as the auxiliary field $\boldsymbol{g}_a$ and state B as a down-going plane vector wavefunction $\boldsymbol{\phi}_j^-(-q',-p';\boldsymbol{r})$. The surface S can be chosen as a plane $z = d$, where $d > z'$. The volume V is then the half-space $z < d$. The surface S should enclose V, but the remaining part at infinity vanishes because both the down-going plane wave and the auxiliary field satisfy radiation conditions. The forces are then $\rho\boldsymbol{f}^{\mathrm{A}} = (\mu/k_{\mathrm{s}})\boldsymbol{a}\,\delta(\boldsymbol{r}-\boldsymbol{r}')$ and $\rho\boldsymbol{f}^{\mathrm{B}} = 0$. This gives

$$\sum_j \int_{-\infty}^{\infty}\int_{-\infty}^{\infty} c_j(q,p)\int_S \Big[\boldsymbol{t}^{(z)}\left(\boldsymbol{\phi}_j^+(q,p;\boldsymbol{r})\right)\cdot\boldsymbol{\phi}_j^-(-q',-p';\boldsymbol{r})$$

$$-\boldsymbol{t}^{(z)}\left(\boldsymbol{\phi}_j^-(-q',-p';\boldsymbol{r})\right)\cdot\boldsymbol{\phi}_j^+(q,p;\boldsymbol{r})\Big]\,\mathrm{d}S\,\mathrm{d}q\,\mathrm{d}p = -\frac{\mu}{k_{\mathrm{s}}}\boldsymbol{a}\cdot\boldsymbol{\phi}_j^+(q',p';\boldsymbol{r})$$

The orthogonality of the plane vector wavefunctions [the last of Eqs. (8.18)] can then be used to obtain

$$c_j(q,p) = \frac{2\mathrm{i}}{k_j h_j}\boldsymbol{a}\cdot\boldsymbol{\phi}_j^-(-q,-p;\boldsymbol{r}')$$

As the vector $\boldsymbol{a}$ is arbitrary this gives for $z > z'$

$$\boldsymbol{G}(\boldsymbol{r};\boldsymbol{r}') = 2\mathrm{i}\sum_j \int_{-\infty}^{\infty}\int_{-\infty}^{\infty} \boldsymbol{\phi}_j^-(-q,-p;\boldsymbol{r}')\boldsymbol{\phi}_j^+(q,p;\boldsymbol{r})\,\frac{\mathrm{d}q\,\mathrm{d}p}{k_j h_j}$$

Due to the symmetries this finally gives the expansion of the Green dyadic in 3D

$$\boldsymbol{G}(\boldsymbol{r};\boldsymbol{r}') = 2\mathrm{i}\sum_j \int_{-\infty}^{\infty}\int_{-\infty}^{\infty} \boldsymbol{\phi}_j^-(-q,-p;\boldsymbol{r}_<)\boldsymbol{\phi}_j^+(q,p;\boldsymbol{r}_>)\,\frac{\mathrm{d}q\,\mathrm{d}p}{k_j h_j} \tag{8.19}$$

where for plane waves $\boldsymbol{r}_> = \boldsymbol{r}$ if $z > z'$ and $\boldsymbol{r}_> = \boldsymbol{r}'$ if $z < z'$ and similarly for $\boldsymbol{r}_<$. Due to the symmetries the order of the wavefunctions can be changed. If the plane wavefunctions are expressed in the spherical angles α and β instead of q and p the expansion becomes

$$\boldsymbol{G}(\boldsymbol{r};\boldsymbol{r}') = 2\mathrm{i}\sum_j \int_0^{2\pi}\int_{C_\pm} \boldsymbol{\phi}_j(\alpha,\beta;\boldsymbol{r})\boldsymbol{\phi}_j^\dagger(\alpha,\beta;\boldsymbol{r}_>)\,\sin\alpha\,\mathrm{d}\alpha\,\mathrm{d}\beta \quad (8.20)$$

where C_+ or C_- (see Figure 8.4) is chosen according to $z > z'$ or $z < z'$. The distinction between up- and down-going waves is treated in a different manner with the integration contour playing an important role.

The transformation between the spherical and plane vector wavefunctions is an important tool when the scattering by a sphere in a half-space is considered. The transformation function needed is [27]

$$B_{\tau\sigma ml,j}(\alpha,\beta) = -\mathrm{i}^{-l}\left(\frac{\epsilon_m(2l+1)(l-m)!}{4\pi(l+m)!}\right)^{1/2}$$

$$\times\begin{bmatrix} \mathrm{i}\Delta_l^m(\alpha)\begin{Bmatrix}\cos m\beta\\ \sin m\beta\end{Bmatrix} & \pi_l^m(\alpha)\begin{Bmatrix}-\sin m\beta\\ \cos m\beta\end{Bmatrix} & 0\\ \pi_l^m(\alpha)\begin{Bmatrix}-\sin m\beta\\ \cos m\beta\end{Bmatrix} & \mathrm{i}\Delta_l^m(\alpha)\begin{Bmatrix}\cos m\beta\\ \sin m\beta\end{Bmatrix} & 0\\ 0 & 0 & -\mathrm{P}_l^m(\cos\alpha)\begin{Bmatrix}\cos m\beta\\ \sin m\beta\end{Bmatrix}\end{bmatrix} \quad (8.21)$$

where

$$\Delta_l^m(\alpha) = [l(l+1)]^{-1/2}\frac{\mathrm{d}}{\mathrm{d}\alpha}\mathrm{P}_l^m(\cos\alpha)$$

$$\pi_l^m(\alpha) = [l(l+1)]^{-1/2}\frac{m\mathrm{P}_l^m(\cos\alpha)}{\sin\alpha}$$

Here the columns correspond to $j = 1,2,3$ and the rows to $\tau = 1,2,3$ and the upper (lower) trigonometric function corresponds to $\sigma = \mathrm{e}$ ($\sigma = \mathrm{o}$). Also the function $B^\dagger_{\tau\sigma ml,j}(\alpha,\beta)$ is needed; it has all explicit "i" changed to "−i".

It is noted that $B_{\tau\sigma ml,j}(\alpha,\beta)$ is essentially a spherical component of the vector spherical harmonics. The transformation from a plane wavefunction to the regular spherical wavefunctions is

$$\boldsymbol{\phi}_j(\alpha,\beta;\boldsymbol{r}) = \sum_{\tau,\sigma,m,l} B^{\dagger}_{\tau\sigma ml,j}(\alpha,\beta)\boldsymbol{\psi}^0_{\tau\sigma ml}(\boldsymbol{r}) \tag{8.22}$$

which is valid without any restrictions. The transformations from a spherical wavefunction to the plane wavefunctions are

$$\boldsymbol{\psi}^0_{\tau\sigma ml}(\boldsymbol{r}) = \sum_j \int_0^{2\pi}\int_0^{\pi} B_{\tau\sigma ml,j}(\alpha,\beta)\boldsymbol{\phi}_j(\alpha,\beta;\boldsymbol{r})\sin\alpha\,\mathrm{d}\alpha\,\mathrm{d}\beta \tag{8.23}$$

which is valid without restrictions, and

$$\boldsymbol{\psi}^+_{\tau\sigma ml}(\boldsymbol{r}) = 2\sum_j \int_0^{2\pi}\int_{C_\pm} B_{\tau\sigma ml,j}(\alpha,\beta)\boldsymbol{\phi}_j(\alpha,\beta;\boldsymbol{r})\sin\alpha\,\mathrm{d}\alpha\,\mathrm{d}\beta \tag{8.24}$$

where the contour C_+ must be chosen for $z > 0$ and C_- for $z < 0$. As noted above these transformations are naturally expressed in the direction of propagation of the plane waves, i.e. the plane waves are viewed as functions of α and β (rather than q and p).

8.2.2 Scattering problem

Consider the scattering by a spherical obstacle in an elastic half-space, see Figure 8.2. For simplicity the obstacle is taken as a cavity with radius a. The distance between the sphere centre and the stress-free surface of the half-space is d. Both Cartesian and spherical coordinates are used with the origin at the centre of the sphere with $z = d$ as the surface of the half-space. The Lamé constants of the half-space are λ and μ and the corresponding wavenumbers are k_s and k_p. For simplicity the incident wave $\boldsymbol{u}^\mathrm{in}$ is supposed to satisfy the boundary condition on the surface of the half-space.

The scattered wave $\boldsymbol{u}^\mathrm{sc}$ is composed of two parts, one that is out-going from the sphere and one that is down-going from the surface of the half-space. As in the previous section this follows from an application of the integral representation with the Green tensor expanded in the spherical and plane vector wavefunctions and thus

$$
\begin{aligned}
\boldsymbol{u}^{\mathrm{sc}}(\boldsymbol{r}) = & \sum_{\tau,\sigma,m,l} f_{\tau\sigma ml}\boldsymbol{\psi}^{+}_{\tau\sigma ml}(\boldsymbol{r}) \\
& + \sum_{j} \int_0^{2\pi} \int_{C_-} g_j(\alpha,\beta)\boldsymbol{\phi}_j(\alpha,\beta;\boldsymbol{r}) \sin\alpha \, \mathrm{d}\alpha \, \mathrm{d}\beta
\end{aligned} \tag{8.25}
$$

To apply the boundary condition on the free surface of the half-space it is best to first transform the spherical wavefunctions. Using Eq. (8.24) in the first term in Eq. (8.25) the result is

$$
\begin{aligned}
\boldsymbol{u}^{\mathrm{sc}}(\boldsymbol{r}) = 2 & \sum_{\tau,\sigma,m,l} f_{\tau\sigma ml} \sum_{j} \int_0^{2\pi} \int_{C_+} B_{\tau\sigma ml,j}(\alpha,\beta)\boldsymbol{\phi}_j(\alpha,\beta;\boldsymbol{r}) \sin\alpha \, \mathrm{d}\alpha \, \mathrm{d}\beta \\
& + \sum_{j} \int_0^{2\pi} \int_{C_-} g_j(\alpha,\beta)\boldsymbol{\phi}_j(\alpha,\beta;\boldsymbol{r}) \sin\alpha \, \mathrm{d}\alpha \, \mathrm{d}\beta
\end{aligned}
$$

When applying the boundary condition it is more convenient to use the Cartesian components of the wavevector and thus write this as

$$
\begin{aligned}
\boldsymbol{u}^{\mathrm{sc}}(\boldsymbol{r}) = 2 & \sum_{\tau,\sigma,m,l} f_{\tau\sigma ml} \sum_{j} \int_{-\infty}^{\infty} \int_{-\infty}^{\infty} B_{\tau\sigma ml,j}(\alpha,\beta)\boldsymbol{\phi}^{+}_j(q,p;\boldsymbol{r}) \frac{\mathrm{d}q \, \mathrm{d}p}{k_j h_j} \\
& + \sum_{j} \int_{-\infty}^{\infty} \int_{-\infty}^{\infty} g_j(q,p)\boldsymbol{\phi}^{-}_j(q,p;\boldsymbol{r}) \frac{\mathrm{d}q \, \mathrm{d}p}{k_j h_j}
\end{aligned}
$$

The α and β in $B_{\tau\sigma ml,j}(\alpha,\beta)$ are determined from $q = k_j \sin\alpha\cos\beta$, $p = k_j \sin\alpha\sin\beta$, and $h_j = k_j\cos\alpha$, where also $h_j = (k_j^2 - s^2)^{1/2}$, $s^2 = q^2 + p^2$, with $\operatorname{Im} h_j \geq 0$. As the incident wave is assumed to satisfy the boundary condition on the surface of the half-space also the scattered wave must satisfy this boundary condition and this gives

$$
\begin{aligned}
2 & \sum_{\tau,\sigma,m,l} f_{\tau\sigma ml} \sum_{j} B_{\tau\sigma ml,j}(\alpha,\beta)\boldsymbol{t}^{(z)}(\boldsymbol{\phi}^{+}_j(q,p;\boldsymbol{r}_0)) \\
& + \sum_{j} g_j(q,p)\boldsymbol{t}^{(z)}(\boldsymbol{\phi}^{-}_j(q,p;\boldsymbol{r}_0)) = 0
\end{aligned}
$$

where $\boldsymbol{r}_0 = \hat{\boldsymbol{z}}d$, i.e. $x = y = 0$. These equations can be solved to obtain

$$g_j(q,p) = 2\sum_{j'} R_{jj'}(q,p) \sum_{\tau,\sigma,m,l} f_{\tau\sigma ml} B_{\tau\sigma ml,j'}(\alpha,\beta) \tag{8.26}$$

where the elements of the reflection matrix are

$$\begin{aligned}
R_{11} &= \mathrm{e}^{2\mathrm{i}h_\mathrm{s}d}, \qquad R_{12} = R_{21} = R_{23} = R_{32} = 0 \\
R_{22} &= \frac{4s^2 h_\mathrm{p} h_\mathrm{s} - \left(2s^2 - k_\mathrm{s}^2\right)^2}{4s^2 h_\mathrm{p} h_\mathrm{s} + \left(2s^2 - k_\mathrm{s}^2\right)^2}\,\mathrm{e}^{2\mathrm{i}h_\mathrm{s}d} \\
R_{33} &= \frac{4s^2 h_\mathrm{p} h_\mathrm{s} - \left(2s^2 - k_\mathrm{s}^2\right)^2}{4s^2 h_\mathrm{p} h_\mathrm{s} + \left(2s^2 - k_\mathrm{s}^2\right)^2}\,\mathrm{e}^{2\mathrm{i}h_\mathrm{p}d} \\
R_{23} &= -\sqrt{\frac{k_\mathrm{p}}{k_\mathrm{s}}}\,\frac{4\mathrm{i}sh_\mathrm{p}\left(2s^2 - k_\mathrm{s}^2\right)^2}{4s^2 h_\mathrm{p} h_\mathrm{s} + \left(2s^2 - k_\mathrm{s}^2\right)^2}\,\mathrm{e}^{\mathrm{i}(h_\mathrm{s}+h_\mathrm{p})d} \\
R_{32} &= -\sqrt{\frac{k_\mathrm{s}}{k_\mathrm{p}}}\,\frac{4\mathrm{i}sh_\mathrm{s}\left(2s^2 - k_\mathrm{s}^2\right)^2}{4s^2 h_\mathrm{p} h_\mathrm{s} + \left(2s^2 - k_\mathrm{s}^2\right)^2}\,\mathrm{e}^{\mathrm{i}(h_\mathrm{s}+h_\mathrm{p})d}
\end{aligned} \tag{8.27}$$

These elements are just the ordinary plane wave reflection coefficients for a stress-free plane. The normalization of the plane wavefunctions is responsible for the unusual normalization of R_{23} and R_{32}.

When the boundary condition on the spherical obstacle is to be applied it is useful to transform the plane waves in Eq. (8.25) to spherical wavefunctions according to Eq. (8.22)

$$\begin{aligned}
\boldsymbol{u}^\mathrm{sc}(\boldsymbol{r}) &= \sum_{\tau,\sigma,m,l} f_{\tau\sigma ml}\boldsymbol{\psi}^+_{\tau\sigma ml}(\boldsymbol{r}) \\
&+ \sum_j \int_0^{2\pi}\int_{C_-} g_j(\alpha,\beta) \sum_{\tau,\sigma,m,l} B^\dagger_{\tau\sigma ml,j}(\alpha,\beta)\boldsymbol{\psi}^0_{\tau\sigma ml}(\boldsymbol{r}) \sin\alpha\,\mathrm{d}\alpha\,\mathrm{d}\beta
\end{aligned}$$

Insertion of $g_j(\alpha,\beta)$ from Eq. (8.26) gives

$$\begin{aligned}
\boldsymbol{u}^\mathrm{sc}(\boldsymbol{r}) &= \sum_{\tau,\sigma,m,l} f_{\tau\sigma ml}\boldsymbol{\psi}^+_{\tau\sigma ml}(\boldsymbol{r}) \\
&\quad + \sum_{\tau,\sigma,m,l}\sum_{\tau',\sigma',m',l'} \boldsymbol{\psi}^0_{\tau\sigma ml}(\boldsymbol{r}) R_{\tau\sigma ml,\tau'\sigma'm'l'} f_{\tau'\sigma'm'l'}
\end{aligned}$$

where the spherical projection of the reflection matrix is

$$R_{\tau\sigma ml,\tau'\sigma'm'l'}$$
$$= \sum_{jj'} \int_0^{2\pi} \int_{C_-} B^{\dagger}_{\tau\sigma ml,j}(\alpha,\beta) R_{jj'}(q,p) B_{\tau\sigma ml,j}(\alpha,\beta) \sin\alpha \,\mathrm{d}\alpha\,\mathrm{d}\beta \tag{8.28}$$

The reflection matrix $R_{jj'}(q,p)$ is seen to be independent of β and this means that the β integral simply becomes the orthogonality integrals for the trigonometric functions and the matrix $R_{\tau\sigma ml,\tau'\sigma'm'l'}$ becomes diagonal in m and m'.

The boundary condition is applied on the total wave so the incident wave is also needed and as usual it is convenient to expand it in the regular spherical wavefunctions

$$\boldsymbol{u}^{\text{in}}(\boldsymbol{r}) = \sum_{\tau,\sigma,m,l} b_{\tau\sigma ml} \boldsymbol{\psi}^0_{\tau\sigma ml}(\boldsymbol{r})$$

The total wave thus has expansions in regular and outgoing spherical wavefunctions exactly as for a spherical obstacle in full space. Therefore, the application of the boundary condition can be reformulated as

$$f_{\tau\sigma ml} - \sum_{\tau''} T^l_{\tau\tau''} \sum_{\tau',\sigma',m',l'} R_{\tau''\sigma ml,\tau'\sigma'm'l'} f_{\tau'\sigma'm'l'} = \sum_{\tau'} T^l_{\tau\tau'} b_{\tau'\sigma ml} \tag{8.29}$$

where the T-matrix elements $T^l_{\tau\tau'}$ for a cavity enters and the fact that the T-matrix is diagonal in all indices except τ and is also independent of m and σ is used. Equation (8.29) determines the coefficients $f_{\tau\sigma ml}$ and then $g_j(\alpha,\beta)$ is obtained from Eq. (8.26). The scattered wave is thus determined.

The present approach is valid for other obstacles than a spherical cavity, it is simply to put in the appropriate T-matrix. This is possible also for nonspherical obstacles (when more sums are needed), at least as long as the circumscribed sphere to the obstacle does not touch the free surface of the half-space. As in the previous section it is also possible to generalize the method to a multi-layered half-space with an obstacle in one of the layers.

9 Scattering by obstacles in waveguides

A waveguide is a structure that forces waves to propagate along a certain direction or along a certain plane. Typical waveguides are plates and bars where the waves are confined to the structure. But also a layered half-space or full space can act as a waveguide for certain types of waves, the most famous of which is the Rayleigh wave. Also a cylindrical hole (empty or filled with another material) acts as a waveguide for certain waves. Typical for a waveguide is that waves can only propagate as certain discrete modes. The scattering by such modes from obstacles inside the waveguide is the topic of this chapter, which takes up a few simple cases.

9.1 Circular obstacle in waveguide: 2D SH waves

As usual the simplest case is for the 2D SH waves. This case is reviewed by Martin and Dalrymple [90], mostly in the setting of water waves (which mathematically is the same problem). The corresponding in-plane elastic problem is considered by Karlsson [74], but this case is not taken up here.

Consider a circular cavity of radius a situated symmetrically in a plate of thickness $2d$, see Figure 9.1. A Cartesian coordinate system has the origin at the centre of the cavity, the x-axis parallel to the plate, and the y-axis perpendicular to the plate. The plate acts as a waveguide and the waveguide modes are the simple separable solutions to Helmholtz equation. To satisfy the stress-free boundary conditions on the surfaces of the plate the y-dependence of a symmetric waveguide mode must be $\cos(n\pi y/d)$, with $n = 0, 1, 2, \ldots$. The waveguide modes are thus

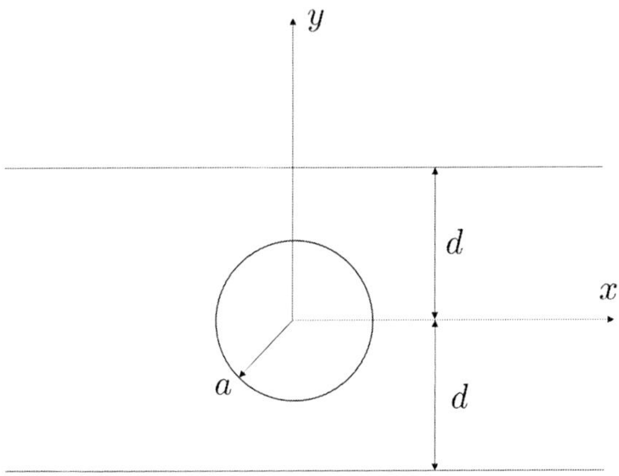

Figure 9.1: The scattering geometry with an obstacles in a layer acting as a waveguide.

$$u_n^{\pm}(x,y) = \mathrm{e}^{\pm \mathrm{i} q_n x} \cos(n\pi y/d), \qquad q_n = \left(k^2 - \left(\frac{n\pi}{d}\right)^2\right)^{1/2}$$

for $n = 0, 1, 2, \ldots$ and with $\operatorname{Im} q_n \geq 0$. It is seen that $q_0 = k$, so this mode has no y-dependence. For $n > kd/\pi$ the wavenumber q_n in the x-direction becomes imaginary and the mode thus decays (or increases) exponentially and is called evanescent. The $+$ $(-)$ gives modes propagating to the right (left). There also exist antisymmetric modes, but it is assumed that the incident wave is a symmetric waveguide mode u_n^+ and due to symmetry the antisymmetric modes are not excited.

The 2D SH problem for the scattering by a circular cavity in a layered structure is treated in Section 8.1.3, and this solution can now be utilized. In the present case the reflection coefficients from the planes become

$$R_1 = R_2 = 1$$

and this gives the plane wave expansion coefficients inside the layer as

$$f_1(q) = f_2(q) = \frac{\mathrm{i}}{4\pi h \sin hd} \sum_m f_m \mathrm{i}^{-m} \sqrt{\epsilon_m} \cos m\beta$$

Here, and everywhere in the following, the σ-index is suppressed as only $\sigma =$ e is needed due to symmetry. The equation that determines the expansion coefficients f_m is Eq. (8.14) which can be written

$$f_m - T_m \sum_{m'=0}^{\infty} R_{mm'} f_{m'} = T_m b_m \tag{9.1}$$

where T_m is the transition matrix element for a circular cavity and

$$R_{mm'} = \frac{\mathrm{i}^{1+m-m'}}{\pi} \sqrt{\epsilon_m \epsilon_{m'}} \int_{-\infty}^{\infty} \frac{\mathrm{e}^{ihd}}{\sin hd} \cos m\beta \cos m'\beta \, \frac{\mathrm{d}q}{h} \tag{9.2}$$

where β is defined through $q = k \cos \beta$ and $h = k \sin \beta$. This integral may look divergent as the integrand has poles whenever $\sin hd = 0$. The integral must, however, be regarded in the limit of vanishing damping. For finite damping the poles and cuts [from the square root in the definition of $h = (k^2 - q^2)^{1/2}$] of the integrand are located in the first and third quadrant in the complex q-plane, as Figure 9.2 shows. When there is no damping the integration contour must deviate from the real axis into the fourth and second quadrant as also shown in Figure 9.2.

The expansion coefficients b_m of the incident waveguide mode are determined from the expansion of a plane wave in the polar wavefunctions as given in Eq. (8.2). The incident wave is taken as propagating in the positive x-direction and is

$$u_{n_0}^{+}(x, y) = \mathrm{e}^{\mathrm{i}q_{n_0} x} \cos n_0 \pi y / d, \qquad q_{n_0} = \sqrt{k^2 - \left(\frac{n_0 \pi}{d}\right)^2}$$

The expansion coefficients thus become

$$b_m = 2\mathrm{i}^m \sqrt{\epsilon_m} \cos m\beta_{n_0} \tag{9.3}$$

where β_{n_0} is determined from $q_{n_0} = k \cos \beta_{n_0}$.

Solving Eq. (9.1) determines the coefficients f_m and thereby also $f_1(q)$ and $f_2(q)$. Eliminating $f_1(q)$ and $f_2(q)$ the scattered wave becomes [using Eq. (8.4)]

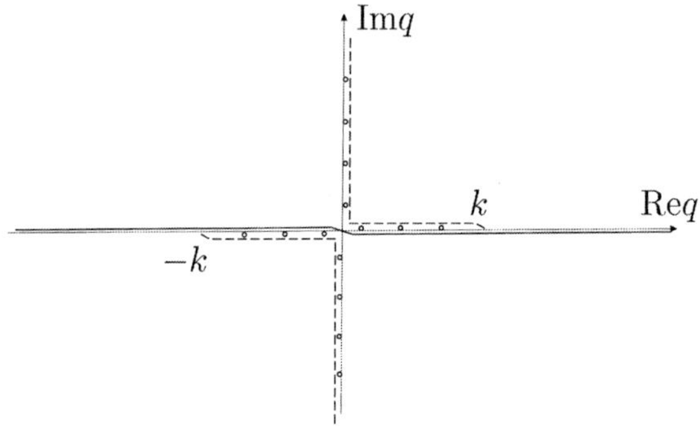

Figure 9.2: The complex q-plane with the poles (small circles) and cuts (dashed), and the integration contour.

$$u^{\rm sc}(x,y) = \sum_m f_m \frac{\sqrt{\epsilon_m}}{2}\left[\mathrm{H}_m(kr)\cos m\varphi\right.$$

$$\left. + \frac{\mathrm{i}^{1-m}}{\pi}\int_{-\infty}^{\infty} \frac{\cos hy}{\sin hd}\cos m\beta\, \mathrm{e}^{\mathrm{i}(qx+hd)}\,\frac{\mathrm{d}q}{h}\right] \tag{9.4}$$

Except close to the obstacle it must be possible to expand the scattered wave in the waveguide modes. For $x > a$ this expansion is in the right-going waves u_n^+ and for $x < -a$ it is in the left-going waves u_n^-. To derive such an expansion, consider $x > a$. The integrand in Eq. (9.4) is an analytic function except for the poles and cuts. The poles appear at $\sin hd = 0$, which in fact defines the waveguide modes. In the complex q-plane these are located along the real axis for $|q| < k$ and along the imaginary axis. Figure 9.2 shows the complex q-plane with the location of the poles, the cuts from the square root defining h, and the integration contour. For materials without damping the integration contour must run below the poles and the cut for $\mathrm{Re}\, q > 0$, because these are located in the first quadrant in the case of damping. It is possible to deform the contour into one running up and down along the cut in the upper half-plane (at the quarter circles at infinity the integrand is exponentially vanishing, so these give no contribution). Along the cut h is real and it is positive on the left side and negative on the right side. Changing to h as integration variable the integral along the cut becomes

$$
\begin{aligned}
I_m(r,\varphi) &= -\frac{\mathrm{i}^{1-m}}{\pi}\int_0^\infty \left[\frac{\cos hy}{\sin hd}\cos m\beta\,\mathrm{e}^{\mathrm{i}(qx+hd)}\right. \\
&\quad \left. +\frac{\cos hy}{-\sin hd}\cos m\beta\,\mathrm{e}^{\mathrm{i}(qx-hd)}\right]\frac{\mathrm{d}q}{h} \\
&= -\frac{2\mathrm{i}^{-m}}{\pi}\int_0^\infty \cos hy\cos m\beta\;\mathrm{e}^{\mathrm{i}qx}\,\frac{\mathrm{d}q}{h}
\end{aligned}
$$

In addition there are of course contributions from the poles. Changing to β as integration variable and switching to polar coordinates the integral becomes

$$
I_m(r,\varphi) = -\frac{\mathrm{i}^{-m}}{\pi}\int_0^{\pi/2-\mathrm{i}\infty}\cos m\beta\left(\mathrm{e}^{\mathrm{i}kr\cos(\beta-\varphi)}+\mathrm{e}^{\mathrm{i}kr\cos(\beta+\varphi)}\right)\mathrm{d}\beta
$$

where the contour starts at the origin, runs along the real axis to $\pi/2$, and then straight down parallel to the imaginary axis. Changing integration variable to $\beta-\varphi$ for the first term and to $\beta+\varphi$ for the second term gives

$$
I_m(r,\varphi) = -\frac{2\mathrm{i}^{-m}}{\pi}\cos m\varphi\int_0^{\pi/2-\mathrm{i}\infty}\mathrm{e}^{\mathrm{i}kr\cos\beta}\cos m\beta\,\mathrm{d}\beta
$$

Finally, $\cos m\beta$ is written as a sum of two exponentials and in the second of these the sign is switched on the integration variable. This gives

$$
\begin{aligned}
I_m(r,\varphi) &= -\frac{\mathrm{i}^{-m}}{\pi}\cos m\varphi\int_{-\pi/2+\mathrm{i}\infty}^{\pi/2-\mathrm{i}\infty}\mathrm{e}^{\mathrm{i}(m\beta+kr\cos\beta)}\,\mathrm{d}\beta \\
&= -\mathrm{H}_m(kr)\cos m\varphi
\end{aligned}
$$

where the second step follows from an integral representation of the Hankel function (Watson [140]). This term exactly cancels the first term in Eq. (9.4). All that remains is the contribution from the poles at $h_n = n\pi/d$ and $q_n = \sqrt{k^2-h_n^2}$, $n=0,1,2,\ldots$. The residues are readily calculated and the result is

$$
u^{\mathrm{sc}}(x,y) = -\sum_m f_m\mathrm{i}^{-m}\sqrt{\epsilon_m}\left[\frac{1}{2kd}\,\mathrm{e}^{\mathrm{i}kx}+\sum_{n=1}^{\infty}\frac{\cos m\beta_n}{q_n d}\,\mathrm{e}^{\mathrm{i}q_n x}\cos h_n y\right]
$$

(9.5)

where β_n is determined from $q_n = k\cos\beta_n$ and $h_n = k\sin\beta_n$. The scattered (transmitted) wave for $x > a$ is thus obtained as a sum over the waveguide modes as is to be expected. For $x < -a$ the procedure is the same, but with the contour closed in the lower complex q-plane.

9.2 Sphere in circular waveguide: torsional waves

Consider next a 3D geometry with a spherical cavity of radius a in an elastic, infinite cylindrical bar with radius d and specialize to the case of an incident torsional waveguide mode. This is a particular case of the work of Olsson [108], who investigates the same geometry for an arbitrary incident waveguide mode. Olsson [109] also generalizes this to an inhomogeneity in a thick-walled pipe. Due to the special case considered here the corresponding acoustic problem considered by Boström [14] becomes very similar.

The surface of the cylinder is traction-free. The cavity is assumed to have its centre on the axis of the cylinder so that the geometry is rotationally symmetric. Due to the geometry both cylindrical coordinates (r, φ, z) and spherical coordinates (R, θ, φ) are used. Note that in this section the radial coordinate in the spherical system is denoted R instead of r to distinguish it from the radius r in the cylindrical coordinates.

To keep things as simple as possible the torsional problem is considered. This means that the displacement only has an azimuthal component that is independent of the azimuthal coordinate. In cylindrical coordinates the displacement is then $\boldsymbol{u} = \hat{\boldsymbol{\varphi}}\, u(r, z)$ and in spherical coordinates it is $\boldsymbol{u} = \hat{\boldsymbol{\varphi}}\, u(R, \theta)$. The only relevant stress components are the tangential component of the traction on the cylinder

$$\sigma_{r\varphi}(r, z) = \mu\left(\frac{\partial u}{\partial r} - \frac{u}{r}\right)$$

and on the sphere

$$\sigma_{\varphi R}(R, \theta) = \mu\left(\frac{\partial u}{\partial R} - \frac{u}{R}\right)$$

To expand the displacement in cylindrical coordinates the vector wavefunctions from Section 2.3.1 are used. For torsional waves only the ones for

$\tau = 1$, $\sigma = \mathrm{e}$, and $m = 0$ are needed and skipping the normalization these simply become (suppressing the vector character and all unnecessary indices)

$$\chi^0(h;r,z) = \mathrm{J}_1(qr)\mathrm{e}^{\mathrm{i}hz}$$

where $q^2 + h^2 = k^2$ and the root is chosen so $\mathrm{Im}\, h \geq 0$ or $\mathrm{Im}\, q \geq 0$ (whichever is relevant). The corresponding outgoing waves are not needed in the present problem. For the spherical coordinates the wavefunctions are given in Section 3.1.3 and only those for $\tau = 1$, $\sigma = \mathrm{e}$, and $m = 0$ are relevant and these can be given as (again skipping the normalization, the vector character, and unnecessary indices; the definition of the associated Legendre functions is used)

$$\psi_l^0(R,\theta) = \mathrm{j}_l(kR)\mathrm{P}_l^1(\cos\theta)$$

for $l = 1, 2, \ldots$. Also the corresponding out-going wavefunctions $\psi_l^+(R,\theta)$ are needed; they of course have the spherical Bessel function replaced by the spherical Hankel function (of the first kind).

The torsional waveguide modes are given by $\chi^0(h_n;r,z)$, where h_n is determined so that the boundary condition $\sigma_{r\varphi}(d,z) = 0$ is satisfied. Using a recurrence relation for Bessel functions the stress is $\sigma_{r\varphi}(r,z) = -\mu q \mathrm{J}_2(qr)$, so $q_n d$, $n = 1, 2, \ldots$, are the roots of the Bessel function of order two, $\mathrm{J}_2(q_n d) = 0$, and then $h_n = (k^2 - q_n^2)^{1/2}$. When $q_n > k$ modes are evanescent with h_n imaginary and decay exponentially.

For the scattering by the cavity in the cylinder it is assumed that the incident wave satisfies the traction-free boundary condition on the cylinder, but it is not further specified. As usual the total wave is decomposed into the incident wave u^{in} and the scattered wave u^{sc}. The scattered wave can be given as one part from the surface of the cylinder and one part from the surface of the sphere (as can be shown by the integral representation)

$$u^{\mathrm{sc}} = \int_{-\infty}^{\infty} g(h)\mathrm{J}_1(qr)\mathrm{e}^{\mathrm{i}hz}\,\frac{\mathrm{d}h}{k} + \sum_{l=1}^{\infty} f_l \mathrm{h}_l(kR)\mathrm{P}_l^1(\cos\theta) \tag{9.6}$$

To be able to satisfy the boundary conditions it is apparently advantageous to be able to transform between the cylindrical and spherical wavefunctions. These are well-known [27] and in the present special case the needed ones are

$$\mathrm{J}_1(qr)\mathrm{e}^{\mathrm{i}hz} = \frac{1}{2}\sum_{l=1}^{\infty} \mathrm{i}^{l-1}\mathrm{P}_l^1(h/k)\mathrm{j}_l(kR)\mathrm{P}_l^1(\cos\theta) \tag{9.7}$$

$$\mathrm{h}_1(kR)\mathrm{P}_l^1(\cos\theta) = \frac{\mathrm{i}^{1-l}}{2k}\int_{-\infty}^{\infty} \mathrm{P}_l^1(h/k)\mathrm{H}_1(qr)\mathrm{e}^{\mathrm{i}hz}\,\mathrm{d}h, \qquad r > 0 \tag{9.8}$$

$$\mathrm{h}_1(kR)\mathrm{P}_l^1(\cos\theta) = \mathrm{i}^{1-l}\int_0^{\infty} \mathrm{P}_l^1(h/k)\mathrm{J}_1(qr)\mathrm{e}^{\mathrm{i}hz}\,\frac{q\,\mathrm{d}q}{kh}, \qquad z > 0 \tag{9.9}$$

There is a counterpart to the last equation with exponential factor $\exp(-\mathrm{i}hz)$ which is valid for $z < 0$.

To apply the stress-free boundary condition on the cylinder the second term in the scattered wave Eq. (9.6) is transformed to cylindrical wavefunctions with the help of Eq. (9.8). Inverting the Fourier transform the result is

$$qg(h)\mathrm{H}_2(qd) + \frac{q}{2}\sum_{l=1}^{\infty} \mathrm{i}^{1-l} f_l \mathrm{P}_l^1(h/k)\mathrm{J}_2(qd)$$

$$= -\frac{1}{2\pi\mu}\int_{-\infty}^{\infty} \sigma_{r\varphi}^{\mathrm{in}}(d,z)\mathrm{e}^{-\mathrm{i}hz}\,\mathrm{d}z = 0$$

where the second equality follows from the assumption that the incident wave satisfies the boundary condition on the cylinder. To apply the stress-free boundary condition on the sphere the first term in the scattered wave Eq. (9.6) is transformed to spherical wavefunctions with the help of Eq. (9.7). Inverting the Legendre series this results in

$$kf_l\left(\mathrm{h}_l'(ka) - \frac{\mathrm{h}_l(ka)}{ka}\right) + \frac{\mathrm{i}^{l-1}}{2}\left(\mathrm{j}_l'(ka) - \frac{\mathrm{j}_l(ka)}{ka}\right)\int_{-\infty}^{\infty} g(h)\mathrm{P}_l^1(h/k)\,\mathrm{d}h$$

$$= -\frac{2l+1}{2\mu l(l+1)}\int_0^{\pi} \sigma_{\varphi R}^{\mathrm{in}}(a,\theta)\mathrm{P}_l^1(\cos\theta)\sin\theta\,\mathrm{d}\theta$$

Eliminate $g(h)$ between the two equations to obtain

$$
\begin{aligned}
& f_l - \sum_{l'=1}^{\infty} T_l R_{ll'} f_{l'} \\
& \quad = -\frac{(2l+1)a}{2\mu l(l+1)\left(ka\mathrm{h}_l'(ka) - \mathrm{h}_l(ka)\right)} \int_0^{\pi} \sigma_{\varphi R}^{\mathrm{in}}(a,\theta) \mathrm{P}_l^1(\cos\theta) \sin\theta \, \mathrm{d}\theta
\end{aligned} \tag{9.10}
$$

Here the T-matrix (diagonal in this case) is introduced

$$
T_l = -\frac{ka\mathrm{j}_l'(ka) - \mathrm{j}_l(ka)}{ka\mathrm{h}_l'(ka) - \mathrm{h}_l(ka)} \tag{9.11}
$$

and the spherical projection of the reflection coefficient from (the inside of) the cylinder is also introduced

$$
R_{ll'} = -\frac{\mathrm{i}^{l-l'}}{2} \int_{-\infty}^{\infty} \frac{\mathrm{H}_2(qd)}{\mathrm{J}_2(qd)} \mathrm{P}_l^1(h/k) \mathrm{P}_{l'}^1(h/k) \, \mathrm{d}h \tag{9.12}
$$

Here the integrand has poles and cuts along the real h-axis so the integration contour must run below the real axis in the fourth quadrant and above the real axis in the second quadrant. Equation (9.10) is a system of equations that determines the coefficients f_l which in turn determines $g(h)$.

As in the previous section the scattered wave can be given as a sum over the waveguide modes. Consider $z > a$ and eliminate $g(h)$ and use the transformation Eq. (9.9) to get

$$
\begin{aligned}
u^{\mathrm{sc}} = \sum_{l=1}^{\infty} \mathrm{i}^{1-l} f_l \Bigg[& \int_0^{\infty} \mathrm{P}_l^1(h/k) \mathrm{J}_1(qr) \mathrm{e}^{\mathrm{i}hz} \frac{q \, \mathrm{d}q}{kh} \\
& - \frac{1}{2} \int_{-\infty}^{\infty} \frac{\mathrm{H}_2(qd)}{\mathrm{J}_2(qd)} \mathrm{P}_l^1(h/k) \mathrm{J}_1(qr) \mathrm{e}^{\mathrm{i}hz} \frac{\mathrm{d}h}{h} \Bigg]
\end{aligned}
$$

In a similar way as in the previous section the second integral can be deformed so that it cancels the first one except for the pole contributions (see Boström [14] for a very similar calculation). The simple poles appear when $\mathrm{J}_2(qd) = 0$ and the residues are readily calculated to yield the result [a recurrence relation for Bessel functions is used to simplify $\mathrm{J}_2'(q_n d)$]

$$
u^{\mathrm{sc}} = \pi \sum_{n=1}^{\infty} \mathrm{J}_1(q_n r) \mathrm{e}^{\mathrm{i}h_n z} \frac{\mathrm{H}_2(q_n d)}{h_n \mathrm{J}_1(q_n d)} \sum_{l=1}^{\infty} \mathrm{i}^{-l} f_l \mathrm{P}_l^1(h_n/k) \tag{9.13}
$$

This solution as a sum over the waveguide modes is thus valid for $z > a$; a similar result applies for $z < -a$.

9.3 Circular hole in thick plate

As a third example of scattering in a waveguide a circular hole in a plate is investigated. The waveguide modes in the plate are the well-known Rayleigh-Lamb modes, but also the simpler SH modes. All these modes are usually expressed in Cartesian coordinates, but here it is natural to employ cylindrical coordinates to express the modes and this is the task of the first subsection. See Achenbach and Xu [4] for more about plate modes in cylindrical coordinates and Diligent *et al.* [44] for the scattering by a hole in a plate.

9.3.1 Rayleigh-Lamb modes

Consider a plate of thickness $2d$ with a circular hole of radius a that runs through the whole thickness of the plate. A cylindrical coordinate system (r, φ, z) with the origin at the centre of the hole is employed, so that the plate surfaces are at $z = \pm d$. Only symmetric modes are considered; these have even horizontal components and an odd vertical component as functions of z. Without loss of generality the incident mode is assumed to propagate in the x-direction and it is, furthermore, assumed that the problem is symmetric around $\varphi = 0$. The σ-index can then be skipped. The other symmetries can be treated in a similar way, but are not further investigated.

The cylindrical vector wavefunctions are defined in Section 2.3. Here they are slightly changed, however, so that the symmetries are taken care of. They are also left without normalization. The following cylindrical vector wavefunctions are defined

$$
\begin{aligned}
\boldsymbol{\chi}_{1m}^{0}(q,\boldsymbol{r}) &= \frac{1}{q}\,\nabla\times\left[\hat{\boldsymbol{z}}\mathrm{J}_m(qr)\sin m\varphi\cos h_{\mathrm{s}}z\right] \\
&= \cos h_{\mathrm{s}}z\left[\hat{\boldsymbol{r}}\,\frac{m}{qr}\mathrm{J}_m(qr)\cos m\varphi-\hat{\boldsymbol{\varphi}}\mathrm{J}_m'(qr)\sin m\varphi\right] \\
\boldsymbol{\chi}_{2m}^{0}(q,\boldsymbol{r}) &= \frac{1}{q}\,\nabla\times\nabla\times\left[\hat{\boldsymbol{z}}\mathrm{J}_m(qr)\cos m\varphi\sin h_{\mathrm{s}}z\right] \\
&= h_{\mathrm{s}}\cos h_{\mathrm{s}}z\left[\hat{\boldsymbol{r}}\,\mathrm{J}_m'(qr)\cos m\varphi-\hat{\boldsymbol{\varphi}}\,\frac{m}{qr}\mathrm{J}_m(qr)\sin m\varphi\right] \\
&\qquad +\hat{\boldsymbol{z}}\,q\mathrm{J}_m(qr)\cos m\varphi\sin h_{\mathrm{s}}z
\end{aligned}
$$

$$
\begin{aligned}
\boldsymbol{\chi}_{3m}^{0}(q,\boldsymbol{r}) &= \nabla\left[\mathrm{J}_m(qr)\cos m\varphi\cos h_{\mathrm{p}}z\right]\\
&= q\cos h_{\mathrm{p}}z\left[\hat{\boldsymbol{r}}\mathrm{J}_m'(qr)\cos m\varphi-\hat{\boldsymbol{\varphi}}\,\frac{m}{qr}\mathrm{J}_m(qr)\sin m\varphi\right]\\
&\quad -\hat{\boldsymbol{z}}\,h_{\mathrm{p}}\mathrm{J}_m(qr)\cos m\varphi\sin h_{\mathrm{p}}z
\end{aligned}
\tag{9.14}
$$

Here the first index $\tau = 1, 2, 3$ corresponds to SH, SV, and P waves. The common radial wavenumber is q and the vertical wavenumbers are $h_{\mathrm{s}} = (k_{\mathrm{s}}^2 - q^2)^{1/2}$ and $h_{\mathrm{p}} = (k_{\mathrm{p}}^2 - q^2)^{1/2}$. The branch of the square root is chosen so that $\operatorname{Im} h_{\mathrm{s}} \geq 0$ and $\operatorname{Im} h_{\mathrm{p}} \geq 0$. The upper index 0 means that the wavefunctions are regular containing Bessel functions. The corresponding out-going wavefunctions are denoted with an upper index $+$ and contain Hankel functions. To apply the boundary conditions on the plate surfaces the corresponding stresses are needed and they are

$$
\begin{aligned}
\boldsymbol{t}^{(z)}\left(\boldsymbol{\chi}_{1m}^{0}(q,\boldsymbol{r})\right) &= \mu\frac{\partial\boldsymbol{\chi}_{1m}^{0}(q,\boldsymbol{r})}{\partial z}\\
&= \mu h_{\mathrm{s}}\sin h_{\mathrm{s}}z\left[\hat{\boldsymbol{r}}\,\frac{m}{qr}\mathrm{J}_m(qr)\cos m\varphi-\hat{\boldsymbol{\varphi}}\mathrm{J}_m'(qr)\sin m\varphi\right]\\
\boldsymbol{t}^{(z)}\left(\boldsymbol{\chi}_{2m}^{0}(q,\boldsymbol{r})\right) &= \mu\frac{\partial\boldsymbol{\chi}_{2m}^{0}(q,\boldsymbol{r})}{\partial z}+\mu\nabla\left(\hat{\boldsymbol{z}}\cdot\chi_{2m}^{0}(q,\boldsymbol{r})\right)\\
&= \mu(2q^2-k_{\mathrm{s}}^2)\sin h_{\mathrm{s}}z\left[\hat{\boldsymbol{r}}\,\mathrm{J}_m'(qr)\cos m\varphi-\hat{\boldsymbol{\varphi}}\,\frac{m}{qr}\mathrm{J}_m(qr)\sin m\varphi\right]\\
&\quad +\hat{\boldsymbol{z}}2qh_{\mathrm{s}}\mathrm{J}_m(qr)\cos m\varphi\cos h_{\mathrm{s}}z\\
\boldsymbol{t}^{(z)}\left(\boldsymbol{\chi}_{3m}^{0}(q,\boldsymbol{r})\right) &= \hat{\boldsymbol{z}}\lambda\nabla\cdot\boldsymbol{\chi}_{3m}^{0}(q,\boldsymbol{r})+2\mu\frac{\partial\boldsymbol{\chi}_{3m}^{0}(q,\boldsymbol{r})}{\partial z}\\
&= -2\mu qh_{\mathrm{p}}\sin h_{\mathrm{p}}z\left[\hat{\boldsymbol{r}}\mathrm{J}_m'(qr)\cos m\varphi-\hat{\boldsymbol{\varphi}}\,\frac{m}{qr}\mathrm{J}_m(qr)\sin m\varphi\right]\\
&\quad +\hat{\boldsymbol{z}}(2q^2-k_{\mathrm{s}}^2)\mathrm{J}_m(qr)\cos m\varphi\cos h_{\mathrm{p}}z
\end{aligned}
\tag{9.15}
$$

The boundary condition on the plate surfaces is the vanishing of the traction. As the horizontal direction of the traction is the same for the SV and P wavefunctions, the waveguide modes can be divided into two groups. For the SH modes the requirement is that $\sin h_{\mathrm{s}}d = 0$, so that $h_{1n} = n\pi/d$,

$n = 1, 2, \ldots$ and correspondingly $q_{1n} = (k_\mathrm{s}^2 - h_{1n}^2)^{1/2}$ (with $\operatorname{Im} q_{1n} \geq 0$). The waveguide mode is simply

$$\boldsymbol{v}_{1mn}^0(\boldsymbol{r}) = \boldsymbol{\chi}_{1m}^0(q_{1n}, \boldsymbol{r}) \tag{9.16}$$

for the regular ones and similarly for the out-going $\boldsymbol{v}_{1mn}^+(\boldsymbol{r})$ (with Hankel functions instead of Bessel functions). For the coupled P-SV modes a linear combination is taken

$$\boldsymbol{v}_{2m}^0(\boldsymbol{r}) = A_m \boldsymbol{\chi}_{2m}^0(q, \boldsymbol{r}) + B_m \boldsymbol{\chi}_{3m}^0(q, \boldsymbol{r})$$

Use of the boundary conditions leads to the classical dispersion relation for the symmetric Rayleigh-Lamb waves [1]

$$\frac{\tan h_\mathrm{s} d}{\tan h_\mathrm{p} d} = -\frac{4q^2 h_\mathrm{s} h_\mathrm{p}}{(2q^2 - k_\mathrm{s}^2)^2}$$

This determines possible values of $q = q_{2n}$, $n = 1, 2, \ldots$ and corresponding $h_{\mathrm{s}2n} = (k_\mathrm{s}^2 - q_{2n}^2)^{1/2}$ and $h_{\mathrm{p}2n} = (k_\mathrm{p}^2 - q_{2n}^2)^{1/2}$ with $\operatorname{Im} h_{\mathrm{s}2n} \geq 0$ and $\operatorname{Im} h_{\mathrm{p}2n} \geq 0$. It also leads to a relation between A_m and B_m so that the regular cylindrical Rayleigh-Lamb waveguide modes can be written

$$\begin{aligned}\boldsymbol{v}_{2mn}^0(\boldsymbol{r}) = {} & 2q_{2n} h_{\mathrm{p}2n} \sin(h_{\mathrm{p}2n} d)\, \boldsymbol{\chi}_{2m}^0(q_{2n}, \boldsymbol{r}) \\ & + (2q_{2n}^2 - k_\mathrm{s}^2) \sin(h_{\mathrm{s}2n} d)\, \boldsymbol{\chi}_{3m}^0(q_{2n}, \boldsymbol{r})\end{aligned} \tag{9.17}$$

and similarly for the out-going ones $\boldsymbol{v}_{2mn}^+(\boldsymbol{r})$.

9.3.2 Scattering problem

When solving for the scattering by a circular hole it is natural to make expansions of the incident and scattered waves in the cylindrical modes of the plate. The incident wave is expanded in the regular modes

$$\boldsymbol{u}^\mathrm{in}(\boldsymbol{r}) = \sum_{j,m,n} b_{jmn} \boldsymbol{v}_{jmn}^0(\boldsymbol{r}) \tag{9.18}$$

where the sum is over $j = 1, 2$, $m = 0, 1, 2, \ldots$, and $n = 1, 2, \ldots$. Here the coefficients b_{jmn} are regarded as known. As an example the incident wave can be a Rayleigh-Lamb mode propagating in the x-direction

$$\boldsymbol{u}^{\text{in}}(\boldsymbol{r}) = 2q_{2n_0}h_{\text{p}2n_0}\sin h_{\text{p}2n_0}d\,\frac{1}{q_{2n_0}}\nabla\times\nabla\times[\hat{\boldsymbol{z}}\exp{(\text{i}q_{2n_0}x)}\sin h_{\text{s}2n_0}z]$$
$$+(2q_{2n_0}^2 - k_\text{s}^2)\sin h_{\text{p}2n_0}d\,\nabla\left[\exp{(\text{i}q_{2n_0}x)}\cos h_{\text{p}2n_0}z\right]$$

for some n_0 and corresponding q_{2n_0}. Using the expansion of a plane wave in polar waves [Eq. (2.17)]

$$\text{e}^{\text{i}kx} = \sum_{m=0}^{\infty}\text{i}^m\epsilon_m\text{J}_m(kr)\cos m\varphi$$

this gives the expansion coefficients

$$b_{jmn} = \delta_{j2}\delta_{nn_0}\,\text{i}^m\epsilon_m$$

It is of course also possible to take an incident SH mode.

The scattered wave is expanded in the out-going waves

$$\boldsymbol{u}^{\text{sc}}(\boldsymbol{r}) = \sum_{j,m,n} f_{jmn}\boldsymbol{v}_{jmn}^{+}(\boldsymbol{r}) \tag{9.19}$$

where the coefficients f_{jmn} are to be determined. The boundary condition on the hole is the vanishing of the traction. For the cylindrical vector wave-functions of the present section the tractions in the r-direction is very similar to those in Section 2.3 [Eq. (2.17)]

$$\boldsymbol{t}^{(r)}\left(\boldsymbol{\chi}_{1m}^0(q,\boldsymbol{r})\right) = 2\mu\frac{\partial}{\partial r}\boldsymbol{\chi}_{1m}^0(h,\boldsymbol{r}) + \mu k_\text{s}\hat{\boldsymbol{r}}\times\left[\nabla\times\boldsymbol{\chi}_{1m}^0(h,\boldsymbol{r})\right]$$
$$= \mu\left[\hat{\boldsymbol{r}}\,2m\frac{\text{d}}{\text{d}r}\left(\frac{\text{J}_m(qr)}{qr}\right)\cos m\varphi\cos h_\text{s}z - \hat{\boldsymbol{\varphi}}\,q\left[2\text{J}_m''(qr)\right.\right.$$
$$\left.\left.+\text{J}_m(qr)\right]\sin m\varphi\cos h_\text{s}z + \hat{\boldsymbol{z}}\,\frac{mh_\text{s}}{qr}\text{J}_m(qr)\cos m\varphi\sin h_\text{s}z\right]$$

$$t^{(r)}\left(\chi^0_{2m}(q,\boldsymbol{r})\right)=2\mu\frac{\partial}{\partial r}\chi^0_{2m}(h,\boldsymbol{r})+\mu k_{\rm s}\hat{\boldsymbol{r}}\left[\times\nabla\times\chi^0_{2m}(h,\boldsymbol{r})\right]$$

$$=\mu\Bigg[\hat{\boldsymbol{r}}\,2qh_{\rm s}\mathrm{J}''_m(qr)\cos m\varphi\cos h_{\rm s}z$$

$$-\hat{\boldsymbol{\varphi}}\,2mh_{\rm s}\frac{\mathrm{d}}{\mathrm{d}r}\left(\frac{\mathrm{J}_m(qr)}{qr}\right)\sin m\varphi\cos h_{\rm s}z$$

$$+\hat{\boldsymbol{z}}\,(2q^2-k_{\rm s}^2)\mathrm{J}'_m(qr)\cos m\varphi\sin h_{\rm s}z\Bigg]$$

$$t^{(r)}\left(\chi^0_{3m}(q,\boldsymbol{r})\right)=\hat{\boldsymbol{r}}\mu\left(2k_{\rm p}^2-k_{\rm s}^2\right)\Phi^0_m(q,\boldsymbol{r})+2\mu\frac{\partial}{\partial r}\chi^0_{3m}(q,\boldsymbol{r})$$

$$=\mu\Bigg[\hat{\boldsymbol{r}}\left((2k_{\rm p}^2-k_{\rm s}^2)\mathrm{J}_m(qr)+2q^2\mathrm{J}''_m(qr)\right)\cos m\varphi\cos h_{\rm p}z$$

$$-\hat{\boldsymbol{\varphi}}\,2mq\frac{\mathrm{d}}{\mathrm{d}r}\left(\frac{\mathrm{J}_m(qr)}{qr}\right)\sin m\varphi\cos h_{\rm p}z$$

$$-\hat{\boldsymbol{z}}\,2qh_{\rm p}\mathrm{J}'_m(qr)\cos m\varphi\sin h_{\rm p}z\Bigg] \tag{9.20}$$

The corresponding tractions for the waveguide modes are obtained by operating with the traction operator on Eqs. (9.16) and (9.17). The boundary condition on the surface of the hole at $r=a$ then becomes

$$t^{(r)}\left(\boldsymbol{u}^{\rm in}(\boldsymbol{r})+\boldsymbol{u}^{\rm sc}(\boldsymbol{r})\right)$$

$$=\sum_{j,m,n}\left[b_{jmn}t^{(r)}\left(\boldsymbol{v}^0_{jmn}(\boldsymbol{r})\right)+f_{jmn}t^{(r)}\left(\boldsymbol{v}^+_{jmn}(\boldsymbol{r})\right)\right]=0 \tag{9.21}$$

This equation has three components in the r-, φ-, and z-directions. The orthogonality properties of the trigonometric functions in φ means that the different m-values decouple. However, this argument does not apply for the z-dependence. Instead the equations are projected onto the mode shapes of the SH waves, meaning that the r- and z-components of the traction are projected onto $\cos n\pi z/2d$ and the φ-component onto $\sin n\pi z/2d$. This leaves a system of equations for the expansion coefficients f_{jmn}, and as this determines the scattered wave the problem is in principle solved.

10 Scattering by cracks in isotropic solids

Scattering by cracks is a subject with important applications and is taken up in this chapter. For a strip-like crack in 2D and for strip-like, rectangular, and circular cracks in 3D it is possible to formulate integral equations over the surface of the crack that are amenable to more or less analytical solutions (although they involve computations of integrals and inversion of a matrix). By expanding the incident and scattered waves in polar or spherical wavefunctions the T-matrix can be determined. For low frequencies it is in some cases possible to obtain explicit expression for the leading order T-matrix elements. The chapter ends with a study of a curved crack (that is part of a circle) for which a similar approach is possible.

The approach used in the present chapter (and the two following chapters) is a hypersingular integral equation method, which dates back to the late 1970s. Neerhoff [101] uses a reciprocity relation and a Green function to derive an integral equation for the crack-opening displacement (henceforth abbreviated COD) for an interface crack in a one-layered half-space for the 2D SH case and solves the equation by expansions and projections with Chebyshev functions. Also for the 2D SH case, van den Berg [130] applies the same method to a crack in full space and obtains the T-matrix. Krenk and Schmidt [77] consider the circular crack, but derives the integral equation by a direct approach. The circular crack is also considered by Martin and Wickham [94] (who also give a short history of the subject), but using the method of dual integral equations, leading to a quite complicated procedure with double integrals that must be computed numerically. Subsequently, the hypersingular integral equation method for crack scattering is applied to a variety of problems as is testified by the present and the following two chapters. Boström [17] gives a review of the method with references. The book by Zhang and Gross [147] considers a number of crack scattering problems and gives numerous references. The hypersingular integral equations lead to mathematical questions that are discussed by Martin and Rizzo [91, 92] and Martin *et al.* [93].

10.1 Strip-like crack: 2D SH waves

It is instructive to start the investigation of scattering by cracks by the scalar 2D problem for SH waves, see van den Berg [130] (this is extended to an arbitrary number of cracks by Gross and Zhang [56]). This involves the formulation of an integral equation and its solution by an expansion of the crack-opening displacement in terms of Chebyshev functions. To obtain the T-matrix an integral representation is used and it is then possible to obtain explicit expressions for the leading order T-matrix elements for low frequencies.

10.1.1 Integral equation

In the 2D SH case there is only an out-of-plane component which satisfies the scalar Helmholtz equation

$$\nabla^2 u + k^2 u = 0$$

where the displacement is only called u (instead of u_z) and the wavenumber similarly is called k (instead of k_s) in this section. As usual the total displacement is divided as $u = u^{\mathrm{in}} + u^{\mathrm{sc}}$, where the incident wave u^{in} is known and the scattered wave u^{sc} is to be determined. This later leads to the determination of the T-matrix for the crack. The strip-like crack is located along the x-axis between $x = -a$ and $x = a$ and is assumed open so that the boundary condition for $y = 0$, $-a < x < a$ is

$$\frac{\partial u}{\partial y} = 0$$

Other boundary conditions are possible.

Taking a Fourier transform in x it is clear that the scattered wave for $y > 0$ can be represented as

$$u^{\mathrm{sc}}(x, y) = \int_{-\infty}^{\infty} f(q) \mathrm{e}^{\mathrm{i}(qx+hy)}\, \mathrm{d}q$$

where $f(q)$ is an unknown function of the Fourier variable q and the wavenumber in the y-direction is $h = (k^2 - q^2)^{1/2}$. To satisfy the radiation condition the square root is chosen so that $\operatorname{Im} h \geq 0$. The scattered wave is then up-going or exponentially decaying (evanescent). As the scattered wave must be odd in y (because an even wave already satisfies the boundary condition), the boundary condition for $y = 0$ can be reformulated as

$$u^{\mathrm{sc}} = 0, \qquad |x| > a$$

$$\frac{\partial u^{\mathrm{sc}}}{\partial y} = -\frac{\partial u^{\mathrm{in}}}{\partial y}, \qquad |x| < a$$

Applying these boundary conditions to the scattered wave the following two equations are obtained

$$\int_{-\infty}^{\infty} f(q)\mathrm{e}^{\mathrm{i}qx}\,\mathrm{d}q = 0, \qquad |x| > a$$

$$\int_{-\infty}^{\infty} \mathrm{i}hf(q)\mathrm{e}^{\mathrm{i}qx}\,\mathrm{d}q = -\frac{\partial u^{\mathrm{in}}(x,0)}{\partial y}, \qquad |x| < a$$

This is a pair of dual integral equations, and these were first solved by representing the unknown f by an integral with an unknown kernel so that the first equation is automatically satisfied and the second gives an equation for the new unknown (see Robertson [117] for an early example for the circular crack). This procedure leads to calculations of double integrals which does not seem to be very attractive from a numerical point of view.

Instead the first equation is rewritten as

$$\int_{-\infty}^{\infty} f(q)\mathrm{e}^{\mathrm{i}qx}\,\mathrm{d}q = \begin{cases} 0, & |x| > a \\ \frac{1}{2}\Delta u(x), & |x| < a \end{cases}$$

where the unknown crack-opening displacement (in the following abbreviated COD) Δu is introduced. The factor $1/2$ comes from the fact that only half of the COD originates from the wave in the upper half-plane. Inverting the Fourier transform this gives the representation

$$f(q) = \frac{1}{4\pi}\int_{-a}^{a} \Delta u(x)\mathrm{e}^{-\mathrm{i}qx}\,\mathrm{d}x$$

Inserting this into the stress boundary condition on the crack yields

$$\frac{\mathrm{i}}{4\pi}\int_{-\infty}^{\infty} h\,\mathrm{e}^{\mathrm{i}qx}\int_{-a}^{a}\Delta u(x')\mathrm{e}^{-\mathrm{i}qx'}\,\mathrm{d}x'\mathrm{d}q = -\frac{\partial u^{\mathrm{in}}(x,0)}{\partial y}, \qquad |x|<a \tag{10.1}$$

This is the desired integral equation for the COD. It is not possible to change the order of the integrations as this would lead to a divergent integral.

Before solving the integral equation it is instructive to derive the same integral equation in a different way, a way that is more amenable to generalizations. Using the stress free boundary condition in the integral representation Eq. (4.18), this equation becomes

$$u^{\mathrm{sc}}(x,y) = \int_{-a}^{a}\Delta u(x')\frac{\partial}{\partial y'}g(x',0;x,y)\,\mathrm{d}x' \tag{10.2}$$

The COD here appears because the normal derivative has different sign on the upper and lower side of the crack. Insert the representation of the Green function in plane waves [Eq. (8.6)], take the derivative with respect to y, use the boundary condition once more, and take the limit $y \to 0^+$ to obtain

$$\begin{aligned}\lim_{y\to 0+}\frac{\mathrm{i}}{4\pi}\int_{-a}^{a}\Delta u(x')\int_{-\infty}^{\infty} h\,\mathrm{e}^{\mathrm{i}(q(x-x')+hy)}\,\mathrm{d}q\,\mathrm{d}x'\\ = -\frac{\partial u^{\mathrm{in}}(x,0)}{\partial y}, \qquad |x|<a\end{aligned}$$

One could as well take the limit $y \to 0^-$ and then the sign on hy in the exponent must be changed. As long as $y > 0$ the q-integral contains an exponential factor $\exp(-|q|y)$ when $|q|$ is large and the order of integrations can be reversed so that the equation becomes identical with Eq. (10.1). The limit in front of the integrals can not be taken inside the integrals as they stand as the inner integral would not exist in any normal sense; instead it must be interpreted as a Hadamard finite-part integral [91]. This type of integral is called hypersingular, and the integral equations for cracks in the following are of this type.

The two ways of deriving the integral equation, by a Fourier representation or by an integral representation with a Green function, are equivalent in this case. However, this is not always so; for situations with plane boundaries or multiple cracks the Green function approach is more general.

10.1.2 Solution of the integral equation

To solve the integral equation some discretization scheme is necessary. In boundary element methods (BEM) this is done by dividing the crack into small elements with a local basis on each element. For a straight crack as considered here a better approach is to use a global (on the crack) basis. The following complete set of Chebyshev functions on the interval $-1 \leq s \leq 1$ is very useful for this purpose

$$\psi_n(s) = \begin{cases} \cos(n \arcsin s), & n = 1, 3, \ldots \\ \mathrm{i}\,\sin(n \arcsin s), & n = 2, 4, \ldots \end{cases} \tag{10.3}$$

where the "i" is inserted for convenience. These functions satisfy the following integral

$$\int_{-1}^{1} \psi_n(s)\, \mathrm{e}^{-\mathrm{i}\gamma s}\, \mathrm{d}s = \frac{n\pi}{\gamma}\, \mathrm{J}_n(\gamma) \tag{10.4}$$

The Chebyshev functions as defined have the form of $(1-s^2)^{1/2}$ times a polynomial. They are related to the conventional Chebyshev polynomials of the second kind and are orthogonal with weight $(1-s^2)^{-1/2}$

$$\int_{-1}^{1} \psi_n^{\star}(s)\psi_{n'}(s) \frac{\mathrm{d}s}{\sqrt{1-s^2}} = \frac{\pi}{2}\, \delta_{nn'}$$

where the star denotes the complex conjugate. As $\psi_1(s) = (1-s^2)^{1/2}$ and $\psi_2(s) = 2\mathrm{i}s(1-s^2)^{1/2}$ this also leads to

$$\int_{-1}^{1} \psi_n(s)\, \mathrm{d}s = \frac{\pi}{2}\, \delta_{n1} \tag{10.5}$$

$$\int_{-1}^{1} \psi_n(s) s\, \mathrm{d}s = \frac{\mathrm{i}\pi}{4}\, \delta_{n2} \tag{10.6}$$

The COD is now expanded in terms of the Chebyshev functions

$$\Delta u(x') = \sum_{n'=1}^{\infty} \alpha_{n'} \psi_{n'}(x'/a) \tag{10.7}$$

This expansion has two very important properties; it satisfies the correct edge condition at the crack tips $x = \pm a$ and it makes possible the use of the integral Eq. (10.4). Insertion of the COD expansion into the integral equation Eq. (10.1) gives

$$\frac{\mathrm{i}}{4}\int_{-\infty}^{\infty} h\,\mathrm{e}^{\mathrm{i}qx}\sum_{n'=1}^{\infty}\alpha_{n'}\frac{n'}{q}\mathrm{J}_{n'}(qa)\,\mathrm{d}q = -\frac{\partial u^{\mathrm{in}}(x,0)}{\partial y}, \qquad |x| < a$$

To get a full discretization a projection onto the Chebyshev functions is also performed. Thus, multiply the equation with $\psi_n(x/a)$ and integrate in x over the crack

$$\sum_{n'=1}^{\infty} Q_{nn'}\alpha_{n'} = -4\mathrm{i}\int_{-a}^{a}\frac{\partial u^{\mathrm{in}}(x,0)}{\partial y}\,\psi_n(x/a)\,\mathrm{d}x \tag{10.8}$$

where

$$Q_{nn'} = \pi nn'(-1)^n\int_{-\infty}^{\infty}\frac{h}{q^2}\,\mathrm{J}_n(qa)\mathrm{J}_{n'}(qa)\,\mathrm{d}q \tag{10.9}$$

In this form the Q-matrix is obviously symmetric (as $Q_{nn'} = 0$ when $n + n'$ is odd, the factor $(-1)^n$ does not destroy the symmetry). The integral is slowly converging and is not suitable for a direct numerical quadrature as it stands. In this simple case the integral can be reduced to a finite range [130]

$$Q_{nn'} = 2\pi nn'(-1)^n\int_0^k\frac{\sqrt{k^2-q^2}}{q^2}\left[\mathrm{J}_>(qa)\mathrm{H}_<(qa) + \frac{\mathrm{i}}{n\pi}\delta_{nn'}\right]\mathrm{d}q \tag{10.10}$$

when $n + n'$ is even. Here $>$ ($<$) denotes the larger (smaller) of n and n' and the Hankel function of the first kind is denoted H_n as in previous chapters. Another way to tackle the integral in Eq. (10.9) is to extract the dominating part for large q and integrate it exactly with the help of the integral

$$\int_0^{\infty}\frac{1}{s}\mathrm{J}_\nu(s)\mathrm{J}_{\nu+2j}(s)\,\mathrm{d}s = \frac{\delta_{j0}}{2\nu} \tag{10.11}$$

where ν can be any complex number such that $\mathrm{Re}\,\nu > -1/2$ and j is a nonnegative integer. The rest of the integrand then behaves as q^{-3} times two

Bessel functions for large q and can therefore be integrated numerically in an efficient way.

Once the expansion coefficients α_n are determined from Eq. (10.8), the scattered wave for $y > 0$ can be obtained as

$$u^{\mathrm{sc}} = \frac{1}{4} \sum_{n=1}^{\infty} n\alpha_n \int_{-\infty}^{\infty} \frac{1}{q} \mathrm{J}_n(qa) \mathrm{e}^{\mathrm{i}(qx+hy)}\, \mathrm{d}q$$

In the far field this integral can be evaluated by the stationary-phase method, see Section 5.2.1. Introducing polar coordinates (r, φ) the stationary point appears for $q = k\cos\varphi$ and the approximation gives

$$u^{\mathrm{sc}} = \frac{1}{4} \sqrt{\frac{2\pi}{kr}}\, \mathrm{e}^{\mathrm{i}(kr+\pi/4)} \tan\varphi \sum_{n=1}^{\infty} n\alpha_n \mathrm{J}_n(ka\cos\varphi)$$

10.1.3 The T-matrix

Expressing the waves in polar wavefunctions the T-matrix relates the coefficients of the scattered wave to those of the incident wave. The even (in φ) wavefunctions already satisfy the boundary condition on the crack, so the incident wave can be taken as an odd wavefunction

$$u^{\mathrm{in}} = \chi^{0}_{\mathrm{o}m}(r, \varphi) = \frac{1}{\sqrt{2}} \mathrm{J}_m(kr) \sin m\varphi$$

On the crack $r = x$ and $\partial/\partial y = (1/x)\partial/\partial\varphi$ for $x > 0$ and $r = -x$ and $\partial/\partial y = (1/x)\partial/\partial\varphi$ (two minus signs cancel) for $x < 0$, so the right-hand side of Eq. (10.8) (except for a factor $-2\mathrm{i}$) becomes

$$\begin{aligned} L_{nm} &= 2 \int_{-a}^{a} \frac{\partial \chi^{0}_{\mathrm{o}m}(r,\varphi)}{\partial y} \psi_n(x/a)\, \mathrm{d}x \\ &= \sqrt{2} \Big(\int_{-a}^{0} \frac{m}{x} (-1)^m \mathrm{J}_m(-kx) \psi_n(x/a)\, \mathrm{d}x \\ &\quad + \int_{0}^{a} \frac{m}{x} \mathrm{J}_m(kx) \psi_n(x/a)\, \mathrm{d}x \Big) \\ &= m\sqrt{2} \int_{-a}^{a} \frac{1}{x} \mathrm{J}_m(kx) \psi_n(x/a)\, \mathrm{d}x \end{aligned} \tag{10.12}$$

for $m = 1, 2, \ldots$ and $n = 1, 2, \ldots$.

To express the scattered wave in the out-going polar wavefunctions the integral representation in terms of the COD [Eq. (10.2)] is used. Expanding the Green function in the polar wavefunctions [Eq. (4.3)] this yields

$$u^{\mathrm{sc}}(r,\varphi) = \mathrm{i}\sum_{\sigma,m} \chi^{+}_{\sigma m}(r,\varphi) \int_{-a}^{a} \Delta u(x') \frac{\partial \chi^{0}_{\sigma m}(r',\varphi')}{\partial y'}\,\mathrm{d}x'$$

As for the incident wave only $\sigma = \mathrm{o}$ contributes and inserting the COD expansion this gives

$$\begin{aligned} u^{\mathrm{sc}}(r,\varphi) &= \mathrm{i}\sum_{m=1}^{\infty} \chi^{+}_{\mathrm{o}m}(r,\varphi) \sum_{n=1}^{\infty} \alpha_n \frac{1}{\sqrt{2}} \int_{-a}^{a} \frac{m}{x'} \mathrm{J}_m(kx')\psi_n(x'/a)\,\mathrm{d}x' \\ &= \mathrm{i}\sum_{m=1}^{\infty} \chi^{+}_{\mathrm{o}m}(r,\varphi) \sum_{n=1}^{\infty} \alpha_n \frac{1}{2} L_{nm} \end{aligned}$$

From this expression the nonzero T-matrix elements can be read off

$$T_{\mathrm{o}m,\mathrm{o}m'} = \sum_{n,n'} L_{nm} \left(Q^{-1}\right)_{nn'} L_{n'm'} \tag{10.13}$$

As the matrix Q is symmetric, so is the T-matrix. The matrix elements $Q_{nn'} = 0$ when $n + n'$ is an odd number and $L_{nm} = 0$ when $n + m$ is an odd number. Therefore, $T_{\mathrm{o}m,\mathrm{o}m'} = 0$ when $m + m'$ is an odd number. All T-matrix elements where at least one of the σ-indices is even are of course vanishing

$$T_{\mathrm{e}m,\mathrm{e}m'} = T_{\mathrm{e}m,\mathrm{o}m'} = T_{\mathrm{o}m,\mathrm{e}m'} = 0$$

10.1.4 Low frequency T-matrix

For low frequencies ka is a small number and it is then possible to obtain explicit expressions. To leading order the matrix Q becomes (for $n + n'$ even; $Q_{nn'} = 0$ for $n + n'$ odd)

$$Q_{nn'} \simeq 2\pi n n'(-1)^n \int_0^{\infty} \frac{\mathrm{i}}{q} \mathrm{J}_n(qa)\mathrm{J}_{n'}(qa)\,\mathrm{d}q = \mathrm{i}\pi n(-1)^n \delta_{nn'}$$

using the integral Eq. (10.11). As this matrix is diagonal it is of course trivial to calculate its inverse. Using the expression for the scattered far field and the incident wave it is then easy to obtain the scattered wave in explicit form (assuming an incident plane wave)

$$u^{\mathrm{sc}} = \frac{1}{4}\sqrt{\frac{2\pi}{kr}}\,\mathrm{e}^{\mathrm{i}(kr+\pi/4)}(ka)^2 \sin\varphi \sin\varphi_0 \tag{10.14}$$

This expression is symmetric in φ and φ_0, which is a manifestation of reciprocity.

The T-matrix can also be obtained explicitly for low frequencies. The other matrix needed becomes

$$\begin{aligned} L_{nm} &= \sqrt{2}m \int_{-1}^{1} \frac{1}{s}\mathrm{J}_m(kas)\psi_n(s)\,\mathrm{d}s \\ &\simeq \frac{\sqrt{2}m(ka)^m}{2^m m!} \int_{-1}^{1} s^{m-1}\psi_n(s)\,\mathrm{d}s \end{aligned}$$

It follows that $m = 1$ gives the leading order contribution and

$$L_{n1} \simeq \frac{ka}{\sqrt{2}} \int_{-1}^{1} \psi_n(s)\,\mathrm{d}s = \frac{\pi ka}{2\sqrt{2}}\,\delta_{n1}$$

using the integral Eq. (10.5). This determines the single leading order T-matrix element

$$T_{\mathrm{o}1,\mathrm{o}1} = \mathrm{i}(ka)^2\frac{\pi}{8} \tag{10.15}$$

From the T-matrix the scattered wave can be computed and in the far field the expression in Eq. (10.14) is then reproduced.

The situation can be compared to the one for a circular cavity, where there exist three T-matrix elements for $m = 0$ (only the even element) and $m = 1$ (both even and odd elements). These elements are given in Eqs. (2.23) and (2.24) (with $\rho' = 0$ and $\mu' = 0$) and each of these elements is exactly twice the crack T-matrix element.

10.2 Strip-like crack: 2D P-SV waves

For the coupled P-SV waves in 2D essentially the same procedure as for the scalar waves is used, see van den Berg [131] and van der Hijden and Neerhoff [133] (who employ the Green function approach). This, in fact, leads to two uncoupled scalar problems, one for the symmetric part and one for the antisymmetric part.

10.2.1 Symmetric and antisymmetric parts

With the open crack situated at the x-axis between $x = -a$ and $x = a$, symmetry with respect to $y = 0$ gives that the scattering problem can be divided into a symmetric and an antisymmetric part, this being the generalization to vector problems from the even and odd parts for scalar problems. For the symmetric part u_x is an even function of y and u_y is an odd function of y and vice versa for the antisymmetric part. It follows that σ_{yy} is even and σ_{xy} is odd for the symmetric part and vice versa for the antisymmetric part. It is then sufficient to consider only the half-plane $y > 0$. For the symmetric part the boundary conditions for $y = 0$ become

$$\begin{aligned} \sigma_{xy}^{\rm sc} &= 0, \qquad \text{all } x \\ u_y^{\rm sc} &= 0, \qquad |x| > a \\ \sigma_{yy}^{\rm sc} &= -\sigma_{yy}^{\rm in}, \qquad |x| < a \end{aligned}$$

and for the antisymmetric part they become

$$\begin{aligned} \sigma_{yy}^{\rm sc} &= 0, \qquad \text{all } x \\ u_x^{\rm sc} &= 0, \qquad |x| > a \\ \sigma_{xy}^{\rm sc} &= -\sigma_{xy}^{\rm in}, \qquad |x| < a \end{aligned}$$

Solutions for $y > 0$ in terms of plane waves can be constructed with the help of Eq. (1.17)

$$\boldsymbol{u} = \nabla \times (\hat{\boldsymbol{z}}\Psi) + \nabla\Phi$$

where Ψ and Φ satisfy scalar Helmholtz equations with wavenumber $k_{\rm s}$ and $k_{\rm p}$, respectively. Thus, it is realized that the scattered wave in the upper half-plane $y > 0$ can be represented by a Fourier transform

$$\begin{aligned} \boldsymbol{u}^{\rm sc}(x,y) &= \int_{-\infty}^{\infty} \left[f_1(q)\nabla \times \left(\hat{\boldsymbol{z}}{\rm e}^{{\rm i}(qx+h_{\rm s}y)}\right) + f_2(q)\nabla\left({\rm e}^{{\rm i}(qx+h_{\rm p}z)}\right)\right] {\rm d}q \\ &= {\rm i}\int_{-\infty}^{\infty} \left[(\hat{\boldsymbol{x}}h_{\rm s} - \hat{\boldsymbol{y}}q)\, f_1(q){\rm e}^{{\rm i}h_{\rm s}y} + (\hat{\boldsymbol{x}}q + \hat{\boldsymbol{y}}h_{\rm p})\, f_2(q){\rm e}^{{\rm i}h_{\rm p}y}\right] {\rm e}^{{\rm i}qx}\, {\rm d}q \end{aligned}$$

Here the wavenumbers in the y-direction are $h_\mathrm{s} = (k_\mathrm{s}^2 - q^2)^{1/2}$ and $h_\mathrm{p} = (k_\mathrm{p}^2 - q^2)^{1/2}$, where $\operatorname{Im} h_\mathrm{s} \geq 0$ and $\operatorname{Im} h_\mathrm{p} \geq 0$. The corresponding traction in the y-direction becomes

$$
\begin{aligned}
\boldsymbol{t}^{(y)} &= \hat{\boldsymbol{y}}\lambda\nabla\cdot\boldsymbol{u}^\mathrm{sc} + \mu\left(\frac{\partial\boldsymbol{u}^\mathrm{sc}}{\partial y} + \nabla u_y^\mathrm{sc}\right) \\
&= \mu\int_{-\infty}^{\infty}\left[\left(\hat{\boldsymbol{x}}\left(2q^2 - k_\mathrm{s}^2\right) + \hat{\boldsymbol{y}}\,2qh_\mathrm{s}\right)f_1(q)\mathrm{e}^{\mathrm{i}h_\mathrm{s}y}\right. \\
&\qquad \left. + \left(-\hat{\boldsymbol{x}}\,2qh_\mathrm{p} + \hat{\boldsymbol{y}}\left(2q^2 - k_\mathrm{s}^2\right)\right)f_2(q)\mathrm{e}^{\mathrm{i}h_\mathrm{p}y}\right]\mathrm{e}^{\mathrm{i}qx}\,\mathrm{d}q
\end{aligned}
$$

10.2.2 Integral equations

The first boundary condition for the symmetric part gives

$$\left(2q^2 - k_\mathrm{s}^2\right)f_1 - 2qh_\mathrm{p}f_2 = 0$$

The second can be written

$$\mathrm{i}\int_{-\infty}^{\infty}\left(-qf_1 + h_\mathrm{p}f_2\right)\mathrm{e}^{\mathrm{i}qx}\,\mathrm{d}q = \begin{cases} 0, & |x| > a \\ \tfrac{1}{2}\Delta u_y(x), & |x| < a \end{cases}$$

Inverting the Fourier transform gives

$$-qf_1 + h_\mathrm{p}f_2 = -\frac{\mathrm{i}}{4\pi}\int_{-\infty}^{\infty}\Delta u_y(x')\,\mathrm{e}^{-\mathrm{i}qx'}\mathrm{d}x'$$

Solving the two equations for f_1 and f_2 gives

$$f_1 = \frac{\mathrm{i}q}{2\pi k_\mathrm{s}^2}\int_{-a}^{a}\Delta u_y(x')\,\mathrm{e}^{-\mathrm{i}qx'}\mathrm{d}x'$$

$$f_2 = \frac{\mathrm{i}\left(2q^2 - k_\mathrm{s}^2\right)}{4\pi k_\mathrm{s}^2 h_\mathrm{p}}\int_{-a}^{a}\Delta u_y(x')\,\mathrm{e}^{-\mathrm{i}qx'}\mathrm{d}x'$$

Insertion into the last boundary condition finally gives

$$\sigma_{yy}^{\mathrm{sc}}(x,0) = \frac{\mathrm{i}\mu}{4\pi k_{\mathrm{s}}^2}\int_{-\infty}^{\infty}\frac{R(q)}{h_{\mathrm{p}}}\,\mathrm{e}^{\mathrm{i}qx}\int_{-a}^{a}\Delta u_y(x')\mathrm{e}^{-\mathrm{i}qx'}\,\mathrm{d}x'\,\mathrm{d}q$$

$$= -\sigma_{yy}^{\mathrm{in}}(x,0), \qquad |x| < a \tag{10.16}$$

where the Rayleigh function is [$R(q) = 0$ determines the wavenumber of the Rayleigh surface wave in an elastic half-space]

$$R(q) = \left(2q^2 - k_{\mathrm{s}}^2\right)^2 + 4q^2 h_{\mathrm{s}} h_{\mathrm{p}}$$

This is the integral equation for the normal COD Δu_y on the crack and it is very similar to the one for the scalar waves in Eq. (10.1).

For the antisymmetric part the first boundary condition gives

$$2qh_{\mathrm{s}}f_1 + \left(2q^2 - k_{\mathrm{s}}^2\right)f_2 = 0$$

The second boundary condition can be written

$$\mathrm{i}\int_{-\infty}^{\infty}\left(h_{\mathrm{s}}f_1 + qf_2\right)\mathrm{e}^{\mathrm{i}qx}\,\mathrm{d}q = \begin{cases} 0, & |x| > a \\ \frac{1}{2}\Delta u_x(x), & |x| < a \end{cases}$$

Inverting the Fourier transform and solving the two equations for f_1 and f_2 give

$$f_1 = \frac{\mathrm{i}\left(2q^2 - k_{\mathrm{s}}^2\right)}{4\pi k_{\mathrm{s}}^2 h_{\mathrm{s}}}\int_{-a}^{a}\Delta u_x(x')\,\mathrm{e}^{-\mathrm{i}qx'}\,\mathrm{d}x'$$

$$f_2 = \frac{-\mathrm{i}q}{2\pi k_{\mathrm{s}}^2}\int_{-a}^{a}\Delta u_x(x')\,\mathrm{e}^{-\mathrm{i}qx'}\,\mathrm{d}x'$$

Insertion into the last boundary condition gives

$$\sigma_{xy}^{\mathrm{sc}}(x,0) = \frac{\mathrm{i}\mu}{4\pi k_{\mathrm{s}}^2}\int_{-\infty}^{\infty}\frac{R(q)}{h_{\mathrm{s}}}\,\mathrm{e}^{\mathrm{i}qx}\int_{-a}^{a}\Delta u_x(x')\,\mathrm{e}^{-\mathrm{i}qx'}\,\mathrm{d}x'\mathrm{d}q$$

$$= -\sigma_{xy}^{\mathrm{in}}(x,0), \qquad |x| < a \tag{10.17}$$

This is the integral equation for the tangential COD Δu_x on the crack and it is almost identical to the one for the normal COD.

10.2.3 The T-matrix

The solution of the integral equations and the calculation of the T-matrix are very similar to the scalar case in the preceding section. Some details are given for the symmetric part and then the results for the antisymmetric part are stated.

To solve the symmetric integral equation Eq. (10.16) the normal COD is expanded in the Chebyshev functions

$$\Delta u_y(x') = \sum_{n'=1}^{\infty} \alpha_{n'} \psi_{n'}(x'/a)$$

This expansion satisfies the correct edge condition at the crack tips. Insertion into the integral equation gives [using Eq. (10.4)]

$$\frac{\mathrm{i}\mu}{4k_\mathrm{s}^2} \int_{-\infty}^{\infty} \frac{R(q)}{h_\mathrm{p}} \sum_{n'=1}^{\infty} \frac{n'}{q} \mathrm{J}_{n'}(qa)\, \mathrm{e}^{\mathrm{i}qx}\, \mathrm{d}q = -\sigma_{yy}^{\mathrm{in}}(x, 0)$$

Projection onto the Chebyshev functions then gives

$$\sum_{n'=1}^{\infty} Q_{nn'} \alpha_{n'} = -2\mathrm{i} L_n \tag{10.18}$$

where

$$Q_{nn'} = \pi n n' (-1)^n \int_{-\infty}^{\infty} \frac{R(q)}{k_\mathrm{s}^2 q^2 h_\mathrm{p}} \mathrm{J}_n(qa) \mathrm{J}_{n'}(qa)\, \mathrm{d}q \tag{10.19}$$

As $R(q)$ contains terms in q^4 this integral may look divergent. However, the q^4-terms cancel for large $|q|$ so the behaviour of the Rayleigh function for large $|q|$ is $R(q) \simeq 2q^2(k_\mathrm{p}^2 - k_\mathrm{s}^2)$ (see the next subsection). As $Q_{nn'} = 0$ when $n + n'$ is odd, the matrix is symmetric [despite the factor $(-1)^n$]. The right-hand side is

$$L_n = \frac{2}{\mu} \int_{-a}^{a} \sigma_{yy}^{\mathrm{in}}(x, 0) \psi_n(x/a)\, \mathrm{d}x \tag{10.20}$$

On the crack it is noticed the normal traction can be calculated as

$$\sigma_{yy}(x, 0) = \sigma_{\varphi\varphi}(r, 0) = \lambda \left(\frac{\partial u_r}{\partial r} + \frac{1}{r} \frac{\partial u_\varphi}{\partial \varphi} \right) + \frac{2\mu}{r} \left(\frac{\partial u_\varphi}{\partial \varphi} + u_r \right)$$

for $x > 0$ (where also $r = x$) and similarly for $x < 0$.

Once the COD coefficients are determined the scattering problem is in principle solved. The scattered wave, and particularly the far field, is obtained as in the SH case.

To obtain the T-matrix the incident wave is taken as a polar wavefunction. The symmetric wavefunction for $\tau = 1$ is for $m = 1, 2, \ldots$ (it does not exist for $m = 0$)

$$\chi^0_{1om}(\boldsymbol{r}) = \frac{1}{\sqrt{2}}\left[\hat{\boldsymbol{r}}\frac{m}{k_\mathrm{s}r}\,\mathrm{J}_m(k_\mathrm{s}r)\cos m\varphi - \hat{\boldsymbol{\varphi}}\mathrm{J}'_m(k_\mathrm{s}r)\sin m\varphi\right]$$

For this incident wave Eq. (10.20) becomes

$$L_{n1om} = 2\sqrt{2}m\int_{-a}^{a}\frac{1}{x}\left(\frac{1}{k_\mathrm{s}x}\,\mathrm{J}_m(k_\mathrm{s}x) - \mathrm{J}'_m(k_\mathrm{s}x)\right)\psi_n(x/a)\,\mathrm{d}x \quad (10.21)$$

where the extra indices $\tau\sigma m$ are added to specify the incident wavefunction. For $m = 1$ this integral may look divergent at $x = 0$, but to leading order the two terms cancel. For $\tau = 2$ the corresponding expressions are for $m = 0, 1, 2, \ldots$

$$\chi^0_{2em}(\boldsymbol{r}) = \frac{\sqrt{\epsilon_m}k_\mathrm{p}}{2k_\mathrm{s}}\left[\hat{\boldsymbol{r}}\mathrm{J}'_m(k_\mathrm{p}r)\cos m\varphi - \hat{\boldsymbol{\varphi}}\frac{m}{k_\mathrm{p}r}\,\mathrm{J}_m(k_\mathrm{p}r)\sin m\varphi\right]$$

and

$$L_{n2em} = \frac{\sqrt{\epsilon_m}}{k_\mathrm{s}}\int_{-a}^{a}\left[\left(2k_p^2 - k_\mathrm{s}^2 - \frac{2m^2}{x^2}\right)\mathrm{J}_m(k_\mathrm{p}x) + \frac{2k_\mathrm{p}}{x}\,\mathrm{J}'_m(k_\mathrm{p}x)\right]\psi_n(x/a)\,\mathrm{d}x \quad (10.22)$$

This may look divergent for $m = 1$, but cancellations save the situation.

To obtain the T-matrix the scattered wave should be expanded in the outgoing wavefunctions. This is accomplished by the integral representation Eq. (4.22) (the 2D version without k_s in front) with the Green dyadic expanded as in Eq. (4.12)

$$\boldsymbol{u}^{\mathrm{sc}}(\boldsymbol{r}) = \frac{\mathrm{i}}{\mu}\sum_{\tau,\sigma,m}\boldsymbol{\chi}^+_{\tau\sigma m}(\boldsymbol{r})\int_{-a}^{a}\Delta u_y(x')\hat{\boldsymbol{y}}\cdot\boldsymbol{t}^{(y)}(\boldsymbol{\chi}^0_{\tau\sigma m}(x',0))\,\mathrm{d}x'$$

which is valid outside the circumscribed circle to the crack. For the symmetric part the sum is only over $\tau\sigma = 1\mathrm{o}$ and $2\mathrm{e}$. Inserting the expansion of the COD this becomes

$$\boldsymbol{u}^{\mathrm{sc}}(\boldsymbol{r}) = \frac{\mathrm{i}}{2} \sum_{\tau,\sigma,m} \boldsymbol{\chi}^{+}_{\tau\sigma m}(\boldsymbol{r}) \sum_{n=1}^{\infty} \alpha_n L_{n\tau\sigma m}$$

From this the T-matrix elements can be read off

$$T_{\tau\sigma m,\tau'\sigma'm'} = \sum_{n,n'} L_{n\tau\sigma m} \left(Q^{-1}\right)_{nn'} L_{n'\tau'\sigma'm'} \tag{10.23}$$

This is of course only valid for the symmetric indices $\tau\sigma = 1\mathrm{o}$ and $2\mathrm{e}$.

For the antisymmetric waves the expression for the T-matrix looks exactly the same

$$T_{\tau\sigma m,\tau'\sigma'm'} = \sum_{n,n'} L_{n\tau\sigma m} \left(P^{-1}\right)_{nn'} L_{n'\tau'\sigma'm'} \tag{10.24}$$

which is now valid for $\tau\sigma = 1\mathrm{e}$ and $2\mathrm{o}$. Here the matrix is

$$P_{nn'} = \pi n n'(-1)^n \int_{-\infty}^{\infty} \frac{R(q)}{k_{\mathrm{s}}^2 q^2 h_{\mathrm{s}}} \mathrm{J}_n(qa)\mathrm{J}_{n'}(qa)\,\mathrm{d}q \tag{10.25}$$

The other quantities are

$$L_{n1\mathrm{e}m} = -k_{\mathrm{s}}\sqrt{\epsilon_m} \int_{-a}^{a} \left(2\mathrm{J}''_m(k_{\mathrm{s}}x) + \mathrm{J}_m(k_{\mathrm{s}}x)\right) \psi_n(x/a)\,\mathrm{d}x \tag{10.26}$$

for $m = 0, 1, 2, \ldots$ and

$$L_{n2\mathrm{o}m} = \frac{2\sqrt{2}mk_{\mathrm{p}}}{k_{\mathrm{s}}} \int_{-a}^{a} \frac{1}{x} \left(\mathrm{J}'_m(k_{\mathrm{p}}x) - \frac{1}{k_{\mathrm{p}}x}\mathrm{J}_m(k_{\mathrm{p}}x)\right) \psi_n(x/a)\,\mathrm{d}x \tag{10.27}$$

for $m = 1, 2, \ldots$.

10.2.4 Low frequency T-matrix

For low frequencies it is possible to obtain explicit expressions for the T-matrix elements. Starting with the symmetric part the Rayleigh function

$R(q) = (2q^2 - k_\mathrm{s}^2)^2 + 4q^2 h_\mathrm{s} h_\mathrm{p}$ is first estimated for large q (which in effect is the same as small $k_\mathrm{s} a$) by first estimating

$$\sqrt{k_\mathrm{s}^2 - q^2} \simeq \mathrm{i}q\left(1 - \frac{k_\mathrm{s}^2}{2q^2}\right), \qquad \sqrt{k_\mathrm{p}^2 - q^2} \simeq \mathrm{i}q\left(1 - \frac{k_\mathrm{p}^2}{2q^2}\right)$$

which leads to

$$R(q) \simeq -2q^2(k_\mathrm{s}^2 - k_\mathrm{p}^2)$$

It follows that [Eq. (10.11)]

$$\begin{aligned} Q_{nn'} &\simeq 4\mathrm{i}\pi n n'(-1)^n \frac{k_\mathrm{s}^2 - k_\mathrm{p}^2}{k_\mathrm{s}^2} \int_0^\infty \frac{1}{s}\,\mathrm{J}_n(s)\mathrm{J}_{n'}(s)\,\mathrm{d}s \\ &= \frac{2\mathrm{i}\pi n(-1)^n(k_\mathrm{s}^2 - k_\mathrm{p}^2)}{k_\mathrm{s}^2}\,\delta_{nn'} \end{aligned}$$

for $n + n'$ even and $Q_{nn'} = 0$ for $n + n'$ odd. The quantities in Eqs. (10.20) and (10.21) can be calculated for small $k_\mathrm{s} a$ by using the series for Bessel functions [Eq. (2.2)] keeping two terms for $m = 0$ and $m = 1$ and one term for $m = 2$. This gives for $\tau = 1$ [using Eqs. (10.5) or (10.6)]

$$L_{n1\mathrm{o}1} = \frac{(k_\mathrm{s} a)^2}{2\sqrt{2}} \int_{-1}^{1} s\,\psi_n(s)\,\mathrm{d}s = \frac{\mathrm{i}\pi}{8\sqrt{2}}(k_\mathrm{s} a)^2 \delta_{n2}$$

$$L_{n1\mathrm{o}2} = -\frac{k_\mathrm{s} a}{\sqrt{2}} \int_{-1}^{1} \psi_n(s)\,\mathrm{d}s = -\frac{\pi}{2\sqrt{2}} k_\mathrm{s} a\,\delta_{n1}$$

Similarly for $\tau = 2$

$$L_{n2\mathrm{e}0} = -\frac{\pi a}{2k_\mathrm{s}}\left(k_\mathrm{s}^2 - k_\mathrm{p}^2\right)\delta_{n1}$$

$$L_{n2\mathrm{e}1} = \frac{\mathrm{i}\pi k_\mathrm{p} a^2}{18\sqrt{2}k_\mathrm{s}}\left(3k_\mathrm{p}^2 - 2k_\mathrm{s}^2\right)\delta_{n2}$$

$$L_{n2\mathrm{e}2} = -\frac{\pi k_\mathrm{p}^2 a}{4\sqrt{2}k_\mathrm{s}}\delta_{n1}$$

It is seen that these quantities behave as $(k_\mathrm{s} a)^2$ for $m = 1$ but as $k_\mathrm{s} a$ for $m = 0$ and $m = 2$. This means that the T-matrix elements for $m = 1$

(including coupling elements to $m = 0, 2$) can be neglected for low frequencies. Physically this can be understood by the fact that for low frequencies $m = 1$ corresponds to rigid body motion of the obstacle. However, the crack has no mass, so the centre of mass of the crack just follows the incident wave and there is no scattering.

Thus, there are leading order T-matrix elements for $m = 0$ and $m = 2$ (monopole and quadrupole). The diagonal element for $m = 0$ becomes

$$T_{2e0,2e0} = \mathrm{i}(k_\mathrm{p}a)^2 \frac{\pi(\lambda + \mu)}{8\mu} \tag{10.28}$$

This element is exactly half of the corresponding element for a circular cavity in Eq. (2.38) (putting $\lambda_1 = \mu_1 = 0$). For $m = 2$ there are four elements

$$T_{1o2,1o2} = \mathrm{i}\left(\frac{k_\mathrm{s}^2 a}{k_\mathrm{p}}\right)^2 \frac{\pi\mu}{16(\lambda + \mu)}$$

$$T_{2e2,2e2} = \mathrm{i}(k_\mathrm{p}a)^2 \frac{\pi\mu}{16(\lambda + \mu)}$$

$$T_{1o2,2e2} = T_{2e2,1o2} = \mathrm{i}(k_\mathrm{s}a)^2 \frac{\pi\mu}{16(\lambda + \mu)} \tag{10.29}$$

These elements are exactly one fourth of the corresponding elements for the cavity in Eqs. (2.40). Finally, there are elements that have no counterpart for the circular cavity, namely elements coupling $m = 0$ and $m = 2$

$$T_{2e0,1o2} = T_{1o2,2e0} = \mathrm{i}(k_\mathrm{s}a)^2 \frac{\pi}{8\sqrt{2}}$$

$$T_{2e0,2e2} = T_{2e2,2e0} = \mathrm{i}(k_\mathrm{p}a)^2 \frac{\pi}{8\sqrt{2}} \tag{10.30}$$

For the antisymmetric part the procedure is the same. The matrix P becomes identical to the matrix Q for low frequencies and the other quantities needed are

$$L_{n1e0} = \frac{(k_\mathrm{s}a)^3}{8} \int_{-1}^{1} s^2 \psi_n(s)\,\mathrm{d}s$$

$$L_{n1e1} = \frac{\mathrm{i}\pi}{8\sqrt{2}} (k_\mathrm{s}a)^2 \delta_{n2}$$

$$L_{n1e2} = -\frac{\pi k_s a}{2\sqrt{2}}\delta_{n1}$$

$$L_{n2o1} = -\frac{i\pi k_p^3 a^2}{8\sqrt{2}k_s}\delta_{n2}$$

$$L_{n2o2} = \frac{\pi k_p^2 a}{2\sqrt{2}k_s}\delta_{n1}$$

The value of the remaining integral does not matter, the result is anyway that to lowest order only $m = 2$ contributes to the scattering. These T-matrix elements are in fact exactly the same as in the symmetric case

$$T_{1e2,1e2} = i\left(\frac{k_s^2 a}{k_p}\right)^2 \frac{\pi\mu}{16(\lambda+\mu)}$$

$$T_{2o2,2o2} = i(k_p a)^2 \frac{\pi\mu}{16(\lambda+\mu)}$$

$$T_{1e2,2o2} = T_{2o2,1e2} = i(k_s a)^2 \frac{\pi\mu}{16(\lambda+\mu)} \tag{10.31}$$

That $m = 0$, which for the antisymmetric part is for $\tau = 1$ (the torsional wave), does not contribute to leading order corresponds to the fact that the torsional wave for the circular cavity does not contribute either.

10.2.5 Other boundary conditions

The crack is assumed open in the previous calculations, this by far being the most common case. It is straightforward to consider spring boundary conditions. A particular spring boundary condition is a fluid-filled crack, for which the boundary conditions can be taken as the slip boundary conditions. This means that the normal displacement and traction are continuous and the tangential tractions are vanishing from both sides. For this case the symmetric part gives no scattering at all, so that the corresponding part of the T-matrix vanishes, while the antisymmetric part is unaffected.

10.3 Strip-like crack: 3D waves

Turning to 3D, the simplest case is the strip-like crack, i.e. a crack that is finite in one direction and infinite in the other. It seems that this problem is not considered in the literature, but it is a special case of a strip-like crack in a half-space (Bövik and Boström [20]). A Cartesian coordinate system (x, y, z) is introduced so the crack is defined by $|x| < a$, $-\infty < y < \infty$, and $z = 0$. The approach is very similar to the previous two sections, the main change is the coupling between SH and P-SV waves, which has the consequence that the antisymmetric part becomes a 2D vector problem.

10.3.1 Symmetric and antisymmetric parts

As for the strip-like crack in 2D it is useful to divide the problem into a symmetric and an antisymmetric part. For the symmetric part u_x and u_y are even functions of z and u_z is an odd function of z and vice versa for the antisymmetric part. It follows that σ_{zz} is even and σ_{zx} and σ_{yz} are odd for the symmetric part and vice versa for the antisymmetric part. It is then sufficient to consider the half-space $z > 0$. For the symmetric part the boundary conditions for $z = 0$ and $-\infty < y < \infty$ become

$$\sigma_{zx}^{\text{sc}} = \sigma_{yz}^{\text{sc}} = 0, \qquad \text{all } x$$

$$u_z^{\text{sc}} = 0, \qquad |x| > a$$

$$\sigma_{zz}^{\text{sc}} = -\sigma_{zz}^{\text{in}}, \qquad |x| < a$$

and for the antisymmetric part they become

$$\sigma_{zz}^{\text{sc}} = 0, \qquad \text{all } x$$

$$u_x^{\text{sc}} = u_y^{\text{sc}} = 0, \qquad |x| > a$$

$$\sigma_{zx} = -\sigma_{zx}^{\text{in}}, \qquad |x| < a$$

$$\sigma_{yz} = -\sigma_{yz}^{\text{in}}, \qquad |x| < a$$

Solutions for $z > 0$ can be constructed with the help of the recipe in Eq. (1.18)

$$\boldsymbol{u} = \nabla \times (\hat{\boldsymbol{z}}\Psi_1) + \nabla \times (\nabla \times (\hat{\boldsymbol{z}}\Psi_2)) + \nabla\Phi$$

where Ψ_1 and Ψ_2 satisfy scalar Helmholtz equations with wavenumber k_{s} and Φ satisfies Helmholtz equation with wavenumber k_{p}. These solutions to

Helmholtz equation are plane waves and it is thus realized that the solution for $z > 0$ can be represented as a double Fourier transform in both x and y

$$\boldsymbol{u}^{\mathrm{sc}} = \frac{1}{2\pi}\int_{-\infty}^{\infty}\int_{-\infty}^{\infty}\Big[f_1\nabla\times\left(\hat{\boldsymbol{z}}\,\mathrm{e}^{\mathrm{i}(qx+py+h_{\mathrm{s}}y)}\right)$$

$$+f_2\nabla\times\left[\nabla\times\left(\hat{\boldsymbol{z}}\,\mathrm{e}^{\mathrm{i}(qx+py+h_{\mathrm{s}}y)}\right)\right]+f_3\nabla\left(\mathrm{e}^{\mathrm{i}(qx+py+h_{\mathrm{p}}z)}\right)\Big]\mathrm{d}q\,\mathrm{d}p$$

Here the wavenumbers in the z-direction are $h_{\mathrm{s}} = (k_{\mathrm{s}}^2 - q^2 - p^2)^{1/2}$ and $h_{\mathrm{p}} = (k_{\mathrm{p}}^2 - q^2 - p^2)^{1/2}$, where $\operatorname{Im} h_{\mathrm{s}} \geq 0$ and $\operatorname{Im} h_{\mathrm{p}} \geq 0$. The unknown amplitudes are functions of the transform variables $f_i = f_i(q, p)$, $i = 1, 2, 3$. As the geometry is independent of the y-coordinate it is possible to perform the whole analysis in the Fourier domain of y and the transform variable p then acts as a parameter [the factor $1/2\pi$ is inserted to conform with the definition of the Fourier inverse transform in Eq. (4.9)]. In this Fourier space the displacement is thus written (omitting the superscript "sc")

$$\tilde{\boldsymbol{u}}(x,p,z) = \int_{-\infty}^{\infty}\left[(\mathrm{i}f_1(\hat{\boldsymbol{x}}p - \hat{\boldsymbol{y}}q)\right.$$

$$\left.+f_2(-\hat{\boldsymbol{x}}qh_{\mathrm{s}} - \hat{\boldsymbol{y}}ph_{\mathrm{s}} + \hat{\boldsymbol{z}}s^2))\,\mathrm{e}^{\mathrm{i}h_{\mathrm{s}}z} + \mathrm{i}f_3(\hat{\boldsymbol{x}}q + \hat{\boldsymbol{y}}p + \hat{\boldsymbol{z}}h_{\mathrm{p}})\,\mathrm{e}^{\mathrm{i}h_{\mathrm{p}}z}\right]\mathrm{e}^{\mathrm{i}qx}\,\mathrm{d}q$$

where $s^2 = q^2 + p^2$. The corresponding normal traction on $z = 0$ is also calculated

$$\tilde{\boldsymbol{t}}^{(z)}(x,p,0) = \mu\int_{-\infty}^{\infty}\left[f_1h_{\mathrm{s}}(-\hat{\boldsymbol{x}}p + \hat{\boldsymbol{y}}q)\right.$$

$$+\mathrm{i}f_2\left(\hat{\boldsymbol{x}}q(k_{\mathrm{s}}^2 - 2h_{\mathrm{s}}^2) + \hat{\boldsymbol{y}}p(k_{\mathrm{s}}^2 - 2h_{\mathrm{s}}^2) + \hat{\boldsymbol{z}}2h_{\mathrm{s}}s^2\right)$$

$$\left.+f_3(-\hat{\boldsymbol{x}}2qh_{\mathrm{p}} - \hat{\boldsymbol{y}}2ph_{\mathrm{p}} + \hat{\boldsymbol{z}}(2k_{\mathrm{p}}^2 - 2h_{\mathrm{p}}^2 - k_{\mathrm{s}}^2))\right]\mathrm{e}^{\mathrm{i}qx}\,\mathrm{d}q$$

10.3.2 Symmetric integral equation

For the symmetric part the first two boundary conditions give

$$f_1 = 0$$

$$f_3 = \frac{\mathrm{i}(k_{\mathrm{s}}^2 - 2h_{\mathrm{s}}^2)}{2h_{\mathrm{p}}}f_2$$

The normal displacement at $z = 0$ can then be written

$$\tilde{u}_z(x,p,0) = \frac{k_{\mathrm{s}}^2}{2}\int_{-\infty}^{\infty} f_2\,\mathrm{e}^{\mathrm{i}qx}\,\mathrm{d}q = \begin{cases} 0, & |x| > a \\ \frac{1}{2}\Delta\tilde{u}_z(x,p), & |x| < a \end{cases}$$

where the second boundary condition is used. Inverting the Fourier transform this gives

$$f_2 = \frac{1}{2\pi k_{\mathrm{s}}^2}\int_{-a}^{a} \Delta\tilde{u}_z(x,p)\,\mathrm{e}^{-\mathrm{i}qx}\,\mathrm{d}x$$

Eliminating f_3 the last boundary condition gives

$$\tilde{\sigma}_{zz}(x,p,0) = \frac{\mathrm{i}\mu}{2}\int_{-\infty}^{\infty} f_2\frac{R(s)}{h_{\mathrm{p}}}\,\mathrm{e}^{\mathrm{i}qx}\,\mathrm{d}q = -\tilde{\sigma}_{zz}^{\mathrm{in}}(x,p,0)$$

where the Rayleigh function $R(s) = 4s^2h_{\mathrm{s}}h_{\mathrm{p}} + (k_{\mathrm{s}}^2 - 2s^2)^2$ enters. With f_2 inserted this is an integral equation for the normal COD.

To solve the integral equation the normal COD is expanded in the Chebyshev functions exactly as in the 2D cases

$$\Delta\tilde{u}_z(x,p) = \sum_{n=1}^{\infty}\alpha_n(p)\psi_n(x/a)$$

This gives

$$f_2 = \frac{1}{2k_{\mathrm{s}}^2}\sum_{n=1}^{\infty}\alpha_n(p)\frac{n}{q}\,\mathrm{J}_n(qa)$$

Inserting this into the integral equation and projecting the integral equation on the Chebychev functions finally give

$$\sum_{n'=1}^{\infty} Q_{nn'}(p)\alpha_{n'}(p) = -\frac{4\mathrm{i}}{\pi\mu}\int_{-a}^{a}\tilde{\sigma}_{zz}^{\mathrm{in}}(x,p,0)\psi_n(x/a)\,\mathrm{d}x \tag{10.32}$$

where

$$Q_{nn'}(p) = nn'(-1)^n\int_{-\infty}^{\infty}\frac{R(s)}{k_{\mathrm{s}}^2q^2h_{\mathrm{p}}}\,\mathrm{J}_n(qa)\mathrm{J}_{n'}(qa)\,\mathrm{d}q \tag{10.33}$$

This solution is almost identical to the corresponding solution for the symmetric integral equation in the 2D P-SV case.

10.3.3 Antisymmetric integral equations

The first boundary condition for the antisymmetric problem gives

$$f_3 = -\frac{2\mathrm{i}h_\mathrm{s}s^2}{2s^2 - k_\mathrm{s}^2}\, f_2$$

The two tangential displacement components at $z = 0$ then become

$$\begin{aligned}
\tilde{u}_x(x,p,0) &= \int_{-\infty}^{\infty} \left[\mathrm{i}f_1 p + f_2\,\frac{k_\mathrm{s}^2 q h_\mathrm{s}}{2s^2 - k_\mathrm{s}^2}\right] \mathrm{e}^{\mathrm{i}qx}\,\mathrm{d}q \\
&= \begin{cases} 0, & |x| > a \\ \frac{1}{2}\Delta\tilde{u}_x(x,p), & |x| < a \end{cases} \\
\tilde{u}_y(x,p,0) &= \int_{-\infty}^{\infty} \left[-\mathrm{i}f_1 q + f_2\,\frac{k_\mathrm{s}^2 p h_\mathrm{s}}{2s^2 - k_\mathrm{s}^2}\right] \mathrm{e}^{\mathrm{i}qx}\,\mathrm{d}q \\
&= \begin{cases} 0, & |x| > a \\ \frac{1}{2}\Delta\tilde{u}_y(x,p), & |x| < a \end{cases}
\end{aligned}$$

where the second boundary condition is used. Inverting the Fourier transforms and solving for f_1 and f_2 give

$$f_1 = -\frac{\mathrm{i}}{4\pi s^2}\left[p\int_{-a}^{a} \Delta\tilde{u}_x(x,p)\mathrm{e}^{-\mathrm{i}qx}\,\mathrm{d}x - q\int_{-a}^{a} \Delta\tilde{u}_y(x,p)\mathrm{e}^{-\mathrm{i}qx}\,\mathrm{d}x\right]$$

$$f_2 = \frac{2s^2 - k_\mathrm{s}^2}{4\pi k_\mathrm{s}^2 h_\mathrm{s} s^2}\left[q\int_{-a}^{a} \Delta\tilde{u}_x(x,p)\mathrm{e}^{-\mathrm{i}qx}\,\mathrm{d}x + p\int_{-a}^{a} \Delta\tilde{u}_y(x,p)\mathrm{e}^{-\mathrm{i}qx}\,\mathrm{d}x\right]$$

Eliminating f_3 the two tangential traction components at $z = 0$ become

$$\tilde{\sigma}_{zx}(x,p,0) = \mu\int_{-\infty}^{\infty} \left[-f_1 p h_\mathrm{s} + \mathrm{i}f_2\,\frac{qR(s)}{k_\mathrm{s}^2 - 2h_\mathrm{s}^2}\right] \mathrm{e}^{\mathrm{i}qx}\,\mathrm{d}q = -\tilde{\sigma}_{zx}^{\mathrm{in}}(x,p,0)$$

$$\tilde{\sigma}_{yz}(x,p,0) = \mu\int_{-\infty}^{\infty} \left[f_1 q h_\mathrm{s} + \mathrm{i}f_2\,\frac{pR(s)}{k_\mathrm{s}^2 - 2h_\mathrm{s}^2}\right] \mathrm{e}^{\mathrm{i}qx}\,\mathrm{d}q = -\tilde{\sigma}_{yz}^{\mathrm{in}}(x,p,0)$$

With f_1 and f_2 inserted this is two coupled integral equations for the two COD components $\Delta\tilde{u}_x$ and $\Delta\tilde{u}_y$.

To solve the integral equations the COD components are expanded in the Chebyshev functions

$$\Delta\tilde{u}_x(x,p) = \sum_{n=1}^{\infty} \beta_n(p)\psi_n(x/a)$$

$$\Delta\tilde{u}_y(x,p) = \sum_{n=1}^{\infty} \gamma_n(p)\psi_n(x/a)$$

Inserting and projecting onto the Chebyshev functions give

$$\begin{aligned}
&\sum_{n'=1}^{\infty} \left(P^1_{nn'}(p)\beta_{n'}(p) + P^3_{nn'}(p)\gamma_{n'}(p)\right) \\
&\qquad = -\frac{4\mathrm{i}}{\pi\mu}\int_{-a}^{a} \tilde{\sigma}^{\mathrm{in}}_{zx}(x,p,0)\psi_n(x/a)\,\mathrm{d}x \\
&\sum_{n'=1}^{\infty} \left(P^3_{nn'}(p)\beta_{n'}(p) + P^2_{nn'}(p)\gamma_{n'}(p)\right) \\
&\qquad = -\frac{4\mathrm{i}}{\pi\mu}\int_{-a}^{a} \tilde{\sigma}^{\mathrm{in}}_{yz}(x,p,0)\psi_n(x/a)\,\mathrm{d}x
\end{aligned} \tag{10.34}$$

where

$$\begin{aligned}
P^1_{nn'}(p) &= nn'(-1)^n \int_{-\infty}^{\infty} \left(\frac{p^2 h_{\mathrm{s}}}{q^2 s^2} + \frac{R}{k_{\mathrm{s}}^2 s^2 h_{\mathrm{s}}}\right) \mathrm{J}_n(qa)\mathrm{J}_{n'}(qa)\,\mathrm{d}q \\
P^3_{nn'}(p) &= nn'(-1)^n \int_{-\infty}^{\infty} \left(-\frac{p h_{\mathrm{s}}}{q s^2} + \frac{pR}{k_{\mathrm{s}}^2 q s^2 h_{\mathrm{s}}}\right) \mathrm{J}_n(qa)\mathrm{J}_{n'}(qa)\,\mathrm{d}q \\
P^2_{nn'}(p) &= nn'(-1)^n \int_{-\infty}^{\infty} \left(\frac{h_{\mathrm{s}}}{s^2} + \frac{p^2 R}{k_{\mathrm{s}}^2 q^2 s^2 h_{\mathrm{s}}}\right) \mathrm{J}_n(qa)\mathrm{J}_{n'}(qa)\,\mathrm{d}q
\end{aligned} \tag{10.35}$$

Solving these equations for β_n and γ_n, $n = 1, 2, \ldots$, in turn determines the f_i, $i = 1, 2, 3$, and thereby the scattered wave. As for the strip-like crack in 2D it is possible to determine the T-matrix (with respect to the cylindrical wavefunctions of Section 2.3.1).

10.4 Rectangular crack

The rectangular crack can be treated in a way very similar to the strip-like one. It is possible to determine the T-matrix, but as this involves integration of spherical vector wavefunctions over the rectangular crack surface, this becomes rather cumbersome. The rectangular crack is treated by Guan and Norris [57] using the Green function approach, see also Itou [68] for some special cases.

10.4.1 Symmetric and antisymmetric parts

As for the strip-like crack in 3D it is useful to divide the problem into a symmetric and an antisymmetric part. A Cartesian coordinate system (x, y, z) is introduced so the crack is defined by $|x| < a$, $|y| < b$, and $z = 0$. For the symmetric part u_x and u_y are even functions of z and u_z is an odd function of z and vice versa for the antisymmetric part. It follows that σ_{zz} is even and σ_{zx} and σ_{yz} are odd for the symmetric part and vice versa for the antisymmetric part. It is then sufficient to consider the half-space $z > 0$. For the symmetric part the boundary conditions for $z = 0$ become

$$\sigma_{zx}^{\mathrm{sc}} = \sigma_{yz}^{\mathrm{sc}} = 0, \qquad \text{all } x \text{ and } y$$

$$u_z^{\mathrm{sc}} = 0, \qquad |x| > a \text{ or } |y| > b$$

$$\sigma_{zz}^{\mathrm{sc}} = -\sigma_{zz}^{\mathrm{in}}, \qquad |x| < a \text{ and } |y| < b$$

and for the antisymmetric part they become

$$\sigma_{zz}^{\mathrm{sc}} = 0, \qquad \text{all } x \text{ and } y$$

$$u_x^{\mathrm{sc}} = u_y^{\mathrm{sc}} = 0, \qquad |x| > a \text{ or } |y| > b$$

$$\sigma_{zx} = -\sigma_{zx}^{\mathrm{in}}, \qquad |x| < a \text{ and } |y| < b$$

$$\sigma_{yz} = -\sigma_{yz}^{\mathrm{in}}, \qquad |x| < a \text{ and } |y| < b$$

Solutions for $z > 0$ can be constructed as in the previous section with the only difference that the representation now is in real space (not in Fourier space in the y-coordinate). The displacement is thus

$$\boldsymbol{u}(x,y,z) = \int_{-\infty}^{\infty}\int_{-\infty}^{\infty} [(\mathrm{i}f_1(\hat{\boldsymbol{x}}p - \hat{\boldsymbol{y}}q) + f_2(-\hat{\boldsymbol{x}}qh_\mathrm{s} - \hat{\boldsymbol{y}}ph_\mathrm{s}$$

$$+\hat{\boldsymbol{z}}s^2))\,\mathrm{e}^{\mathrm{i}h_\mathrm{s}z} + \mathrm{i}f_3(\hat{\boldsymbol{x}}q + \hat{\boldsymbol{y}}p + \hat{\boldsymbol{z}}h_\mathrm{p})\mathrm{e}^{\mathrm{i}h_\mathrm{p}z}\big]\,\mathrm{e}^{\mathrm{i}(qx+py)}\,\mathrm{d}q\,\mathrm{d}p$$

where $s^2 = q^2 + p^2$, $h_\mathrm{s} = (k_\mathrm{s}^2 - s^2)^{1/2}$, and $h_\mathrm{p} = (k_\mathrm{p}^2 - s^2)^{1/2}$ with $\operatorname{Im} h_\mathrm{s} \geq 0$ and $\operatorname{Im} h_\mathrm{p} \geq 0$. The corresponding normal traction on $z = 0$ is also calculated

$$\boldsymbol{t}^{(z)}(x,y,0) = \mu \int_{-\infty}^{\infty}\int_{-\infty}^{\infty} [f_1 h_\mathrm{s}(-\hat{\boldsymbol{x}}p + \hat{\boldsymbol{y}}q)$$

$$+\mathrm{i}f_2\left(-\hat{\boldsymbol{x}}q(k_\mathrm{s}^2 - 2h_\mathrm{s}^2) - \hat{\boldsymbol{y}}p(k_\mathrm{s}^2 - 2h_\mathrm{s}^2) + \hat{\boldsymbol{z}}2h_\mathrm{s}s^2\right)$$

$$+f_3(-\hat{\boldsymbol{x}}2qh_\mathrm{p} - \hat{\boldsymbol{y}}2ph_\mathrm{p} + \hat{\boldsymbol{z}}(2k_\mathrm{p}^2 - 2h_\mathrm{p}^2 - k_\mathrm{s}^2))\big]\,\mathrm{e}^{\mathrm{i}(qx+py)}\,\mathrm{d}q\,\mathrm{d}p$$

10.4.2 Integral equations

The derivation and solution of the symmetric and antisymmetric integral equations follow the same steps as in the previous section. The symmetric integral equation thus becomes

$$\frac{\mathrm{i}\mu}{2}\int_{-\infty}^{\infty}\int_{-\infty}^{\infty} f_2\frac{R(s)}{h_\mathrm{p}}\,\mathrm{e}^{\mathrm{i}(qx+py)}\,\mathrm{d}q\,\mathrm{d}p = -\sigma_{zz}^{\mathrm{in}}(x,y,0)$$

where

$$f_2 = \frac{1}{4\pi^2 k_\mathrm{s}^2}\int_{-a}^{a}\int_{-b}^{b} \Delta u_z(x,y)\mathrm{e}^{-\mathrm{i}(qx+py)}\,\mathrm{d}x\,\mathrm{d}y$$

To solve the integral equation the normal COD is expanded in the Chebyshev functions

$$\Delta u_z(x,p) = \sum_{n=1}^{\infty}\sum_{j=1}^{\infty} \alpha_{nj}\psi_n(x/a)\psi_j(y/b)$$

It should noted that these expansions have the expected behaviour along the edges of the crack, but no measures are taken for the stronger singularities at the corners of the crack, see Helsing *et al.* [63] for the corner singularity for the static mode I case and further references. Inserting this expansion and projecting on the Chebyshev functions give

$$\sum_{n'=1}^{\infty}\sum_{j'=1}^{\infty} Q_{njn'j'}(p)\alpha_{n'j'}$$
$$= \frac{8\mathrm{i}k_{\mathrm{s}}}{\pi^2\mu}\int_{-a}^{a}\int_{-b}^{b}\sigma_{zz}^{\mathrm{in}}(x,y,0)\psi_n(x/a)\psi_j(y/b)\,\mathrm{d}x\,\mathrm{d}y \qquad (10.36)$$

where

$$Q_{njn'j'}(p) = nn'(-1)^{n+j}$$
$$\times\int_{-\infty}^{\infty}\int_{-\infty}^{\infty}\frac{R(s)}{k_{\mathrm{s}}q^2p^2h_{\mathrm{p}}}\,\mathrm{J}_n(qa)\mathrm{J}_{n'}(qa)\mathrm{J}_j(pb)\mathrm{J}_{j'}(pb)\,\mathrm{d}q\,\mathrm{d}p \quad (10.37)$$

For the antisymmetric problem the integral equations become

$$\mu\int_{-\infty}^{\infty}\int_{-\infty}^{\infty}\left(-f_1ph_{\mathrm{s}}+\mathrm{i}f_2\frac{qR(s)}{k_{\mathrm{s}}^2-2h_{\mathrm{s}}^2}\right)\mathrm{e}^{\mathrm{i}(qx+py)}\,\mathrm{d}q\,\mathrm{d}p = -\sigma_{zx}^{\mathrm{in}}(x,y,0)$$

$$\mu\int_{-\infty}^{\infty}\int_{-\infty}^{\infty}\left(f_1qh_{\mathrm{s}}+\mathrm{i}f_2\frac{pR(s)}{k_{\mathrm{s}}^2-2h_{\mathrm{s}}^2}\right)\mathrm{e}^{\mathrm{i}(qx+py)}\,\mathrm{d}q\,\mathrm{d}p = -\sigma_{yz}^{\mathrm{in}}(x,y,0)$$

where

$$f_1 = -\frac{\mathrm{i}}{8\pi^2s^2}\int_{-a}^{a}\int_{-b}^{b}(p\Delta u_x(x,y)-q\Delta u_y(x,y))\,\mathrm{e}^{-\mathrm{i}(qx+py)}\,\mathrm{d}x\,\mathrm{d}y$$

$$f_2 = -\frac{k_{\mathrm{s}}^2-2h_{\mathrm{s}}^2}{8\pi^2k_{\mathrm{s}}^2s^2h_{\mathrm{s}}}\int_{-a}^{a}\int_{-b}^{b}(q\Delta u_x(x,y)+p\Delta u_y(x,y))\,\mathrm{e}^{-\mathrm{i}(qx+py)}\,\mathrm{d}x\,\mathrm{d}y$$

To solve the integral equations the tangential CODs are expanded in the Chebyshev functions

$$\Delta u_x(x,y) = \sum_{n=1}^{\infty}\sum_{j=1}^{\infty}\beta_{nj}\psi_n(x/a)\psi_j(y/b)$$

$$\Delta u_y(x,y) = \sum_{n=1}^{\infty}\sum_{j=1}^{\infty}\gamma_{nj}\psi_n(x/a)\psi_j(y/b)$$

Insertion and projection then gives the system of equations

$$\sum_{n'=1}^{\infty}\sum_{j'=1}^{\infty}\left(P^3_{njn'j'}\beta_{n'j'}+P^2_{njn'j'}\gamma_{n'j'}\right)$$
$$=\frac{8\mathrm{i}k_{\mathrm{s}}}{\pi^2\mu}\int_{-a}^{a}\int_{-b}^{b}\sigma^{\mathrm{in}}_{yz}(x,y,0)\psi_n(x/a)\psi_j(y/b)\,\mathrm{d}x\,\mathrm{d}y$$
$$\sum_{n'=1}^{\infty}\sum_{j'=1}^{\infty}\left(P^1_{njn'j'}\beta_{n'j'}+P^3_{njn'j'}\gamma_{n'j'}\right)$$
$$=\frac{8\mathrm{i}k_{\mathrm{s}}}{\pi^2\mu}\int_{-a}^{a}\int_{-b}^{b}\sigma^{\mathrm{in}}_{zx}(x,y,0)\psi_n(x/a)\psi_j(y/b)\,\mathrm{d}x\,\mathrm{d}y \quad (10.38)$$

where

$$P^1_{njn'j'}=nn'(-1)^{n+j}\int_{-\infty}^{\infty}\int_{-\infty}^{\infty}\left(\frac{k_{\mathrm{s}}h_{\mathrm{s}}}{q^2s^2}+\frac{R}{k_{\mathrm{s}}p^2s^2h_{\mathrm{s}}}\right)I_{njn'j'}\,\mathrm{d}q\,\mathrm{d}p$$
$$P^2_{njn'j'}=nn'(-1)^{n+j}\int_{-\infty}^{\infty}\int_{-\infty}^{\infty}\left(\frac{k_{\mathrm{s}}h_{\mathrm{s}}}{p^2s^2}+\frac{R}{k_{\mathrm{s}}q^2s^2h_{\mathrm{s}}}\right)I_{njn'j'}\,\mathrm{d}q\,\mathrm{d}p$$
$$P^3_{njn'j'}=nn'(-1)^{n+j}\int_{-\infty}^{\infty}\int_{-\infty}^{\infty}\left(-\frac{k_{\mathrm{s}}h_{\mathrm{s}}}{qps^2}+\frac{R}{k_{\mathrm{s}}qps^2h_{\mathrm{s}}}\right)I_{njn'j'}\,\mathrm{d}q\,\mathrm{d}p$$

where the product of four Bessel functions is

$$I_{njn'j'}=\mathrm{J}_n(qa)\mathrm{J}_{n'}(qa)\mathrm{J}_j(pb)\mathrm{J}_{j'}(pb)$$

All the expansion coefficients for the CODs can now be determined and this in turn determines the scattered wave.

10.5 Circular crack

Another crack in 3D that is possible to consider with a singular integral equation approach is the circular crack. The crack is assumed open, but as for the 2D P-SV strip-like crack it is possible to obtain the solution for the fluid-filled crack by letting the symmetric part of the T-matrix vanish. The derivation here follows that of Krenk and Schmidt [77]. Their work is extended to spring boundary conditions on the crack and the calculation of the T-matrix by Boström and Eriksson [23].

The solution procedure uses the Hankel transform (sometimes called the Fourier-Hankel transform). The Hankel transform of order m of a function $f(r)$ of the radial coordinate $0 \leq r < \infty$ is defined as

$$\hat{f}_m(\xi) = \int_0^\infty f(r) J_m(\xi r) r \, \mathrm{d}r \tag{10.39}$$

It is symmetric so its inverse is

$$f(r) = \int_0^\infty \hat{f}_m(\xi) J_m(\xi r) \xi \, \mathrm{d}\xi \tag{10.40}$$

This transform is useful when a transform is needed for the radial coordinate in polar and cylindrical coordinates.

10.5.1 Symmetric and antisymmetric parts

Due to symmetry the scattering by a circular crack with radius a can be divided into a symmetric and an antisymmetric part. It is natural to employ cylindrical coordinates (r, φ, z) with the crack situated at $z = 0$, $r < a$. For the symmetric part u_z is an odd function of z and u_r and u_φ are even functions of z and vice versa for the antisymmetric part. It follows that σ_{zz} is even and σ_{zr} and $\sigma_{\varphi z}$ are odd for the symmetric part and vice versa for the antisymmetric part. It is then sufficient to consider the half-space $z > 0$. For the symmetric part the boundary conditions for the scattered wave for $z = 0$ become

$$\begin{aligned}
&\sigma_{zr}^{\mathrm{sc}} = \sigma_{\varphi z}^{\mathrm{sc}} = 0, \qquad \text{all } r \\
&u_z^{\mathrm{sc}} = 0, \qquad r > a \\
&\sigma_{zz}^{\mathrm{sc}} = -\sigma_{zz}^{\mathrm{in}}, \qquad r < a
\end{aligned}$$

and for the antisymmetric part they become

$$\begin{aligned}
&\sigma_{zz}^{\mathrm{sc}} = 0, \qquad \text{all } r \\
&u_r^{\mathrm{sc}} = u_\varphi^{\mathrm{sc}} = 0, \qquad r > a \\
&\sigma_{zr}^{\mathrm{sc}} = -\sigma_{zr}^{\mathrm{in}}, \qquad r < a
\end{aligned}$$

$$\sigma_{\varphi z}^{\mathrm{sc}} = -\sigma_{\varphi z}^{\mathrm{in}}, \qquad r < a$$

The symmetric part is a scalar problem determined by the normal traction of the incident wave, whereas the antisymmetric part is a vector problem determined by the two tangential traction components of the incident wave.

The cylindrical vector wavefunctions are defined in Section 2.3.1 and as commented there it can be relevant to take a common radial wavenumber q and correspondingly two different wavenumbers in the z-direction. This is the case here and this in effect corresponds to a Hankel transform. The following ansatz is thus made for the scattered wave

$$\begin{aligned}\boldsymbol{u}^{\mathrm{sc}}(\boldsymbol{r}) = \sum_{m=0}^{\infty} \int_0^{\infty} \Bigg[& f_{1m} \nabla \times \left(\hat{\boldsymbol{z}} \frac{1}{q} \mathrm{J}_m(qr) \sin m\varphi \, \mathrm{e}^{\mathrm{i}h_{\mathrm{s}} z} \right) \\ & + f_{2m} \nabla \times \nabla \times \left(\hat{\boldsymbol{z}} \frac{\mathrm{i}}{q} \mathrm{J}_m(qr) \cos m\varphi \, \mathrm{e}^{\mathrm{i}h_{\mathrm{s}} z} \right) \\ & + f_{3m} \nabla \left(\mathrm{J}_m(qr) \cos m\varphi \, \mathrm{e}^{\mathrm{i}h_{\mathrm{p}} z} \right) \Bigg] q \, \mathrm{d}q\end{aligned}$$

Here the three unknowns f_{1m}, f_{2m}, and f_{3m} depend on the variable q and $h_{\mathrm{s}} = (k_{\mathrm{s}}^2 - q^2)^{1/2}$ and $h_{\mathrm{p}} = (k_{\mathrm{p}}^2 - q^2)^{1/2}$ with $\operatorname{Im} h_{\mathrm{s}} \geq 0$ and $\operatorname{Im} h_{\mathrm{p}} \geq 0$. Compared to the wavefunctions in Section 2.3.1 the normalization has been left out and only the symmetric part in φ is included. The normalization is not needed when the T-matrix is calculated (as T is given with respect to the spherical vector wavefunctions) and the antisymmetric part in φ only gives some sign changes, which are stated at the end.

When the action of the nabla operator is written out, there appear derivatives of the Bessel functions and this prevents a direct inversion of the Hankel transform. However, taking the sum and difference of the r- and φ-components and using recurrence relations for Bessel functions the expansion can be written as

$$\boldsymbol{u}^{\mathrm{sc}}(\boldsymbol{r}) = \sum_{m=0}^{\infty} \left[(\hat{\boldsymbol{r}} U_{rm} + \hat{\boldsymbol{z}} U_{zm}) \cos m\varphi + \hat{\boldsymbol{\varphi}} U_{\varphi m} \sin m\varphi \right]$$

where

$$U_{rm} \pm U_{\varphi m} = \int_0^{\infty} \left[(f_{1m} \pm h_{\mathrm{s}} f_{2m}) \, \mathrm{e}^{\mathrm{i}h_{\mathrm{s}} z} \mp q f_{3m} \mathrm{e}^{\mathrm{i}h_{\mathrm{p}} z} \right] \mathrm{J}_{m\pm 1}(qr) q \, \mathrm{d}q$$

$$U_{zm} = \mathrm{i}\int_0^\infty \left[q f_{2m} \mathrm{e}^{\mathrm{i}h_\mathrm{s} z} + h_\mathrm{p} f_{3m} \mathrm{e}^{\mathrm{i}h_\mathrm{p} z} \right] \mathrm{J}_m(qr) q \, \mathrm{d}q$$

The corresponding traction on $z = 0$ is straightforward to calculate

$$\boldsymbol{t}^{\mathrm{sc}}(\boldsymbol{r}) = \sum_{m=0}^{\infty} \left[(\hat{\boldsymbol{r}} T_{rm} + \hat{\boldsymbol{z}} T_{zm}) \cos m\varphi + \hat{\boldsymbol{\varphi}} T_{\varphi m} \sin m\varphi \right]$$

where

$$T_{rm} \pm T_{\varphi m} = \mathrm{i}\mu \int_0^\infty \left[h_\mathrm{s} f_{1m} \mp (2q^2 - k_\mathrm{s}^2) f_{2m} \mp 2q h_\mathrm{p} f_{3m} \right] \mathrm{J}_{m\pm 1}(qr) q \, \mathrm{d}q$$

$$T_{zm} = \mu \int_0^\infty \left[-2q h_\mathrm{s} f_{2m} + (2q^2 - k_\mathrm{s}^2) f_{3m} \right] \mathrm{J}_m(qr) q \, \mathrm{d}q$$

10.5.2 Symmetric integral equation

In this subsection the integral equation for the simpler symmetric part is derived and solved. The first boundary conditions immediately give

$$f_{1m} = 0$$

$$f_{3m} = \frac{k_\mathrm{s}^2 - 2q^2}{2q h_\mathrm{p}} f_{2m}$$

The z-component of the partial wave displacement at $z = 0$ can then be written

$$U_{zm} = \mathrm{i}\int_0^\infty \frac{k_\mathrm{s}^2}{2q} f_{2m} \mathrm{J}_m(qr) q \, \mathrm{d}q = \begin{cases} 0, & r > a \\ \frac{1}{2}\Delta U_{zm}(r), & r < a \end{cases}$$

where the second boundary condition is used. The normal partial wave COD $\Delta U_{zm}(r)$ is here introduced. Inversion of the Hankel transform gives

$$f_{2m} = -\frac{\mathrm{i}q}{k_\mathrm{s}^2} \int_0^a \Delta U_{zm}(r) \mathrm{J}_m(qr) r \, \mathrm{d}r$$

Putting everything together the last boundary condition gives

$$
\begin{aligned}
-\mathrm{i}\int_0^\infty \frac{R(q)}{2k_\mathrm{s}^3 h_\mathrm{p}} \int_0^a \Delta U_{zm}(r')\mathrm{J}_m(qr')r'\,\mathrm{d}r'\,\mathrm{J}_m(qr)q\,\mathrm{d}q \\
= \frac{1}{\mu k_\mathrm{s}} T_{zm}^{\mathrm{in}}(r), \qquad r < a
\end{aligned} \tag{10.41}
$$

where $R(q) = \left(2q^2 - k_\mathrm{s}^2\right)^2 + 4q^2 h_\mathrm{s} h_\mathrm{p}$ is the Rayleigh function and $T_{zm}^{\mathrm{in}}(r)$ is the normal partial wave traction of the incident wave on the crack. This is the desired integral equation for the normal COD.

To solve the integral equation it is advantageous to use the following expansion functions

$$
p_j^m(s) = (-1)^j \frac{\mathrm{P}_{m+2j+1}^m(\sqrt{1-s^2})}{\mathrm{P}_{m+2j+1}^{m+1}(0)} \tag{10.42}
$$

with $j = 0, 1, \ldots$ and where $\mathrm{P}_{m+2j+1}^m(x)$ is an associated Legendre function. With $s = r/a$ these functions have the correct square root behaviour $\sqrt{1 - r^2/a^2}$ at the crack edge $r = a$ and they also have the correct r^m behaviour at $r = 0$. It is possible to calculate the following integral analytically

$$
\int_0^a p_j^m(s)\mathrm{J}_m(\gamma s)s\,\mathrm{d}s = \frac{1}{\gamma}\,\mathrm{j}_{m+2j+1}(\gamma) \tag{10.43}
$$

where $\mathrm{j}_{m+2j+1}(\gamma)$ is a spherical Bessel function. The orthogonality of the Legendre functions can be written in the following way

$$
\int_0^1 p_j^m(s)p_{j'}^m(s)\frac{s\,\mathrm{d}s}{\sqrt{1-s^2}} = \frac{\delta_{jj'}(2m+2j+1)!}{(2m+4j+3)(2j+1)!(\mathrm{P}_{m+2j+1}^{m+1}(0))^2}
$$

For $m = j = 0$ the first of these functions is $p_0^0(s) = (1 - s^2)^{1/2}$ and this gives

$$
\int_0^1 p_j^0(s)s\,\mathrm{d}s = \frac{1}{3}\delta_{j0} \tag{10.44}
$$

This integral is useful later on.

The partial wave COD is thus expanded as

$$
\Delta U_{zm}(r) = \sum_{j=0}^\infty \gamma_j^m p_j^m(r/a)
$$

This expansion is now inserted into Eq. (10.41) which is also multiplied by $rp_j^m(r/a)$ and integrated in r from 0 to a to get

$$\sum_{j'=0}^{\infty} Q_{jj'}^m \gamma_{j'}^m = -\frac{\mathrm{i}}{\mu k_{\mathrm{s}} a^2} \int_0^a T_{zm}^{\mathrm{in}}(r) p_j^m(r/a) r \, \mathrm{d}r \tag{10.45}$$

where

$$Q_{jj'}^m = -\int_0^\infty \frac{R(q)}{2k_{\mathrm{s}}^3 q h_{\mathrm{p}}} \mathrm{j}_{m+2j+1}(qa) \mathrm{j}_{m+2j'+1}(qa) \, \mathrm{d}q \tag{10.46}$$

The matrix $Q_{jj'}^m$ is apparently symmetric and it can be reduced to a finite range [77]. It is seen that the matrix elements for higher values of m can be obtained from those for $m = 0$ and $m = 1$. Solving Eq. (10.45) gives the expansion coefficients of the COD and this in turn determines f_{1m}, f_{2m}, and f_{3m} and thereby the scattered wave.

10.5.3 Antisymmetric integral equations

Turning to the antisymmetric part the first boundary condition gives

$$f_{2m} = \frac{2q^2 - k_{\mathrm{s}}^2}{2qh_{\mathrm{s}}} f_{3m}$$

By the second boundary condition the tangential partial wave displacement on $z = 0$ can then be written

$$U_{rm} \pm U_{\varphi m} = \int_0^\infty \left[f_{1m} \mp \frac{k_{\mathrm{s}}^2}{2q} f_{3m} \right] \mathrm{J}_{m\pm1}(qr) q \, \mathrm{d}q$$
$$= \begin{cases} 0, & r > a \\ \frac{1}{2}\left(\Delta U_{rm} \pm \Delta U_{\varphi m}\right), & r < a \end{cases}$$

where the tangential partial wave CODs $\Delta U_{rm}(r)$ and $\Delta U_{\varphi m}(r)$ are introduced. Inverting the Hankel transform and solving for f_{1m} and f_{3m} give

$$f_{1m} = \frac{1}{4} \int_0^a \left(\Delta U_{rm} + \Delta U_{\varphi m}\right) \mathrm{J}_{m+1}(qr) r \, \mathrm{d}r$$
$$+ \frac{1}{4} \int_0^a \left(\Delta U_{rm} - \Delta U_{\varphi m}\right) \mathrm{J}_{m-1}(qr) r \, \mathrm{d}r$$

$$f_{3m} = -\frac{q}{2k_s^2}\int_0^a (\Delta U_{rm} + \Delta U_{\varphi m})\,\mathrm{J}_{m+1}(qr)r\,\mathrm{d}r$$
$$+\frac{q}{2k_s^2}\int_0^a (\Delta U_{rm} - \Delta U_{\varphi m})\,\mathrm{J}_{m-1}(qr)r\,\mathrm{d}r$$

With f_{2m} eliminated the last two boundary conditions give

$$\mathrm{i}\int_0^\infty \left(-\frac{h_s}{k_s}\,f_{1m} \pm \frac{R(q)}{2k_s q h_s}\,f_{3m}\right)\mathrm{J}_{m\pm1}(qr)q\,\mathrm{d}q$$
$$= \frac{1}{\mu k_s}\left(T_{rm}^{\mathrm{in}}(r) \pm T_{\varphi m}^{\mathrm{in}}(r)\right) \tag{10.47}$$

where T_{rm}^{in} and $T_{\varphi m}^{\mathrm{in}}$ are the tangential partial wave tractions of the incident wave on the crack. With f_{1m} and f_{3m} inserted this becomes two coupled integral equations for the tangential CODs.

As $\mathrm{J}_{-1}(qr) = -\mathrm{J}_1(qr)$ a decoupling into two independent integral equations happens for $m = 0$. For the torsional part the integral equation for $m = 0$ is

$$-\mathrm{i}\int_0^\infty \frac{h_s}{k_s}\,f_{10}\mathrm{J}_1(qr)q\,\mathrm{d}q = \frac{1}{\mu k_s}T_{\varphi 0}^{\mathrm{in}}(r) \tag{10.48}$$

where

$$f_{10} = \frac{1}{2}\int_0^a \Delta U_{\varphi 0}\mathrm{J}_1(qr)r\,\mathrm{d}r$$

The azimuthal COD is expanded in the Legendre functions

$$\Delta U_{\varphi 0}(r) = \sum_{j=0}^{\infty} \beta_j^0 p_j^1(r/a)$$

After insertion and projection as for the symmetric part this gives the discretized equation

$$\sum_{j'=0}^{\infty} P_{jj'}^1 \beta_{j'}^0 = -\frac{\mathrm{i}}{2\mu k_s a^2}\int_0^a T_{\varphi 0}^{\mathrm{in}}(r)p_j^1(r/a)r\,\mathrm{d}r \tag{10.49}$$

where

$$P_{jj'}^m = -\frac{1}{4k_s}\int_0^\infty \frac{h_s}{q}\,\mathrm{j}_{m+2j+1}(qa)\mathrm{j}_{m+2j+1}(qa)\,\mathrm{d}q \tag{10.50}$$

This matrix is introduced for general m for later use. As for the symmetric part the integration can be reduced to a finite interval.

For the radial part for $m = 0$ the integral equation is

$$\mathrm{i}\int_0^\infty \frac{R(q)}{2k_\mathrm{s}qh_\mathrm{s}}\, f_{30}\mathrm{J}_1(qr)q\,\mathrm{d}q = \frac{1}{\mu k_\mathrm{s}}T_{r0}^{\mathrm{in}}(r) \tag{10.51}$$

where

$$f_{30} = -\frac{q}{k_\mathrm{s}}\int_0^a \Delta U_{r0}\mathrm{J}_1(qr)r\,\mathrm{d}r$$

The radial COD is expanded

$$\Delta U_{r0}(r) = \sum_{j=0}^\infty \alpha_j^0 p_j^1(r/a)$$

and proceeding as for the torsional part this results in

$$\sum_{j'=0}^\infty O_{jj'}^1\alpha_{j'}^0 = \frac{\mathrm{i}}{2\mu k_\mathrm{s}a^2}\int_0^a T_{r0}^{\mathrm{in}}(r)p_j^1(r/a)r\,\mathrm{d}r \tag{10.52}$$

where

$$O_{jj'}^m = \int_0^\infty \frac{R(q)}{4k_\mathrm{s}^3qh_\mathrm{s}}\,\mathrm{j}_{m+2j+1}(qa)\mathrm{j}_{m+2j+1}(qa)\,\mathrm{d}q \tag{10.53}$$

This matrix is introduced for general m for later use. The integration can be reduced to a finite interval.

Turning to $m = 1, 2, \ldots$ for the coupled antisymmetric integral equations Eq. (10.47), the CODs are expanded in the Chebyshev functions

$$\Delta U_{rm}(r) + \Delta U_{\varphi m}(r) = \sum_{j=0}^\infty \alpha_j^m p_j^{m+1}(r/a)$$

$$\Delta U_{rm}(r) - \Delta U_{\varphi m}(r) = \sum_{j=0}^\infty \beta_j^m p_j^{m-1}(r/a)$$

Inserting the expansions and projecting as for the symmetric part give the system of equations

$$\sum_{j'=0}^\infty \left[\left(P_{jj'}^{m+1} + O_{jj'}^{m+1}\right)\alpha_{j'}^m + \left(P_{j,j'-1}^{m+1} - O_{j,j'-1}^{m+1}\right)\beta_{j'}^m\right] = S_j^{m+}$$

$$\sum_{j'=0}^{\infty}\left[\left(P_{j,j'+1}^{m-1}-O_{j,j'+1}^{m-1}\right)\alpha_{j'}^{m}+\left(P_{jj'}^{m-1}+O_{jj'}^{m-1}\right)\beta_{j'}^{m}\right]=S_{j}^{m-} \quad (10.54)$$

The right-hand sides are

$$S_j^{m\pm}=-\frac{\mathrm{i}}{\mu k_{\mathrm{s}}a^2}\int_0^a\left(T_{rm}^{\mathrm{in}}(r)\pm T_{\varphi m}^{\mathrm{in}}(r)\right)p_j^{m\pm 1}(r/a)r\,\mathrm{d}r \quad (10.55)$$

Solving Eqs. (10.54) gives the expansion coefficients of the CODs and this in turn determines f_{1m}, f_{2m}, and f_{3m} and thereby the scattered wave.

10.5.4 The T-matrix

To obtain the T-matrix, which is expressed with respect to the spherical vector wavefunctions, the incident wave is taken as one of the regular spherical vector wavefunctions $\boldsymbol{\psi}_{\tau\sigma ml}^{0}(\boldsymbol{r})$. The index σ is suppressed, at present only the group $\tau\sigma=1\mathrm{o},2\mathrm{e},3\mathrm{e}$ is considered, giving the same azimuthal dependence as the expansion of the scattered wave in the cylindrical vector wavefunctions in the preceding subsections. At the end it is stated how the T-matrix for the other group $\tau\sigma=1\mathrm{e},2\mathrm{o},3\mathrm{o}$ is obtained.

Starting with the symmetric part, the right-hand side in Eq. (10.45) contains the partial wave traction $T_{zm}^{\mathrm{in}}(r)$. On the crack $z=0$, so it is trivial to transform between the cylindrical and spherical systems there. The radius r and azimuthal angle φ are the same and $z=0$ corresponds to $\theta=\pi/2$. The normal traction component is $\sigma_{zz}=-\sigma_{\theta\theta}=-\lambda\nabla\cdot\boldsymbol{u}-2\mu(\partial u_\theta/\partial\theta+u_r)/r$. Calculating this normal traction gives the spherical partial wave normal tractions $t_{z\tau l}^{m}(r)$ on the crack

$$t_{z1l}^{m}=\frac{2m\mu\gamma_{ml}}{r\sqrt{l(l+1)}}\mathrm{P}_l^{m\prime}(0)\mathrm{j}_l(k_{\mathrm{s}}r)$$

$$t_{z2l}^{m}=\frac{2\mu\gamma_{ml}}{r\sqrt{l(l+1)}}\left[\mathrm{P}_l^{m\prime\prime}(0)\left(\mathrm{j}_l'(k_{\mathrm{s}}r)+\frac{\mathrm{j}_l(k_{\mathrm{s}}r)}{k_{\mathrm{s}}r}\right)\right.$$
$$\left.+l(l+1)\mathrm{P}_l^{m}(0)\frac{\mathrm{j}_l(k_{\mathrm{s}}r)}{k_{\mathrm{s}}r}\right]$$

$$t_{z3l}^{m} = -\frac{\mu\gamma_{ml}}{k_{\mathrm{p}}}\left(\frac{k_{\mathrm{p}}}{k_{\mathrm{s}}}\right)^{3/2}\left[\left(2k_{\mathrm{p}}^{2}-k_{\mathrm{s}}^{2}\right)\mathrm{P}_{l}^{m}(0)\mathrm{j}_{l}(k_{\mathrm{p}}r)\right.$$
$$\left.+2k_{\mathrm{p}}\mathrm{P}_{l}^{m}(0)\frac{\mathrm{j}_{l}'(k_{\mathrm{p}}r)}{r}+2\mathrm{P}_{l}^{m\prime\prime}(0)\frac{\mathrm{j}_{l}(k_{\mathrm{p}}r)}{r^{2}}\right]$$

where thus the azimuthal factor $\cos m\varphi$ or $\sin m\varphi$ is removed and $\theta = \pi/2$ is inserted. The normalization constant γ_{ml} is given in Eq. (3.11). For an incident wavefunction the right-hand side in Eq. (10.45) becomes (except for a constant factor)

$$L_{zj\tau l}^{m} = \frac{1}{\mu k_{\mathrm{s}}a^{2}}\int_{0}^{a} t_{z\tau l}^{m}(r)p_{j}^{m}(r/a)r\,\mathrm{d}r \tag{10.56}$$

To obtain the scattered wave the integral representation Eq. (4.21) is used (where the traction vanishes due to the boundary condition)

$$\boldsymbol{u}^{\mathrm{sc}}(\boldsymbol{r}) = \frac{k_{\mathrm{s}}}{\mu}\int_{S}\boldsymbol{\Delta u}(\boldsymbol{r}')\cdot(\hat{\boldsymbol{z}}\cdot\boldsymbol{\Sigma}(\boldsymbol{r}';\boldsymbol{r}))\,\mathrm{d}S'$$

where the integration is over the area of the crack and the COD appears due to the difference in normal on the top and bottom sides of the crack. The Green dyadic is expanded in the spherical vector wavefunctions according to Eq. (4.16). Assuming that the field point $\boldsymbol{r}$ lies outside the circumscribing sphere to the crack this gives for the Green stress triadic

$$\hat{\boldsymbol{z}}\cdot\boldsymbol{\Sigma}(\boldsymbol{r}';\boldsymbol{r}) = \mathrm{i}\sum_{\tau,\sigma,m,l}\boldsymbol{t}^{(z)}(\boldsymbol{\psi}_{\tau\sigma ml}^{0}(\boldsymbol{r}'))\boldsymbol{\psi}_{\tau\sigma ml}^{+}(\boldsymbol{r})$$

where $\boldsymbol{t}^{(z)}$ is the traction operator on the crack. This gives the desired expansion of the scattered wave in the out-going spherical vector wavefunctions with expansion coefficients

$$f_{\tau ml} = \frac{\mathrm{i}k_{\mathrm{s}}}{\mu}\int_{S}\boldsymbol{\Delta u}(\boldsymbol{r}')\cdot\boldsymbol{t}^{(z)}(\boldsymbol{\psi}_{\tau ml}^{0}(\boldsymbol{r}'))\mathrm{d}S'$$

where the σ-index is suppressed as it is given implicitly by $\tau\sigma = 1\mathrm{o}, 2\mathrm{e}, 3\mathrm{e}$. For the symmetric part there is only a z-component of the COD, so after performing the azimuthal integration there remains

$$f_{\tau ml} = \frac{\mathrm{i}k_{\mathrm{s}}}{\mu}\frac{2\pi}{\epsilon_{m}}\sum_{j=0}^{\infty}\gamma_{j\tau'l'}^{m}\int_{0}^{a}t_{z\tau l}^{m}(r)p_{j}^{m}(r/a)r\,\mathrm{d}r = \mathrm{i}k_{\mathrm{s}}a\,\frac{2\pi}{\epsilon_{m}}\sum_{j=0}^{\infty}L_{zj\tau l}^{m}\gamma_{j\tau'l'}^{m}$$

where the extra indices $\tau'l'$ on the expansion coefficients of the COD specify the incident wave (as a regular spherical vector wavefunction). Altogether this means that the symmetric part of the T-matrix becomes

$$T_{\tau ml,\tau' m' l'} = \delta_{mm'} \frac{2\pi}{\epsilon_m} \sum_{j=0}^{\infty} \sum_{j'=0}^{\infty} L^m_{zj\tau l} \left(Q^m\right)^{-1}_{jj'} L^m_{zj'\tau' l'} \tag{10.57}$$

As stated above the σ-index is here given implicitly by the τ-index, so that $\tau\sigma = 1\mathrm{o}, 2\mathrm{e}, 3\mathrm{e}$.

For the antisymmetric part the procedure is the same. Then the spherical partial wave tangential traction components on the crack are needed. The tangential tractions are

$$\sigma_{zr} = -\sigma_{r\theta} = -\mu \left(\frac{\partial u_\theta}{\partial r} - \frac{u_\theta}{r} + \frac{1}{r}\frac{\partial u_r}{\partial \theta} \right)$$

$$\sigma_{\varphi z} = -\sigma_{\theta\varphi} = -\frac{\mu}{r} \left(\frac{\partial u_\varphi}{\partial \theta} - \cot\theta \, u_\varphi + \frac{1}{\sin\theta} \frac{\partial u_\theta}{\partial \varphi} \right)$$

The spherical partial wave tangential traction components become

$$t^m_{r1l}(r) = -\frac{m\mu k_\mathrm{s} \gamma_{ml}}{\sqrt{l(l+1)}} \mathrm{P}^m_l(0) \left(\mathrm{j}'_l(k_\mathrm{s} r) - \frac{\mathrm{j}_l(k_\mathrm{s} r)}{k_\mathrm{s} r} \right)$$

$$t^m_{\varphi 1l}(r) = \frac{\mu \gamma_{ml}}{\sqrt{l(l+1)}} \frac{\mathrm{j}_l(k_\mathrm{s} r)}{r} \left[\mathrm{P}^{m\prime\prime}_l(0) + m^2 \mathrm{P}^m_l(0) \right]$$

$$t^m_{r2l}(r) = \frac{\mu k_\mathrm{s} \gamma_{ml}}{\sqrt{l(l+1)}} \mathrm{P}^{m\prime}_l(0) \left(\mathrm{j}''_l(k_\mathrm{s} r) + \frac{l(l+1)-2}{k_\mathrm{s}^2 r^2} \mathrm{j}_l(k_\mathrm{s} r) \right)$$

$$t^m_{\varphi 2l}(r) = -\frac{2m\mu\gamma_{ml}}{r\sqrt{l(l+1)}} \mathrm{P}^{m\prime}_l(0) \left(\mathrm{j}'_l(k_\mathrm{s} r) + \frac{\mathrm{j}_l(k_\mathrm{s} r)}{k_\mathrm{s} r} \right)$$

$$t^m_{r3l} = \frac{2\mu\gamma_{ml}}{r} \left(\frac{k_\mathrm{p}}{k_\mathrm{s}} \right)^{3/2} \mathrm{P}^{m\prime}_l(0) \left(\mathrm{j}'_l(k_\mathrm{p} r) - \frac{\mathrm{j}_l(k_\mathrm{p} r)}{k_\mathrm{p} r} \right)$$

$$t^m_{\varphi 3l} = -2m\mu\gamma_{ml} \left(\frac{k_\mathrm{p}}{k_\mathrm{s}} \right)^{3/2} \mathrm{P}^{m\prime}_l(0) \frac{\mathrm{j}_l(k_\mathrm{p} r)}{k_\mathrm{p} r^2}$$

where again the azimuthal factor $\cos m\varphi$ or $\sin m\varphi$ is removed and $\theta = \pi/2$. These are of course also valid for the special case $m = 0$.

For the torsional waves for $m = 0$ the right-hand side in Eq. (10.49) becomes

$$L^0_{\varphi j1l} = \frac{1}{\mu a} \int_0^a t^0_{\varphi 1l}(r) p^1_j(r/a) r \, \mathrm{d}r$$

As for the symmetric part the expansion coefficients of the scattered wave become

$$f_{10l} = \mathrm{i}k_\mathrm{s} a \, 2\pi \sum_{j=0}^{\infty} L^0_{\varphi j1l} \beta^m_{j\tau' l'}$$

and the T-matrix becomes

$$T_{10l,10l'} = 2\pi \sum_{j=0}^{\infty} \sum_{j'=0}^{\infty} L^0_{\varphi j1l} \left(P^1\right)^{-1}_{jj'} L^0_{\varphi j1l'} \tag{10.58}$$

As noted the σ-index is suppressed and is given implicitly by the τ-index, so that for $\tau = 1$ it is $\sigma = \mathrm{o}$. However, for $m = 0$ there is no wavefunction for $\sigma = \mathrm{o}$, so for just this case ($\tau = 1$ and $m = 0$) the suppressed σ-index is instead $\sigma = \mathrm{e}$.

For the radial waves for $m = 0$ the right-hand side in Eq. (10.51) becomes

$$L^0_{rj\tau l} = \frac{1}{\mu a} \int_0^a t^0_{r\tau l}(r) p^1_j(r/a) r \, \mathrm{d}r$$

for $\tau = 2, 3$. The T-matrix becomes

$$T_{\tau 0l,\tau' 0l'} = 2\pi \sum_{j=0}^{\infty} \sum_{j'=0}^{\infty} L^0_{rj\tau l} \left(O^1\right)^{-1}_{jj'} L^0_{rj\tau' l'} \tag{10.59}$$

where $\tau = 2, 3$ and $\tau' = 2, 3$. There is thus coupling between the SV and P wavefunctions in this case.

For the antisymmetric part and $m = 1, 2, \ldots$ the right-hand sides in the integral equations Eq. (10.54) become

$$L^{m\pm}_{j\tau l} = \frac{1}{\mu a} \int_0^a \left[t^m_{r\tau l}(r) \pm t^m_{\varphi\tau l}(r)\right] p^{m\pm 1}_j(r/a) r \, \mathrm{d}r$$

The expansion coefficients of the scattered wave become

$$f_{\tau ml} = \mathrm{i}(k_{\mathrm{s}}a)^2 \frac{\pi}{\epsilon_m} \sum_{j=0}^{\infty} \left(L^m_{+j\tau l} \alpha^m_{j\tau l} + L^m_{-j\tau l} \beta^m_{j\tau l} \right)$$

Solving the two Eq. (10.54) for $\alpha^m_{j\tau l}$ and $\beta^m_{j\tau l}$ and inserting gives the T-matrix elements, but this can not be written as compactly as for the symmetric case.

As stated in the beginning of this subsection the σ-index is given implicitly by τ, so that it is the T-matrix elements for $\tau\sigma = 1\mathrm{o}, 2\mathrm{e}, 3\mathrm{e}$ that are given. There is one exception and that is the torsional part that is stated for $\tau\sigma = 1\mathrm{e}$ as it does not exist for $\tau\sigma = 1\mathrm{o}$. The elements for the other group $\tau\sigma = 1\mathrm{e}, 2\mathrm{o}, 3\mathrm{o}$ are in fact the same as for the first group except that there is a change of sign in elements that couple $\tau = 1$ to $\tau = 2, 3$, i.e. in the elements $\tau\tau' = 12, 13, 21, 31$.

10.5.5 Low frequency T-matrix

For low frequencies it is possible to obtain explicit expressions for the leading order $T-$matrix elements. As the dimensionless parameter $k_{\mathrm{s}}a$ is small the matrices in Eqs. (10.45) and Eq. (10.49) are evaluated with the help of Eq. (10.11) exactly as for the 2D crack in the P-SV case. This gives

$$Q^m_{jj'} \simeq -\frac{k_{\mathrm{s}}^2 - k_{\mathrm{p}}^2}{k_{\mathrm{s}}^3 a} \frac{\mathrm{i}\pi\delta_{jj'}}{2(2m+4j+3)}$$

$$P^m_{jj'} \simeq -\frac{1}{k_{\mathrm{s}}a} \frac{\mathrm{i}\pi\delta_{jj'}}{4(2m+4j+1)}$$

$$O^m_{jj'} \simeq -\frac{k_{\mathrm{s}}^2 - k_{\mathrm{p}}^2}{k_{\mathrm{s}}^3 a} \frac{\mathrm{i}\pi\delta_{jj'}}{4(2m+4j+1)}$$

For the symmetric part the quantity $L^m_{zj\tau l}$ defined in Eq. (10.56) must be estimated for small $k_{\mathrm{s}}a$. Using the asymptotic form of the spherical Bessel functions this is straightforward and gives that only the following three elements are of leading order [using also Eq. (10.44)]

$$L^0_{zj22} = -k_{\mathrm{s}}a\sqrt{\frac{2}{15\pi}} \frac{\delta_{j0}}{3}$$

$$L^0_{zj30} = -k_\mathrm{p}a\left(\frac{k_\mathrm{p}}{k_\mathrm{s}}\right)^{3/2}\left(\frac{4}{3}-\frac{k_\mathrm{s}^2}{k_\mathrm{p}^2}\right)\frac{\delta_{j0}}{3\sqrt{4\pi}}$$

$$L^0_{zj32} = -k_\mathrm{p}a\left(\frac{k_\mathrm{p}}{k_\mathrm{s}}\right)^{3/2}\frac{2\delta_{j0}}{9\sqrt{5\pi}}$$

As the T-matrix is diagonal in m it is natural to write (the σ-indices are suppressed as in the previous subsection)

$$T_{\tau ml,\tau' m' l'} = T^m_{\tau l,\tau' l'}$$

The leading order T-matrix elements for the symmetric part then become

$$T^0_{30,30} = \mathrm{i}(k_\mathrm{p}a)^3\frac{(3\lambda+2\mu)^2}{27\pi\mu(\lambda+\mu)}$$

$$T^0_{22,22} = \mathrm{i}(k_\mathrm{s}a)^3\frac{8(\lambda+2\mu)}{45\pi(\lambda+\mu)}$$

$$T^0_{32,32} = \mathrm{i}(k_\mathrm{p}a)^3\frac{16\mu}{135\pi(\lambda+\mu)}$$

$$T^0_{22,32} = T^0_{32,22} = \mathrm{i}(k_\mathrm{s}k_\mathrm{p})^{3/2}a^3\frac{8\sqrt{2\mu(\lambda+2\mu)}}{45\sqrt{3}\pi(\lambda+\mu)}$$

$$T^0_{22,30} = T^0_{30,22} = -\mathrm{i}(k_\mathrm{s}k_\mathrm{p})^{3/2}a^3\frac{4(3\lambda+2\mu)\sqrt{\lambda+2\mu}}{9\sqrt{30}\pi(\lambda+\mu)\sqrt{\mu}}$$

$$T^0_{32,30} = T^0_{30,32} = -\mathrm{i}(k_\mathrm{p}a)^3\frac{4(3\lambda+2\mu)}{27\sqrt{5}\pi(\lambda+\mu)} \tag{10.60}$$

It is seen that the leading order elements appear only for $m = 0$ and only for the $\tau = 2$ and $\tau = 3$ parts (SV and P waves; and the suppressed value is $\sigma = \mathrm{e}$). Furthermore, only $l = 0$ and $l = 2$ are involved (but not $l = 0$ for $\tau = 2$ as it does not exist), i.e. the monopole and quadrupole waves. Thus, it is only the combinations $\tau l = 30, 22, 32$ that give leading order contributions.

For the antisymmetric part the procedure is the same. Neither the torsional nor the radial part for $m = 0$ gives any leading order contributions. The only leading order contributions come from

$$L^{1-}_{j22} = \frac{2k_\mathrm{s}a\delta_{j0}}{3\sqrt{10\pi}}$$

$$L^{1-}_{j32} = \left(\frac{k_\mathrm{p}}{k_\mathrm{s}}\right)^{3/2} \frac{2k_\mathrm{p} a\delta_{j0}}{3\sqrt{15\pi}}$$

The leading order T-matrix elements for the antisymmetric part then becomes

$$\begin{aligned} T^1_{22,22} &= -\mathrm{i}(k_\mathrm{s}a)^3 \frac{16(\lambda+2\mu)}{15\pi\lambda} \\ T^1_{32,32} &= -\mathrm{i}(k_\mathrm{p}a)^3 \frac{32\mu}{45\pi\lambda} \\ T^0_{22,32} &= T^0_{32,22} = -\mathrm{i}(k_\mathrm{p}k_\mathrm{s})^{3/2} a^3 \frac{32\sqrt{\mu(\lambda+2\mu)}}{15\sqrt{6}\pi\lambda} \end{aligned} \tag{10.61}$$

The leading order elements thus appear only for $m = 1$ and $l = 2$ and only for the $\tau = 2$ and $\tau = 3$ parts.

Comparing the scattering by a sphere and a circular crack it is noted that there are many more nonzero elements for the sphere for low frequencies, where $l = 0, 1, 2$ all contribute for all relevant values of m and σ (the T-matrix for a sphere is independent of m and σ). On the other hand, $\tau = 1$ (SH waves) does not contribute in either case.

10.6 Curved crack: 2D SH waves

The present type of integral equation method is also well suited to solve the scattering problem for a curved crack that is part of a circle. This problem is thus considered in this section for the simplest case, i.e. 2D SH waves. The method can be used also for the 2D P-SV case. The problem with a curved crack is a special case of a crack at a fiber interface, a problem that is solved by Yang and Norris [145], Norris and Yang [105], and Wang and Wang [138] for SH waves and by Yang and Norris [146] for P-SV waves.

Consider the scattering by a curved crack that is part of a circle with radius a. Polar coordinates (r, φ) are used and the crack is defined by $r = a$ and $|\varphi| < \delta$ (assuming $-\pi < \varphi \leq \pi$). As usual the anti-plane displacement component is simply called u and the shear wavenumber k. The crack is assumed open so that the boundary condition is that the normal derivative of u vanishes.

The total displacement u is divided into the incident wave u^{in} and the scattered wave u^{sc}. Note that the incident wave is defined everywhere, of course outside the circle $r = a$, but also for $r < a$. The incident wave is left arbitrary for the moment. The scattered wave admits one expansion for $r > a$ and another for $r < a$. It is natural to divide the problem into even and odd parts (with respect to φ) and starting with the even part the expansions of the scattered wave are

$$u^{\mathrm{sc}} = \begin{cases} \sum_{m=0}^{\infty} f_m \chi_{\mathrm{e}m}^{+}(r,\varphi), & r > a \\ \sum_{m=0}^{\infty} c_m \chi_{\mathrm{e}m}^{0}(r,\varphi), & r < a \end{cases}$$

where the wavefunctions are defined in Eq. (2.9). In this simple case they are just a product of a Hankel or Bessel function and a cosine function.

The normal stress is continuous everywhere on the circle $r = a$, both on the crack and on the rest. As the incident wave has a continuous stress this directly gives

$$f_m \mathrm{H}_m'(ka) = c_m \mathrm{J}_m'(ka)$$

For $r = a$ the displacement is, furthermore, continuous outside the crack, but has a jump Δu (the crack-opening displacement, COD) on the crack, thus

$$\sum_{m=0}^{\infty} \frac{\sqrt{\epsilon_m}}{2} \left[f_m \mathrm{H}_m(ka) - c_m \mathrm{J}_m(ka) \right] \cos m\varphi = \begin{cases} \Delta u(\varphi), & |\varphi| < \delta \\ 0, & |\varphi| > \delta \end{cases}$$

Inverting the Fourier series gives

$$f_m \mathrm{H}_m(ka) - c_m \mathrm{J}_m(ka) = \frac{\sqrt{\epsilon_m}}{\pi} \int_{-\delta}^{\delta} \Delta u(\varphi) \cos m\varphi \, \mathrm{d}\varphi$$

Solving for f_m gives

$$f_m = \frac{\mathrm{i}\sqrt{\epsilon_m}}{2} ka \mathrm{J}_m'(ka) \int_{-\delta}^{\delta} \Delta u(\varphi) \cos m\varphi \, \mathrm{d}\varphi$$

To proceed the COD is expanded in the even Chebyshev functions

$$\Delta u(\varphi) = \sum_{n=1,3,\ldots} \alpha_n \psi_n(\varphi/\delta)$$

This gives the correct square root behaviour at the crack tips. Note that it is Chebyshev functions of odd order $n = 1, 3, \ldots$ that are even functions. Equation (10.6) then gives

$$\int_{-\delta}^{\delta} \psi_n(\varphi/\delta) \cos m\varphi \, \mathrm{d}\varphi = \frac{n\pi}{m} \mathrm{J}_n(m\delta)$$

for $n = 1, 3, \ldots$. Note that $m = 0$ must be taken as the limit. This gives the scattered wave for $r > a$ in the form

$$u^{\mathrm{sc}} = \sum_{m=0}^{\infty} \sum_{n=1,3,\ldots} \alpha_n \frac{\mathrm{i}\pi n \epsilon_m ka}{4m} \mathrm{J}_n(m\delta) \mathrm{J}'_m(ka) \mathrm{H}_m(kr) \cos m\varphi$$

The only remaining boundary condition is the vanishing of the normal traction on the crack and using this from the outside gives

$$\sum_{m=0}^{\infty} \sum_{n=1,3,\ldots} \alpha_n \frac{\mathrm{i}\pi^2 n n' \epsilon_m ka}{4m} \mathrm{J}_m(m\delta) \mathrm{J}'_m(ka) \mathrm{H}'_m(ka) \cos m\varphi$$

$$= -\frac{1}{k} \frac{\partial u^{\mathrm{in}}(a, \varphi)}{\partial r}$$

This equation is finally projected on the Chebyshev functions resulting in

$$\sum_{n'=1,3,\ldots} Q_{nn'} \alpha_{n'} = -\frac{1}{k} \int_{-\delta}^{\delta} \frac{\partial u^{\mathrm{in}}(a, \varphi)}{\partial r} \psi_n(\varphi/\delta) \, \mathrm{d}\varphi$$

where

$$Q_{nn'} = nn' \sum_{m=0}^{\infty} \frac{\epsilon_m}{m^2} \mathrm{J}'_m(ka) \mathrm{H}'_m(ka) \mathrm{J}_n(m\delta) \mathrm{J}_{n'}(m\delta) \tag{10.62}$$

where the $m = 0$ term must be treated as a limit. Solving this equation gives the expansion coefficients of the COD and thereby of the scattered field.

To obtain the T-matrix the incident wave is chosen as an even wavefunction and it is then easily seen that the result is

$$T_{\mathrm{e}m,\mathrm{e}m'} = \frac{\sqrt{\epsilon_m \epsilon_{m'}}}{mm'} \mathrm{J}'_m(ka)\mathrm{J}'_{m'}(ka) \times \sum_{n=1,3,\ldots} \sum_{n'=1,3,\ldots} nn'\mathrm{J}_n(m\delta)\left(Q^{-1}\right)_{nn'} \mathrm{J}_{n'}(m'\delta) \quad (10.63)$$

For the odd part the solution is almost identical. The differences are that the functions $\cos m\varphi$ are exchanged to $\sin m\varphi$ (and then $m = 0$ does not exist) and that the COD expansion is in the odd Chebyshev functions. The matrix Q remains the same (but n and n' run over the even positive integers) and the equation for the COD expansion coefficients is the same except for a sign change on the right-hand side and that the sum is over $n = 2, 4, \ldots$. The T-matrix then also looks the same except for an overall sign change and the change in summation from odd to even integers.

11 Scattering by cracks in anisotropic solids

In the previous chapter scattering by a crack of simple shape in an isotropic solid is considered. Most of the methods used there can be extended to the case of an anisotropic solid and in this chapter a strip-like crack in 2D and a rectangular crack in 3D are taken up. It is also possible to consider strip-like cracks in 3D [96, 47] and a circular crack in the isotropy plane of a transversely isotropic solid [81].

11.1 Strip-like crack: 2D SH waves

Consider first the simplest case with 2D SH waves and a straight open crack situated at $y = 0$, $|x| < a$ (see Mattsson and Niklasson [95] for the more general case with an interface crack).

11.1.1 Integral equation

Let the stiffness tensor be the most general one admitting 2D SH waves, for which the two relevant stress-strain relations are

$$\sigma_{yz} = C_{44}\frac{\partial u}{\partial y} + C_{45}\frac{\partial u}{\partial x}$$

$$\sigma_{zx} = C_{45}\frac{\partial u}{\partial y} + C_{55}\frac{\partial u}{\partial x}$$

where the anti-plane displacement is denoted u, without an index "z". This means that the equation of motion becomes

$$C_{55}\frac{\partial^2 u}{\partial x^2} + 2C_{45}\frac{\partial^2 u}{\partial x \partial y} + C_{44}\frac{\partial^2 u}{\partial y^2} + \rho\omega^2 u = 0$$

As opposed to the situation in an isotropic material there is no mirror symmetry in the present case (except for $C_{45}=0$), so the solution must proceed in the whole xy-plane. Taking a Fourier transform in x it is realized that the scattered wave can be represented as

$$u^{\rm sc} = \int_{-\infty}^{\infty} f_1(q) {\rm e}^{{\rm i}(qx+h_1 y)}\,{\rm d}q, \qquad y>0$$

$$u^{\rm sc} = \int_{-\infty}^{\infty} f_2(q) {\rm e}^{{\rm i}(qx-h_2 y)}\,{\rm d}q, \qquad y<0$$

Here f_1 and f_2 are coefficients to be determined and the wavenumbers in the y-direction are

$$h_1 = \sqrt{k^2 - c_1 q^2} - c_2 q$$

$$h_2 = \sqrt{k^2 - c_1 q^2} + c_2 q$$

where the branches are chosen so $\operatorname{Im} h_1 \geq 0$ and $\operatorname{Im} h_2 \geq 0$, and where $k = \omega\sqrt{\rho/C_{44}}$, $c_1 = C_{55}/C_{44} - (C_{45}/C_{44})^2$ (so that $c_1 > 0$), and $c_2 = C_{45}/C_{44}$. The wavenumbers h_1 and h_2 are chosen so that the energy flow is upwards or downwards, respectively. If there is damping in the material, then k has a positive imaginary part and the solution decays exponentially upwards and downwards. If $c_2 \neq 0$ it may happen that $h_1 < 0$ for some q, so that the wave vector is downwards although the energy flow is upwards.

The boundary conditions on $y=0$ are that σ_{yz} is continuous for all x, that u is continuous for $|x|>a$, and that $\sigma_{yz}^{\rm sc} = -\sigma_{yz}^{\rm in}$ for $|x|<a$. The last can be written as

$$\frac{\partial u^{\rm sc}}{\partial y} + c_2 \frac{\partial u^{\rm sc}}{\partial x} = -\frac{\partial u^{\rm in}}{\partial y} - c_2 \frac{\partial u^{\rm in}}{\partial x}$$

The first of the boundary conditions gives

$$(h_1 + c_2 q) f_1(q) = (-h_2 + c_2 q) f_2(q)$$

which can be simplified to $f_2(q) = -f_1(q)$. The continuity of u can be written

$$\int_{-\infty}^{\infty} (f_1(q) - f_2(q)) {\rm e}^{{\rm i}qx}\,{\rm d}q = \begin{cases} 0, & |x|>a \\ \Delta u(x), & |x|<a \end{cases}$$

where Δu is the crack-opening displacement (COD). Inverting the Fourier transform gives

$$f_1(q) = \frac{1}{4\pi}\int_{-a}^{a} \Delta u(x)\mathrm{e}^{-\mathrm{i}qx}\,\mathrm{d}x$$

Inserting this into the last boundary condition gives

$$\int_{-\infty}^{\infty} \sqrt{k^2 - c_1 q^2}\,\mathrm{e}^{\mathrm{i}qx}\int_{-a}^{a} \Delta u(x')\mathrm{e}^{-\mathrm{i}qx'}\,\mathrm{d}x'\mathrm{d}q$$
$$= -4\pi\mathrm{i}\left(\frac{\partial u^{\mathrm{in}}}{\partial y} + c_2\frac{\partial u^{\mathrm{in}}}{\partial x}\right), \qquad |x| < a \tag{11.1}$$

This is the desired integral equation for the COD. It is not possible to change the order of the integrations as this would lead to a divergent integral. This integral equation is almost identical to the corresponding one [Eq. (10.1)] in the case of isotropy.

As in the case of isotropy the COD is expanded in terms of the Chebyshev functions

$$\Delta u(x') = \sum_{n'=1}^{\infty} \alpha_{n'}\psi_{n'}(x'/a)$$

This expansion has the correct square root behaviour at the crack tips. Inserting this expansion and projecting on the Chebyshev functions give

$$\sum_{n'=1}^{\infty} Q_{nn'}\alpha_{n'} = -4\mathrm{i}\int_{-a}^{a}\left(\frac{\partial u^{\mathrm{in}}}{\partial y} + c_2\frac{\partial u^{\mathrm{in}}}{\partial x}\right)\psi_n(x/a)\,\mathrm{d}x \tag{11.2}$$

where

$$Q_{nn'} = \pi n n'(-1)^n\int_{-\infty}^{\infty}\frac{\sqrt{k^2 - c_1 q^2}}{q^2}\mathrm{J}_n(qa)\mathrm{J}_{n'}(qa)\,\mathrm{d}q \tag{11.3}$$

As in the case of isotropy this integral can be reduced to a finite range.

It is now possible to solve for the expansion coefficients for the COD and then determine f_1 and f_2 and thereby the scattered wave. As there are no polar wavefunctions for anisotropic media it is of course not possible to determine a T-matrix.

11.1.2 Far field

The scattered field for $y > 0$ is

$$u^{\mathrm{sc}} = \frac{1}{4}\sum_{n=1}^{\infty} \alpha_n \int_{-\infty}^{\infty} \mathrm{J}_n(qa)\, \mathrm{e}^{\mathrm{i}(qx+h_1 y)}\, \frac{\mathrm{d}q}{q}$$

Introduce polar coordinates (r, φ) and change integration variable to $t = q/k$ so that the field can be written

$$u^{\mathrm{sc}} = \frac{1}{4}\sum_{n=1}^{\infty} \alpha_n \int_{-\infty}^{\infty} \mathrm{J}_n(kat)\, \mathrm{e}^{\mathrm{i}kr\psi(t)}\, \frac{\mathrm{d}t}{t}$$

where the phase function is

$$\psi(t) = t\cos\varphi + \left(\sqrt{1 - c_1 t^2} - c_2 t\right)\sin\varphi$$

This integral can be calculated in the far field with the stationary-phase method with kr as the large parameter. Putting $\psi'(t) = 0$ determines the only stationary point

$$t_s = \frac{\cos\varphi - c_2\sin\varphi}{\sqrt{c_1^2\sin^2\varphi + c_1(\cos\varphi - c_2\sin\varphi)^2}}$$

The second derivative is

$$\begin{aligned}\psi''(t_s) &= -\frac{\sin\varphi}{(1 - c_1 t_s^2)^{3/2}} \\ &= -\frac{1}{\sin^2\varphi}\left((\cos\varphi - c_2\sin\varphi)^2/c_1 + \sin^2\varphi\right)^{3/2}\end{aligned}$$

and the value of the phase function at the stationary point becomes

$$\psi(t_s) = \sqrt{(\cos\varphi - c_2\sin\varphi)^2/c_1 + \sin^2\varphi}$$

The scattered wave in the far field then becomes

$$\begin{aligned}u^{\mathrm{sc}} &= \frac{1}{4}\sqrt{\frac{2\pi c_1}{kr}}\, \mathrm{e}^{\mathrm{i}(kr\psi(t_s) - \pi/4)} \\ &\quad \times \frac{\sin\varphi}{(\psi''(t_s))^{1/2}(\cos\varphi - c_2\sin\varphi)} \sum_{n=1}^{\infty} \alpha_n \mathrm{J}_n(kat_s)\end{aligned}$$

which is thus valid for $0 < \varphi < \pi$. An almost identical result is valid for $\pi < \varphi < 2\pi$.

11.2 Strip-like crack: 2D P-SV waves

The 2D problem for the scattering of in-plane P-SV waves by a strip-like crack can be solved in a similar way as the scalar problem, see Dhawan [42], who employs a Fredholm integral equation approach, and Mattsson *et al.* [96] who use the hypersingular integral equation approach for the 3D strip-like crack. The problem is also very similar to the corresponding one when the material is isotropic (Section 10.2).

11.2.1 Symmetric and antisymmetric parts

The crack is located along the x-axis between $x = -a$ and $x = a$. For simplicity it is assumed that the material is orthotropic with the coordinate planes as symmetry planes. The 2D stress-displacement relations are then

$$\sigma_{xx} = C_{11}\frac{\partial u_x}{\partial x} + C_{12}\frac{\partial u_y}{\partial y}$$

$$\sigma_{yy} = C_{12}\frac{\partial u_x}{\partial x} + C_{22}\frac{\partial u_y}{\partial y}$$

$$\sigma_{xy} = C_{66}\left(\frac{\partial u_x}{\partial y} + \frac{\partial u_y}{\partial x}\right)$$

Inserting into the equations of motion gives

$$C_{11}\frac{\partial^2 u_x}{\partial x^2} + C_{12}\frac{\partial^2 u_y}{\partial x \partial y} + C_{66}\left(\frac{\partial^2 u_x}{\partial y^2} + \frac{\partial^2 u_y}{\partial x \partial y}\right) + \rho\,\omega^2 u_x = 0$$

$$C_{22}\frac{\partial^2 u_y}{\partial y^2} + C_{12}\frac{\partial^2 u_x}{\partial x \partial y} + C_{66}\left(\frac{\partial^2 u_y}{\partial x^2} + \frac{\partial^2 u_x}{\partial x \partial y}\right) + \rho\,\omega^2 u_y = 0$$

Exactly as in the corresponding problem in the case of isotropy in Section 10.2, it is possible to divide the problem into symmetric and antisymmetric parts. For the symmetric part the boundary conditions for $y = 0$ become

$$\sigma_{xy}^{\mathrm{sc}} = 0, \qquad \text{all } x$$

$$u_y^{\mathrm{sc}} = 0, \qquad |x| > a$$

$$\sigma_{yy}^{\mathrm{sc}} = -\sigma_{yy}^{\mathrm{in}}, \qquad |x| < a$$

and for the antisymmetric part they become

$$\sigma_{yy}^{\mathrm{sc}} = 0, \qquad \text{all } x$$

$$u_x^{\mathrm{sc}} = 0, \qquad |x| > a$$

$$\sigma_{xy}^{\mathrm{sc}} = -\sigma_{xy}^{\mathrm{in}}, \qquad |x| < a$$

Taking a Fourier transform in x with transform parameter q the equations of motion become

$$-q^2 C_{11}\tilde{u}_x + \mathrm{i}qC_{12}\frac{\partial \tilde{u}_y}{\partial y} + C_{66}\left(\frac{\partial^2 \tilde{u}_x}{\partial y^2} + \mathrm{i}q\frac{\partial \tilde{u}_y}{\partial y}\right) + \rho\,\omega^2\tilde{u}_x = 0$$

$$C_{22}\frac{\partial^2 \tilde{u}_y}{\partial y^2} + \mathrm{i}qC_{12}\frac{\partial \tilde{u}_x}{\partial y} + C_{66}\left(-q^2\tilde{u}_y + \mathrm{i}q\frac{\partial \tilde{u}_x}{\partial y}\right) + \rho\,\omega^2\tilde{u}_y = 0$$

These equations are solved by assuming a solution in the form

$$\tilde{u}_j = A_j\,\mathrm{e}^{\mathrm{i}hy}, \qquad j = x, y$$

This gives the following equations

$$A_x(\rho\,\omega^2 - C_{11}q^2 - C_{66}h^2) - A_y qh(C_{12} + C_{66}) = 0$$

$$A_y(\rho\,\omega^2 - C_{22}h^2 - C_{66}q^2) - A_x qh(C_{12} + C_{66}) = 0$$

The determinant of this homogeneous system gives two values for h^2, then h_i are determined by $\operatorname{Im} h_i \geq 0$, $i = 1, 2$ (a real h_i should be positive). The system also determines the quotient between A_{xi} and A_{yi}, so that

$$A_{ix} = \alpha_i f_i, \qquad \alpha_i = \frac{\rho\,\omega^2 - C_{22}h_i^2 - C_{66}q^2}{qh_i(C_{12} + C_{66})}$$

$$A_{iy} = f_i$$

where $f_i = f_i(q)$, $i = 1, 2$, are to be determined.

The scattered displacement field for $y > 0$ can thus be given as

$$u_x^{\mathrm{sc}} = \sum_{i=1}^{2}\int_{-\infty}^{\infty} \alpha_i f_i(q)\,\mathrm{e}^{\mathrm{i}(qx + h_i y)}\,\mathrm{d}q$$

$$u_y^{\mathrm{sc}} = \sum_{i=1}^{2} \int_{-\infty}^{\infty} f_i(q)\, \mathrm{e}^{\mathrm{i}(qx+h_i y)}\, \mathrm{d}q$$

(the factor 2π in the inverse transform is ignored as it can be incorporated into f_i). The stress components needed for the boundary conditions become

$$\sigma_{xy}^{\mathrm{sc}} = \mathrm{i}C_{66} \sum_{i=1}^{2} \int_{-\infty}^{\infty} (\alpha_i h_i + q) f_i(q)\, \mathrm{e}^{\mathrm{i}(qx+h_i y)}\, \mathrm{d}q$$

$$\sigma_{yy}^{\mathrm{sc}} = \mathrm{i} \sum_{i=1}^{2} \int_{-\infty}^{\infty} (\alpha_i C_{12} q + C_{22} h_i) f_i(q)\, \mathrm{e}^{\mathrm{i}(qx+h_i y)}\, \mathrm{d}q$$

11.2.2 Integral equations

To derive the integral equations the symmetric part is first considered. The first boundary condition is valid for all x and therefore gives

$$\sum_{i=1}^{2} (\alpha_i h_i + q) f_i(q) = 0$$

The second boundary condition can be written

$$\int_{-\infty}^{\infty} (f_1(q) + f_2(q)) \mathrm{e}^{\mathrm{i}qx}\, \mathrm{d}q = \begin{cases} 0, & |x| > a \\ \frac{1}{2}\Delta u_y(x), & |x| < a \end{cases}$$

Inverting the Fourier transform gives

$$f_1(q) + f_2(q) = \frac{1}{4\pi} \int_{-\infty}^{\infty} \Delta u_y(x') \mathrm{e}^{-\mathrm{i}qx'}\, \mathrm{d}x'$$

Solving the two equations for f_1 and f_2 gives

$$f_1 = \frac{\alpha_2 h_2 + q}{4\pi(\alpha_2 h_2 - \alpha_1 h_1)} \int_{-a}^{a} \Delta u_y(x') \mathrm{e}^{-\mathrm{i}qx'}\, \mathrm{d}x'$$

$$f_2 = -\frac{\alpha_1 h_1 + q}{4\pi(\alpha_2 h_2 - \alpha_1 h_1)} \int_{-a}^{a} \Delta u_y(x') \mathrm{e}^{-\mathrm{i}qx'}\, \mathrm{d}x'$$

Insertion into the last boundary condition finally gives

$$\sigma_{yy}^{\mathrm{sc}}(x,0) = \frac{1}{4\pi}\int_{-\infty}^{\infty} E(q)\,\mathrm{e}^{\mathrm{i}qx}\int_{-a}^{a}\Delta u_y(x')\mathrm{e}^{-\mathrm{i}qx'}\,\mathrm{d}x'\,\mathrm{d}q$$

$$= -\sigma_{yy}^{\mathrm{in}}(x,0), \qquad |x| < a \tag{11.4}$$

where

$$E(q) =$$

$$\frac{C_{12}q(\alpha_1\alpha_2(h_2-h_1)+q(\alpha_1-\alpha_2))+C_{22}(\alpha_1h_1^2-\alpha_2h_2^2+q(h_1-h_2))}{\alpha_2h_2-\alpha_1h_1}$$

This integral equation is almost the same as in the case of isotropy [Eq. (10.16)]; the only difference is that $E(q)$ is more complicated in the case of anisotropy. The solution procedure is identical, the same COD expansion and projection is used, so this is not further elaborated on. However, one difference is that it is not possible to determine a T-matrix (as there are no polar wavefunctions for anisotropic solids). It is, on the other hand, possible to calculate the scattered wave in the far field by the stationary-phase method as done in the previous section for the corresponding scalar problem. As this becomes quite complicated it is not further pursued.

For the antisymmetric part the first boundary condition is valid for all x and therefore gives

$$\sum_{i=1}^{2}(\alpha_iC_{12}q + C_{22}h_i)f_i(q) = 0$$

The second boundary condition can be written

$$\int_{-\infty}^{\infty}(\alpha_1f_1(q)+\alpha_2f_2(q))\mathrm{e}^{\mathrm{i}qx}\,\mathrm{d}q = \begin{cases} 0, & |x| > a \\ \frac{1}{2}\Delta u_x(x), & |x| < a \end{cases}$$

Inverting the Fourier transform gives

$$\alpha_1f_1(q)+\alpha_2f_2(q) = \frac{1}{4\pi}\int_{-a}^{a}\Delta u_x(x')\mathrm{e}^{-\mathrm{i}qx'}\,\mathrm{d}x'$$

Solving the two equations for f_1 and f_2 gives

$$f_1 = \frac{\alpha_2 C_{12} q + C_{22} h_2}{4\pi C_{22}(\alpha_1 h_2 - \alpha_2 h_1)} \int_{-a}^{a} \Delta u_y(x') \mathrm{e}^{-\mathrm{i}qx'} \, \mathrm{d}x'$$

$$f_2 = -\frac{\alpha_1 C_{12} q + C_{22} h_1}{4\pi C_{22}(\alpha_1 h_2 - \alpha_2 h_1)} \int_{-a}^{a} \Delta u_y(x') \mathrm{e}^{-\mathrm{i}qx'} \, \mathrm{d}x'$$

Insertion into the last boundary condition finally gives for $|x| < a$

$$\begin{aligned} \sigma_{yy}^{\mathrm{sc}}(x,0) &= \frac{1}{4\pi} \int_{-\infty}^{\infty} D(q)\, \mathrm{e}^{\mathrm{i}qx} \int_{-a}^{a} \Delta u_y(x') \mathrm{e}^{-\mathrm{i}qx'} \, \mathrm{d}x' \, \mathrm{d}q \\ &= -\sigma_{yy}^{\mathrm{in}}(x,0), \qquad |x| < a \end{aligned} \tag{11.5}$$

where

$$D(q) = \frac{g_1 \left(C_{12}\alpha_2 q + C_{22} h_2\right) - g_2 \left(C_{12}\alpha_1 q + C_{22} h_1\right)}{(\alpha_1 h_2 - \alpha_2 h_1)}$$

where

$$g_i = \rho\,\omega^2 - C_{22} h_i^2 + C_{66} q^2, \qquad i1, 2$$

Exactly the same remarks as for the symmetric part apply.

11.3 Rectangular crack

A rectangular crack is next investigated, see Boström *et al.* [24] for the more general case with a rectangular crack in a half-space. The anisotropy is allowed to be arbitrary so all the elements of the stiffness tensor c_{ijkl} can be nonzero. It is convenient to use Cartesian tensors and the Cartesian coordinate system is thus (x_1, x_2, x_3). In this system the crack is located at $|x_1| < a$, $|x_2| < b$, and $x_3 = 0$. The crack is assumed open, but there are no particular difficulties to consider other boundary conditions.

11.3.1 Green tensor

Using the symmetries of the strain and stiffness tensors, the stress-displacement relation in the general case can be written

$$\sigma_{ij} = c_{ijkl} \partial_k u_l$$

The general equation of motion is

$$\partial_k \sigma_{ki} + \rho\omega^2 u_i = 0$$

The standard way of solving these equations in Cartesian coordinates is to eliminate the stress tensor and make a plane wave ansatz [9]. This leads to the Christoffel equation for determining possible wave vectors. However, a slightly different approach is preferred. Taking a double Fourier transform in x_1 and x_2 with transform variables k_1 and k_2 the two equations above can be viewed as a generalized eigenvalue problem

$$\begin{pmatrix} a_{11} & a_{12} & a_{13} & \mathrm{i} & 0 & 0 \\ a_{21} & a_{22} & a_{23} & 0 & \mathrm{i} & 0 \\ a_{31} & a_{32} & a_{33} & 0 & 0 & \mathrm{i} \\ a_{41} & a_{42} & a_{43} & 0 & 0 & 0 \\ a_{51} & a_{52} & a_{53} & 0 & 0 & 0 \\ 0 & 0 & a_{63} & a_{64} & a_{65} & 0 \end{pmatrix} \begin{pmatrix} U_{n1} \\ U_{n2} \\ U_{n3} \\ T_{n1} \\ T_{n2} \\ T_{n3} \end{pmatrix}$$
$$= k_{n3} \begin{pmatrix} b_{11} & b_{12} & b_{13} & 0 & 0 & 0 \\ b_{21} & b_{22} & b_{23} & 0 & 0 & 0 \\ b_{31} & b_{32} & b_{33} & 0 & 0 & 0 \\ b_{41} & b_{42} & b_{43} & \mathrm{i} & 0 & 0 \\ b_{51} & b_{52} & b_{53} & 0 & \mathrm{i} & 0 \\ 0 & 0 & 0 & 0 & 0 & \mathrm{i} \end{pmatrix} \begin{pmatrix} U_{n1} \\ U_{n2} \\ U_{n3} \\ T_{n1} \\ T_{n2} \\ T_{n3} \end{pmatrix} \tag{11.6}$$

The elements are

$$a_{11} = C_{15}k_1 + C_{56}k_2, \quad a_{12} = C_{56}k_1 + C_{25}k_2, \quad a_{13} = C_{55}k_1 + C_{45}k_2,$$

$$a_{21} = C_{14}k_1 + C_{46}k_2, \quad a_{22} = C_{46}k_1 + C_{24}k_2, \quad a_{23} = C_{45}k_1 + C_{44}k_2,$$

$$a_{31} = C_{13}k_1 + C_{36}k_2, \quad a_{32} = C_{36}k_1 + C_{23}k_2, \quad a_{33} = C_{35}k_1 + C_{34}k_2,$$

$$a_{41} = C_{11}k_1^2 + 2C_{16}k_1k_2 + C_{66}k_2^2 - \rho\omega^2,$$

$$a_{42} = C_{16}k_1^2 + (C_{12} + C_{66})k_1k_2 + C_{26}k_2^2,$$

$$a_{43} = C_{15}k_1^2 + (C_{14} + C_{56})k_1k_2 + C_{46}k_2^2,$$

$$a_{52} = C_{66}k_1^2 + 2C_{26}k_1k_2 + C_{22}k_2^2 - \rho\omega^2,$$

$$a_{53} = C_{56}k_1^2 + (C_{25} + C_{46})k_1k_2 + C_{24}k_2^2,$$

$$a_{63} = -\rho\,\omega^2, \quad a_{64} = -\mathrm{i}k_1, \quad a_{65} = -\mathrm{i}k_2$$

and

$$b_{11} = -C_{55}, \quad b_{12} = -C_{45}, \quad b_{13} = -C_{35},$$

$$b_{22} = -C_{44}, \quad b_{23} = -C_{34}, \quad b_{33} = -C_{33},$$

$$b_{41} = -C_{15}k_1 - C_{56}k_2, \quad b_{42} = -C_{14}k_1 - C_{46}k_2,$$

$$b_{43} = -C_{13}k_1 - C_{36}k_2, \quad b_{51} = -C_{56}k_1 - C_{25}k_2,$$

$$b_{52} = -C_{46}k_1 - C_{24}k_2, \quad b_{53} = -C_{36}k_1 - C_{23}k_2$$

Here the abbreviated notation for the elastic constants is used [9]. This generalized eigenvalue problem has six solutions for k_{n3}, $n = 1,2,3,4,5,6$, and corresponding eigenvectors U_{nj} (the displacement polarization vectors) and T_{nj}, $j = 1,2,3$ (the traction polarization vectors). These are written in vector notation as $\boldsymbol{k}_n = (k_1, k_2, k_{n3})$ and $\boldsymbol{U}_n$ and $\boldsymbol{T}_n$. The wave vectors are sorted into three propagating in the positive x_3-direction and three in the negative x_3-direction. This is most easily done by having (small) damping in the material, because then an up-going wave must have a wavenumber with a positive imaginary part and a down-going wave must have a wavenumber with a negative imaginary part. If no damping is present the group velocity must be examined. The up-going and down-going waves are denoted $\boldsymbol{k}_{\boldsymbol{n}}^+$, $\boldsymbol{U}_n^+$, $\boldsymbol{T}_n^+$ and $\boldsymbol{k}_{\boldsymbol{n}}^-$, $\boldsymbol{U}_n^-$, $\boldsymbol{T}_n^-$, respectively, where now $n = 1,2,3$.

To proceed the integral equation for the crack scattering is derived by the integral representation. For this the Green tensor for the anisotropic solid is needed. This tensor is the solution of

$$\partial_k \Sigma_{kij}(\boldsymbol{r};\boldsymbol{r}') + \rho\,\omega^2 G_{ij}(\boldsymbol{r};\boldsymbol{r}') = -\delta_{ij}\delta(\boldsymbol{r} - \boldsymbol{r}')$$

where Σ_{kij} is the Green stress tensor and has the same relation to the Green tensor as the stress tensor has to the displacement

$$\Sigma_{ijk}(\boldsymbol{r};\boldsymbol{r}') = c_{ijlm}\partial_l G_{mk}(\boldsymbol{r};\boldsymbol{r}')$$

Note that the derivatives in these two equations are with respect to the first argument $\boldsymbol{r}$; the second argument $\boldsymbol{r}'$ acts as a parameter in the equations. Taking a double Fourier transform in x_1 and x_2, it is clear that the Green tensor admits the following representation (using also the symmetry of the Green tensor in arguments and indices)

$$G_{ij}(\boldsymbol{r};\boldsymbol{r}') = \int_{-\infty}^{\infty}\int_{-\infty}^{\infty}\sum_{n=1}^{3} D_n^{\pm}U_{ni}^{\pm}U_{nj}^{\pm}\exp\left(\mathrm{i}\boldsymbol{k}_n^{\pm}\cdot(\boldsymbol{r}-\boldsymbol{r}')\right)\mathrm{d}k_1\,\mathrm{d}k_2 \tag{11.7}$$

with superscript "+" for $x_3 > x_3'$ and "−" for $x_3 < x_3'$. The corresponding Green stress tensor is

$$\Sigma_{3ij}(\boldsymbol{r};\boldsymbol{r}') = \int_{-\infty}^{\infty}\int_{-\infty}^{\infty}\sum_{n=1}^{3} D_n^{\pm}T_{ni}^{\pm}U_{nj}^{\pm}\exp\left(\mathrm{i}\boldsymbol{k}_n^{\pm}\cdot(\boldsymbol{r}-\boldsymbol{r}')\right)\mathrm{d}k_1\,\mathrm{d}k_2 \tag{11.8}$$

There remains to determine the integration constants $D_n^{\pm} = D_n^{\pm}(k_1,k_2)$. These are determined by the jump conditions coming from the delta function

$$[G_{ij}(\boldsymbol{r};\boldsymbol{r}')]_{x_3=x_3'} = 0, \qquad [\Sigma_{ij}(\boldsymbol{r};\boldsymbol{r}')]_{x_3=x_3'} = -\delta_{ij}\delta(x_1-x_1')\delta(x_2-x_2')$$

where $[f(x_3;x_3')]_{x_3=x_3'} = \lim_{\epsilon\to 0^+}(f(x_3+\epsilon;x_3') - f(x_3-\epsilon;x_3'))$ is the jump at $x_3 = x_3'$. These conditions lead to

$$\sum_{n=1}^{3}\left(D_n^{+}U_{ni}^{+}U_{nj}^{+} - D_n^{-}U_{ni}^{-}U_{nj}^{-}\right) = 0$$

$$\sum_{n=1}^{3}\left(D_n^{+}T_{ni}^{+}U_{nj}^{+} - D_n^{-}T_{ni}^{-}U_{nj}^{-}\right) = -\frac{\delta_{ij}}{4\pi^2}$$

where $i,j = 1,2,3$. It is noted that there are in all 18 equations for the six unknown integration constants. However, the equations are not linearly independent and the six equations obtained when $i = j$ are used to determine the constants.

11.3.2 Integral equation

The integral representation for an anisotropic solid is derived in the same way as for an isotropic solid. The results looks almost the same, the difference stems from the fact that the normalization of the Green tensor is different in the two cases. The integral representation becomes

$$u_j^{\mathrm{sc}}(\boldsymbol{r}) = \int_S n_k \left[u_i(\boldsymbol{r}')\Sigma_{kij}(\boldsymbol{r}', \boldsymbol{r}) - \sigma_{ki}(\boldsymbol{r}')G_{ij}(\boldsymbol{r}', \boldsymbol{r})\right] \mathrm{d}S' \tag{11.9}$$

where $\boldsymbol{r}$ must not be on the crack and S must mean both sides of the crack. In this case the normal to the crack is $\boldsymbol{e}_3$ on the top side and $-\boldsymbol{e}_3$ on the bottom side and using this and the boundary condition on the crack results in

$$u_j^{\mathrm{sc}}(\boldsymbol{r}) = \int_{S_0} \Delta u_i(\boldsymbol{r}')\Sigma_{3ij}(\boldsymbol{r}', \boldsymbol{r}) \, \mathrm{d}S'$$

where Δu_j is the COD and the integration is over the crack surface. Apply the normal traction operator to this equation, let the field point approach the crack, and use the boundary condition once more to end up with

$$-\sigma_{3j}^{\mathrm{in}}(x_1, x_2, 0)$$

$$= \lim_{x_3 \to 0+} \int_{-a}^{a} \int_{-b}^{b} \Delta u_i(x_1', x_2') S_{ij}(x_1', x_2', 0; x_1, x_2, x_3) \, \mathrm{d}x_1' \, \mathrm{d}x_2' \tag{11.10}$$

where $\sigma_{3j}^{\mathrm{in}}$ is the traction vector due to the incident wave on the crack. The kernel in the integral equation is obtained by operating twice with the traction operator (once on each argument) on the Green tensor

$$S_{ij}(\boldsymbol{r}; \boldsymbol{r}') = c_{3ikl} c_{3jmn} \frac{\partial^2 G_{ln}(\boldsymbol{r}; \boldsymbol{r}')}{\partial x_k \partial x_m'}$$

where the derivatives are written out explicitly to avoid confusion. Equation (11.10) is a hypersingular integral equation for the COD, and this is why the limit in front of the integrals cannot be moved into the integrand. The solution procedure makes it possible at a later stage to take the limit inside the integrals.

To solve the integral equation a double expansion in Chebyshev functions is performed to get the right edge condition, thus

$$\Delta u_i(x_1', x_2') = \sum_{l'=1}^{\infty} \sum_{j'=1}^{\infty} \beta_{il'j'} \psi_{l'}(x_1'/a) \psi_{j'}(x_2'/b)$$

The expansion is inserted into Eq. (11.10) and the equation is multiplied by $\psi_l(x_1/a)\psi_j(x_2/b)$ and integrated over the crack

$$\sum_{i'=1}^{3}\sum_{l'=1}^{\infty}\sum_{j'=1}^{\infty} Q_{ilj,i'l'j'}\beta_{i'l'j'}(k_2) = M_{ilj} \tag{11.11}$$

where the matrix is

$$Q_{ilj,i'l'j'} = (-1)^{l'+j'}\int_{-\infty}^{\infty}\int_{-\infty}^{\infty}$$

$$\times \sum_{n=1}^{3} D_n^+ T_{ni}^+ T_{ni'}^+ \mathrm{J}_l(k_1 a)\mathrm{J}_{l'}(k_1 a)\mathrm{J}_j(k_2 b)\mathrm{J}_{j'}(k_2 b)\,\frac{\mathrm{d}k_1\,\mathrm{d}k_2}{k_1^2 k_2^2} \tag{11.12}$$

and the right-hand side is

$$M_{ilj} = -\int_{-a}^{a}\int_{-b}^{b} \sigma_{3i}^{\mathrm{in}}(x_1, x_2, 0)\psi_l(x_1/a)\psi_j(x_2/b)\,\mathrm{d}x_1\,\mathrm{d}x_2 \tag{11.13}$$

The system Eq. (11.11) determines the COD. With the COD determined the integral representation can be used to obtain the scattered wave everywhere.

From the previous chapter it is seen that the solution procedure for the strip-like and rectangular cracks are very alike. This is true also in the case of anisotropy, so the strip-like crack is not considered.

12 Scattering by cracks in structures

In the previous two chapters the scattering by a crack of simple shape in infinite space (isotropic or anisotropic) is considered. In the present chapter this is extended to the scattering by a crack when there are more (infinite) scattering surfaces present. Typically this can be an infinite plane between two elastic half-spaces. As a special case the crack may then be an interface crack. In the following an interface crack and a tilted crack in two half-planes for 2D SH waves and a strip-like crack in a half-space in 3D are considered. There are more similar cases that can be solved by essentially the same method, for instance one (or more) crack in a (layered) plate or half-space.

12.1 Interface crack: 2D SH waves

To start as simple as possible, 2D SH waves are first considered. An example, which is quite similar to the simple straight crack in infinite space considered in Section 10.1, is the interface crack between two half-planes and this case is therefore taken up first, although more general boundary conditions are considered. Boström [15] investigates the corresponding open interface crack and Chevalier *et al.* [37] the interface crack with spring boundary conditions. Neerhoff [101] and Kundu [78, 79, 80] considers interface cracks in a layered plate or half-plane.

12.1.1 Integral equations

Let the x-axis be the interface between the two half-planes $y < 0$ and $y > 0$. For $y > 0$ the density is ρ_1, the shear modulus is μ_1, and the shear wavenumber is k_1. In the half-plane $y < 0$ the corresponding quantities are

ρ_2, μ_2, and k_2. The (out-of-plane) displacement in the two half-planes are u_1 and u_2. The crack is situated along the interface for $|x| < a$. To show how this is treated, more general boundary conditions are assumed at the "crack" (simply called crack in the following), assuming a jump in both displacement and traction. Thus, the boundary conditions at $y = 0$, $|x| < a$ are taken as

$$u_1 - u_2 = \alpha_1 \frac{\partial u_1}{\partial y} + \alpha_2 \frac{\partial u_2}{\partial y}$$

$$\mu_1 \frac{\partial u_1}{\partial y} - \mu_2 \frac{\partial u_2}{\partial y} = \beta_1 u_1 + \beta_2 u_2 \tag{12.1}$$

These boundary conditions on the crack can be used as a model for a thin interface layer and in this case the parameters should be chosen according to

$$\alpha_1 = \frac{d}{2}\left(\frac{\mu_1}{\mu_0} - 1\right), \qquad \alpha_2 = \frac{d}{2}\left(\frac{\mu_2}{\mu_0} - 1\right)$$

$$\beta_1 = \frac{\omega^2 d}{2}\left(\rho_1 - \rho_0\right), \qquad \beta_2 = \frac{\omega^2 d}{2}\left(\rho_2 - \rho_0\right)$$

Here ρ_0, μ_0, and d are the density, shear modulus, and thickness, respectively, of the thin interface layer. It is noted that the jump in displacement is due to stiffness effects and the jump in stress is due to inertia effects and is frequency dependent. These boundary conditions are discussed fully by Bövik and Olsson [32] and they show that there should also be second order derivatives terms in x of the displacements in the second boundary condition for these conditions to be exact to order d. Such second order derivatives do, however, lead to extra singularities at the crack tips and are therefore avoided. By choosing the parameters differently the boundary conditions could presumably be used also to model other situations, e.g. a damaged interface.

At the interface between the two half-planes and outside the crack the waves satisfy welded boundary conditions

$$u_1 = u_2, \qquad \mu_1 \frac{\partial u_1}{\partial y} = \mu_2 \frac{\partial u_2}{\partial y}$$

The displacements satisfy Helmholtz equations with the appropriate wavenumber. As usual the displacements are divided into incident and scattered parts. The incident part is taken as a wave that is excited anywhere and satisfies the welded boundary conditions along the whole interface as if there was no crack. Thus the waves are written

$$u_1 = u_1^{\text{in}} + U_1, \qquad u_2 = u_2^{\text{in}} + U_2$$

where the incident parts u_1^{in} and u_2^{in} are assumed known and satisfy the welded boundary conditions for $y = 0$ and all x. The scattered parts due to the presence of the crack are denoted U_1 and U_2, and for simplicity of notation the index "sc" is not included. As an example, an incident plane wave from the lower half-plane leads to the following incident waves

$$u_1^{\text{in}} = T_{21}\mathrm{e}^{\mathrm{i}(q_0x+h_{10}y)}, \qquad y > 0$$

$$u_2^{\text{in}} = \mathrm{e}^{\mathrm{i}(q_0x+h_{20}y)} + R_{21}\mathrm{e}^{\mathrm{i}(q_0x-h_{20}y)}, \qquad y < 0$$

where the wave incident on the interface has unit amplitude and the transmission and reflection coefficients are

$$T_{21} = \frac{2\mu_2 h_{20}}{\mu_1 h_{10} + \mu_2 h_{20}}$$

$$R_{21} = \frac{\mu_2 h_{20} - \mu_1 h_{10}}{\mu_1 h_{10} + \mu_2 h_{20}}$$

The horizontal wavenumber q_0 is given and is related to the angle of incidence of the plane wave. The wavenumbers in the y-direction are

$$h_{i0} = \sqrt{k_i^2 - q_0^2}, \qquad i = 1, 2$$

and if a wave is evanescent the root is chosen according to $\operatorname{Im} h_{i0} \geq 0$, $i = 1, 2$.

To formulate integral equations the scattered waves are represented as Fourier integrals

$$U_1 = \int_{-\infty}^{\infty} f_1(q)\, \mathrm{e}^{\mathrm{i}(qx+h_1y)}\, \mathrm{d}q$$

$$U_2 = \int_{-\infty}^{\infty} f_2(q)\, \mathrm{e}^{\mathrm{i}(qx-h_2y)}\, \mathrm{d}q$$

where

$$h_i = \sqrt{k_i^2 - q^2}, \qquad i = 1, 2$$

where the branches are chosen so that $\operatorname{Im} h_i \geq 0$, $i = 1, 2$. The boundary conditions outside the crack then leads to

$$\int_{-\infty}^{\infty} (f_1(q) - f_2(q))\, \mathrm{e}^{\mathrm{i}qx}\, \mathrm{d}q = \begin{cases} 0, & |x| > a \\ \Delta U, & |x| < a \end{cases}$$

$$\mathrm{i} \int_{-\infty}^{\infty} (\mu_1 h_1 f_1(q) + \mu_2 h_2 f_2(q))\, \mathrm{e}^{\mathrm{i}qx}\, \mathrm{d}q = \begin{cases} 0, & |x| > a \\ \Delta T, & |x| < a \end{cases}$$

where both the COD ΔU and the jump in traction ΔT are introduced. Inverting the Fourier transforms determines the Fourier kernels

$$f_1(q) = \frac{1}{2\pi(\mu_1 h_1 + \mu_2 h_2)} \int_{-a}^{a} (\mu_2 h_2 \Delta U - \mathrm{i}\Delta T)\, \mathrm{e}^{-\mathrm{i}qx}\, \mathrm{d}x$$

$$f_2(q) = -\frac{1}{2\pi(\mu_1 h_1 + \mu_2 h_2)} \int_{-a}^{a} (\mu_1 h_1 \Delta U + \mathrm{i}\Delta T)\, \mathrm{e}^{-\mathrm{i}qx}\, \mathrm{d}x$$

Inserting f_1 and f_2 in the Fourier representations of U_1 and U_2 and applying the two boundary conditions on the crack [Eq. (12.1)] finally give

$$\begin{aligned} &\int_{-\infty}^{\infty} \frac{\mu_1 h_1 + \mu_2 h_2 - \mathrm{i} h_1 h_2 (\alpha_1 \mu_2 + \alpha_2 \mu_1)}{2\pi(\mu_1 h_1 + \mu_2 h_2)}\, \mathrm{e}^{\mathrm{i}qx} \int_{-a}^{a} \Delta U(x') \mathrm{e}^{-\mathrm{i}qx'}\, \mathrm{d}x' \mathrm{d}q \\ &\qquad + \int_{-\infty}^{\infty} \frac{\alpha_2 h_2 - \alpha_1 h_1}{2\pi(\mu_1 h_1 + \mu_2 h_2)}\, \mathrm{e}^{\mathrm{i}qx} \int_{-a}^{a} \Delta T(x') \mathrm{e}^{-\mathrm{i}qx'}\, \mathrm{d}x' \mathrm{d}q \\ &= \alpha_1 \left. \frac{\partial u^{\mathrm{in},1}(x,y)}{\partial y} \right|_{y=0} + \alpha_2 \left. \frac{\partial u^{\mathrm{in},2}(x,y)}{\partial y} \right|_{y=0} \end{aligned}$$

$$\begin{aligned} &\int_{-\infty}^{\infty} \frac{\mu_1 h_1 + \mu_2 h_2 + \mathrm{i}(\beta_1 + \beta_2)}{2\pi(\mu_1 h_1 + \mu_2 h_2)}\, \mathrm{e}^{\mathrm{i}qx} \int_{-a}^{a} \Delta T(x') \mathrm{e}^{-\mathrm{i}qx'}\, \mathrm{d}x' \mathrm{d}q \\ &\qquad + \int_{-\infty}^{\infty} \frac{\beta_2 \mu_1 h_1 - \beta_1 \mu_2 h_2}{2\pi(\mu_1 h_1 + \mu_2 h_2)}\, \mathrm{e}^{\mathrm{i}qx} \int_{-a}^{a} \Delta U(x') \mathrm{e}^{-\mathrm{i}qx'}\, \mathrm{d}x' \mathrm{d}q \\ &= \beta_1 u^{\mathrm{in},1}(x,0) + \beta_2 u^{\mathrm{in},2}(x,0) \end{aligned} \tag{12.2}$$

This is two coupled integral equations for the two unknowns ΔU and ΔT.

12.1.2 Solution of the integral equations

To solve the integral equations the jumps in displacement and traction are expanded in suitable systems. For the COD the same Chebyshev functions as used in the previous two chapters [as defined in Eq. (10.3)] are used, thus

$$\Delta U(x) = \sum_{n=1}^{\infty} c_n \psi_n(x/a)$$

This expansion has the expected square root behaviour at the crack edges. However, these expansion functions are not suitable for the jump in traction, because this jump should have a square root singularity at the edges. Instead the following expansion is made

$$\Delta T(x) = \frac{\mu_1}{\sqrt{a^2 - x^2}} \sum_{n=1}^{\infty} d_n \phi_n(x/a)$$

Here μ_1 is inserted for dimensional reasons and the Chebyshev polynomials of the first kind are

$$\phi_n(s) = \begin{cases} \cos(n \arcsin s), & n = 0, 2, 4, \ldots \\ \mathrm{i}\sin(n \arcsin s), & n = 1, 3, \ldots \end{cases} \tag{12.3}$$

These functions have the convenient property

$$\int_{-1}^{1} \frac{\phi_n(s)}{\sqrt{1-s^2}} \, \mathrm{e}^{-\mathrm{i}\gamma s} \, \mathrm{d}s = \mathrm{J}_n(\gamma) \tag{12.4}$$

Insert the two expansions into the integral equations Eq. (12.2) and multiply the first integral equation with $\psi_n(x/a)$ and the second with $\phi_n(x/a)(a^2 - x^2)^{-1/2}$ and integrate from $-a$ to a in x to obtain the discretized equations

$$\sum_{n'=1}^{\infty} B_{nn'} c_{n'} + \sum_{n'=0}^{\infty} C_{nn'} d_{n'}$$

$$= - \int_{-a}^{a} \left(\alpha_1 \left. \frac{\partial u^{\mathrm{in},1}(x,y)}{\partial y} \right|_{y=0} + \alpha_2 \left. \frac{\partial u^{\mathrm{in},2}(x,y)}{\partial y} \right|_{y=0} \right) \psi_n(x/a) \, \mathrm{d}x$$

$$\sum_{n'=1}^{\infty} E_{nn'}c_{n'} + \sum_{n'=0}^{\infty} F_{nn'}d_{n'}$$
$$= \int_{-a}^{a} \left(\beta_1 u^{\mathrm{in},1}(x,0) + \beta_2 u^{\mathrm{in},2}(x,0)\right) \frac{\phi_n(x/a)}{\sqrt{a^2-x^2}}\,\mathrm{d}x$$

Here the matrices are

$$B_{nn'} = nn'(-1)^n \int_{-\infty}^{\infty} \frac{\mu_1 h_1 + \mu_2 h_2 - \mathrm{i}h_1 h_2(\alpha_1\mu_2 + \alpha_2\mu_1)}{2\pi(\mu_1 h_1 + \mu_2 h_2)}$$
$$\times \mathrm{J}_n(qa)\mathrm{J}_{n'}(qa)\,\frac{\mathrm{d}q}{q^2}$$

$$C_{nn'} = n\mu_1(-1)^n \int_{-\infty}^{\infty} \frac{\alpha_2 h_2 - \alpha_1 h_1}{2\pi(\mu_1 h_1 + \mu_2 h_2)}\,\mathrm{J}_n(qa)\mathrm{J}_{n'}(qa)\,\frac{\mathrm{d}q}{q}$$

$$E_{nn'} = n' \int_{-\infty}^{\infty} \frac{\beta_2\mu_1 h_1 - \beta_1\mu_2 h_2}{2\pi(\mu_1 h_1 + \mu_2 h_2)}\,\mathrm{J}_n(qa)\mathrm{J}_{n'}(qa)\,\frac{\mathrm{d}q}{q}$$

$$F_{nn'} = \mu_1 \int_{-\infty}^{\infty} \frac{\mu_1 h_1 + \mu_2 h_2 + \mathrm{i}(\beta_1 + \beta_2)}{2\pi(\mu_1 h_1 + \mu_2 h_2)}\,\mathrm{J}_n(qa)\mathrm{J}_{n'}(qa)\,\mathrm{d}q$$

12.1.3 Scattered waves

Once the equations are solved for c_n, $n = 1, 2, \ldots$, and d_n, $n = 0, 1, \ldots$, the crack-scattered waves can be obtained from

$$U_1 = \sum_{n=1}^{\infty} nc_n \frac{\mu_2 h_2}{2\pi(\mu_1 h_1 + \mu_2 h_2)} \mathrm{J}_n(qa)\mathrm{e}^{\mathrm{i}(qx+h_1 y)}$$
$$-\mathrm{i}\mu_1 \sum_{n=0}^{\infty} d_n \frac{1}{2\pi(\mu_1 h_1 + \mu_2 h_2)} \mathrm{J}_n(qa)\mathrm{e}^{\mathrm{i}(qx+h_1 y)}, \qquad y > 0$$

$$U_2 = -\sum_{n=1}^{\infty} nc_n \frac{\mu_1 h_1}{2\pi(\mu_1 h_1 + \mu_2 h_2)} \mathrm{J}_n(qa)\mathrm{e}^{\mathrm{i}(qx+h_2 y)}$$
$$-\mathrm{i}\mu_1 \sum_{n=0}^{\infty} d_n \frac{1}{2\pi(\mu_1 h_1 + \mu_2 h_2)} \mathrm{J}_n(qa)\mathrm{e}^{\mathrm{i}(qx+h_2 y)}, \qquad y < 0$$

If the scattered waves are of interest in the far field these q-integrals can be calculated by the method of stationary-phase. Polar coordinates (r, φ) are then introduced, where thus r is large (compared to both the crack size and the wavelengths). The stationary points appear at $q_s = k_1 \cos\varphi$ and $q_s = k_2 \cos\varphi$ for $y > 0$ and $y < 0$, respectively, and the waves in the far field becomes

$$U_1 = \sqrt{\frac{2}{\pi k_1 r}}\, \mathrm{e}^{\mathrm{i}(k_1 r - \pi/4)} F_1(\varphi)$$

$$U_2 = \sqrt{\frac{2}{\pi k_2 r}}\, \mathrm{e}^{\mathrm{i}(k_2 r - \pi/4)} F_2(\varphi)$$

The far field amplitudes are

$$F_1(\varphi) = \frac{\mu_1 k_1 \sin\varphi}{2\left(\mu_1 k_1 \sin\varphi + \mu_2 \sqrt{k_2^2 - k_1^2 \cos^2\varphi}\right)}$$

$$\times \left[\frac{\mu_2 \sqrt{k_2^2 - k_1^2 \cos^2\varphi}}{\mu_1 k_1 \cos\varphi} \sum_{n=1}^{\infty} n c_n \mathrm{J}_n(k_1 a \cos\varphi) - \mathrm{i} \sum_{n=0}^{\infty} d_n \mathrm{J}_n(k_1 a \cos\varphi) \right]$$

$$F_2(\varphi) = \frac{-\mu_1 k_2 \sin\varphi}{2\left(\mu_2 k_2 \sin\varphi - \mu_1 \sqrt{k_1^2 - k_2^2 \cos^2\varphi}\right)}$$

$$\times \left[\frac{\sqrt{k_1^2 - k_2^2 \cos^2\varphi}}{k_2 \cos\varphi} \sum_{n=1}^{\infty} n c_n \mathrm{J}_n(k_2 a \cos\varphi) - \mathrm{i} \sum_{n=0}^{\infty} d_n \mathrm{J}_n(k_2 a \cos\varphi) \right]$$

Here all square roots are taken with an imaginary part that is zero or positive.

12.2 Crack in layered media: 2D SH waves

When the crack is situated in a medium that is layered with plane parallel infinite interfaces, there are two cases. One is when the crack is parallel with the interfaces and then Fourier representations can be used in each layer separately, including one directly above and one directly below the crack. Applying all the boundary conditions it is still possible to obtain a hypersingular integral equation for the COD; the only difference is that the kernel is more complicated.

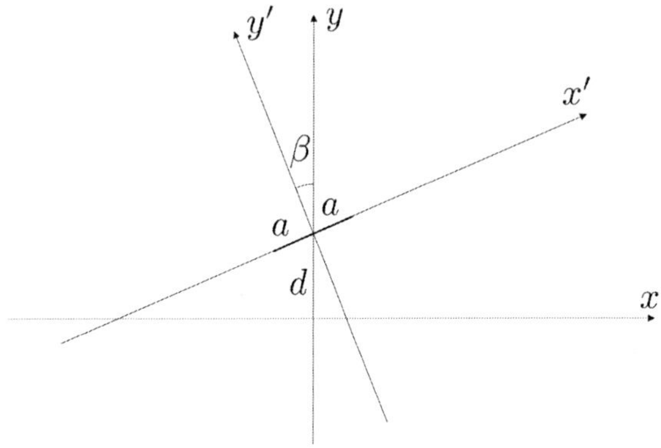

Figure 12.1: The scattering geometry with a tilted crack in two half-planes.

If the crack is not parallel with the layering this approach is no longer possible. By using the Green function approach with the integral representation, however, a hypersingular integral equation for the COD can be derived as shown in the present section.

12.2.1 Green function

Consider a scattering configuration with a crack in a space consisting of two half-planes with different properties. The crack has width $2a$ and is located in the upper half-plane with its centre at the distance d from the interface and is titled counterclockwise with the angle β relative the interface, see Figure 12.1. Two coordinate systems are introduced, the xy-system has the x-axis at the interface between the two layers and the y-axis through the centre of the crack. The $x'y'$-system is fixed in the crack with the x'-axis along the crack and the y'-axis through the centre of the crack. In the upper half-plane the shear modulus is μ_1 and the shear wavenumber is k_1 and in the lower half-plane the corresponding quantities are μ_2 and k_2. The transformation between the coordinate systems is

$$x = x' \cos\beta - y' \sin\beta$$

$$y = d + x' \sin\beta + y' \cos\beta$$

The boundary conditions between the two half-planes are that the displacement and traction are continuous

$$u_1 = u_2$$

$$\mu_1 \frac{\partial u_1}{\partial y} = \mu_2 \frac{\partial u_2}{\partial y}$$

for $y = 0$ and all x. Here u_1 and u_2 are the displacements in the upper and lower half-plane, respectively.

Contrary to previous chapters the Green function needed in this case is not the free space one. Instead the Green function is the one for the two half-planes with the source in the upper half-plane (because the crack is located there). Thus the Green function should satisfy the respective Helmholtz equations

$$\nabla^2 g_1(x, y; x_0, y_0) + k_1^2 g_1(x, y, x_0, y_0) = -\delta(x - x_0)\delta(y - y_0)$$

for $y > 0$ and

$$\nabla^2 g_2(x, y; x_0, y_0) + k_2^2 g_2(x, y, x_0, y_0) = 0$$

for $y < 0$. In addition welded boundary conditions for $y = 0$ and radiation conditions are to be satisfied. The solution is most easily obtained by starting with the free space Green function for the upper half-plane [Eq. (8.6)] to satisfy the source term and then add terms to satisfy the boundary conditions

$$\begin{aligned} g_1(x, y; x_0, y_0) &= \frac{\mathrm{i}}{4\pi} \int_{-\infty}^{\infty} \frac{1}{h_1} \mathrm{e}^{\mathrm{i}(q(x-x_0)+h_1|y-y_0|)}\, \mathrm{d}q \\ &+ \frac{\mathrm{i}}{4\pi} \int_{-\infty}^{\infty} \frac{R}{h_1} \mathrm{e}^{\mathrm{i}(q(x-x_0)+h_1(y+y_0))}\, \mathrm{d}q \end{aligned}$$

for $y > 0$ and

$$g_2(x, y; x_0, y_0) = \frac{\mathrm{i}}{4\pi} \int_{-\infty}^{\infty} \frac{T}{h_1} \mathrm{e}^{\mathrm{i}(q(x-x_0)+h_1 y_0 - h_2 y)}\, \mathrm{d}q$$

for $y < 0$. Here $h_i = (k_i^2 - q^2)^{1/2}$ with $\operatorname{Im} h_i \geq 0$, $i = 1, 2$. The solution is chosen to fulfil the radiation condition and the extra factors from the free part are incorporated explicitly. The boundary conditions on $y = 0$ give that R and T are the ordinary plane wave reflection and transmission coefficients

$$R = \frac{\mu_1 h_1 - \mu_2 h_2}{\mu_1 h_1 + \mu_2 h_2}$$

$$T = \frac{2\mu_1 h_1}{\mu_1 h_1 + \mu_2 h_2}$$

To be useful the Green function should be expressed in the crack coordinate system (x', y'). This can of course be performed with the transformation of the coordinates and this is useful for the reflected and transmitted parts but for the free part it gives a useless expression due to the absolute values. However, the free part of the Green function has the same appearance in all coordinate systems so the Green function in the upper half-plane can be expressed in the following form in the (x', y') system (note that also the source coordinates must be transformed)

$$g_1(x', y'; x'_0, y'_0) = \frac{\mathrm{i}}{4\pi} \int_{-\infty}^{\infty} \frac{1}{h_1} \left(\mathrm{e}^{\mathrm{i}(q(x'-x'_0)+h_1|y'-y'_0|)} \right.$$

$$\left. + R\,\mathrm{e}^{\mathrm{i}(h_1 \sin\beta(x'+x'_0)+q\cos\beta(x'-x'_0))+h_1\cos\beta(y'+y'_0)+q\sin\beta(y'-y'_0)+2dh_1)} \right) \mathrm{d}q \tag{12.5}$$

12.2.2 Integral equation

To derive the integral representation in this case it is first useful to divide the scattering problem into two parts as usual

$$u_1 = u_1^{\mathrm{in}} + u_1^{\mathrm{sc}}$$

$$u_2 = u_2^{\mathrm{in}} + u_2^{\mathrm{sc}} \tag{12.6}$$

However, here the incident waves are the ones that exist in the absence of the crack and the incident waves thus satisfy the boundary condition at the interface. The sources of the incident waves can be located in either half-plane (or in both).

Integral representations are now derived for the scattered waves (not the total waves). These have no sources outside the crack, so there is no "incident wave" for the scattered waves. The integral representation for the upper half-plane becomes (repeating the steps in Section 4.2.1)

$$\int_{-\infty}^{\infty}\left[u_1^{\mathrm{sc}}(x_0,0)\frac{\partial}{\partial y_0}g_1(x_0,0;x,y)-g_1(x_0,0;x,y)\frac{\partial u_1^{\mathrm{sc}}(x_0,0)}{\partial y_0}\right]\mathrm{d}x_0$$

$$+\int_S\left[u_1^{\mathrm{sc}}(\boldsymbol{r}')\nabla' g_1(\boldsymbol{r}';\boldsymbol{r})-(\nabla' u_1^{\mathrm{sc}}(\boldsymbol{r}'))g_1(\boldsymbol{r}';\boldsymbol{r})\right]\cdot\hat{\boldsymbol{n}}\,\mathrm{d}S'$$

$$=\begin{cases} u_1^{\mathrm{sc}}(x,y), & y>0 \text{ and } \boldsymbol{r} \text{ outside } S \\ 0, & y<0 \text{ or } \boldsymbol{r} \text{ inside } S \end{cases}$$

where S is any surface enclosing the crack and $\hat{\boldsymbol{n}}$ its outward-pointing normal. For the lower half-plane there is no crack, so the integral representation becomes

$$-\int_{-\infty}^{\infty}\left[u_2^{\mathrm{sc}}(x_0,0)\frac{\partial}{\partial y_0}g_2(x_0,0;x,y)-g_2(x_0,0;x,y)\frac{\partial u_2^{\mathrm{sc}}(x_0,0)}{\partial y_0}\right]\mathrm{d}x_0$$

$$=\begin{cases} u_2^{\mathrm{sc}}(x,y), & y<0 \\ 0, & y>0 \end{cases}$$

where the minus sign in front of the integral comes from the normal direction which is now downwards. If the lower line in this integral representation is used together with the boundary conditions for both the displacement and Green function the result is

$$\frac{\mu_1}{\mu_2}\int_{-\infty}^{\infty}\left[u_1^{\mathrm{sc}}(x_0,0)\frac{\partial}{\partial y_0}g_1(x_0,0;x,y)\right.$$
$$\left.-g_1(x_0,0;x,y)\frac{\partial u_1^{\mathrm{sc}}(x_0,0)}{\partial y_0}\right]\mathrm{d}x_0=0$$

which is thus valid for $y>0$. This says that the first integral in the integral representation in the upper half-plane vanishes for $y>0$. Using the first integral representation for $y>0$, shrinking S to the crack (so the crack-opening displacement appears), applying the boundary condition on the crack in the integrand, and applying the boundary condition on the crack once more, the result is the hypersingular integral equation

$$\lim_{y'\to 0}\int_{-a}^{a}\Delta u(x_0',0)\frac{\partial^2}{\partial y'\partial y_0'}g_1(x_0',0;x',y')\,\mathrm{d}x_0'=-\frac{\partial u_1^{\mathrm{in}}(x',0)}{\partial y'} \quad (12.7)$$

As usual the limit cannot be taken inside the integral because of the singularity of the Green function. This integral equation is very similar to the one for a single crack, the difference is that the Green function and the incident wave include the effects due to the two half-planes.

To solve the integral equation the COD is expanded in terms of the Chebyshev functions

$$\Delta u(x') = \sum_{n'=1}^{\infty} \alpha_{n'} \psi_{n'}(x'/a) \tag{12.8}$$

Inserting this expansion into Eq. (12.7) and projecting on the Chebyshev functions give

$$\sum_{n'=1}^{\infty} Q_{nn'} \alpha_{n'} = -4\mathrm{i} \int_{-a}^{a} \frac{\partial u^{\mathrm{in}}(x',0)}{\partial y} \psi_n(x'/a)\, \mathrm{d}x' \tag{12.9}$$

where

$$\begin{aligned} Q_{nn'} &= \pi(-1)^n n n' \int_{-\infty}^{\infty} \frac{h_1}{q^2} \mathrm{J}_n(qa) \mathrm{J}_{n'}(qa)\, \mathrm{d}q \\ &\quad - \pi n n' \int_{-\infty}^{\infty} \frac{R}{h_1} \left(h_1^2 \cos^2\beta - q^2 \sin^2\beta\right) \frac{\mathrm{J}_n(a(q\cos\beta - h_1 \sin\beta))}{q\cos\beta - h_1\sin\beta} \\ &\quad \times \frac{\mathrm{J}_{n'}(a(q\cos\beta + h_1\sin\beta))}{q\cos\beta + h_1\sin\beta}\, \mathrm{e}^{2\mathrm{i}h_1 d}\, \mathrm{d}q \end{aligned} \tag{12.10}$$

This in principle solves the scattering problem. It is possible to write down the scattered waves, in particular in the far field, as in the case with a crack in full space.

The present problem is straightforward to generalize to a layered plane (or half-plane) with an arbitrary number of layers. It is also possible to consider the corresponding waveguide problem as done in Section 9.2.

12.3 Crack in layered media: 2D P-SV waves

2D P-SV problems with a crack in the presence of plane interfaces can be treated in the same way as the corresponding 2D SH problems. For interior cracks it seems that this has not been pursued in the literature. Interface

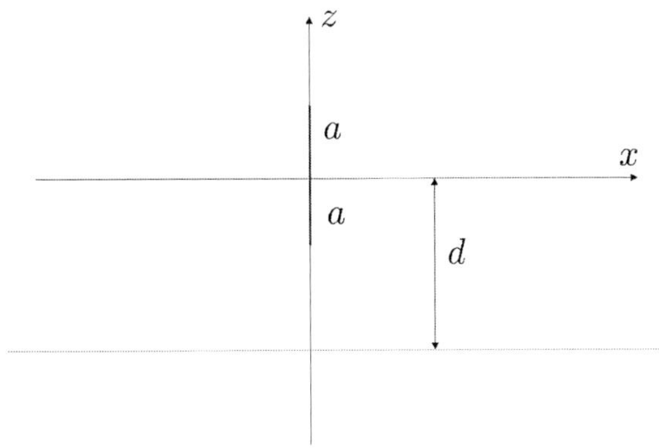

Figure 12.2: The 3D scattering geometry with a crack in a half-space.

cracks in a layered half-plane (also fluid-loaded) are treated by Yang and Bogy [143] and Gracewski and Bogy [53, 54]. They perform a careful analysis of the edge singularities, reproducing the result of Dundurs [45] that the COD has oscillations very close to the crack tip. With similar methods an interface crack between two half-planes is investigated by Itou [69] and Qu [115, 116].

12.4 Strip-like crack in half-space

Turn next to the 3D problem of a strip-like crack in a half-space, see Boström and Bövik [20]. For simplicity the crack is assumed to be perpendicular to the surface of the half-space. Introduce a coordinate system (x, y, z) so the crack is defined by $|z| < a$, $-\infty < y < \infty$, and $x = 0$. The surface of the half-space is $z = -d$, see Figure 12.2. For $d > a$ the crack is an interior one and for $d = a$ it is surface-breaking; both cases are treated in the following. For simplicity the incident wave $\boldsymbol{u}^{\text{in}}$ is assumed to satisfy the boundary condition on the surface of the half-space.

12.4.1 Interior crack

As opposed to the strip-like crack in full space, the integral equation is derived from the integral representation with the Green function; in this case the half-space Green function. This is in fact the only possible way as it is impossible to use Fourier representations in the present case.

Recall the integral representation from Section 4.2.2 [Eq. (4.22)]

$$\boldsymbol{u}^{\text{in}}(\boldsymbol{r}') + \frac{k_{\text{s}}}{\mu} \int_S \Big[\boldsymbol{u}(\boldsymbol{r}) \cdot (\hat{\boldsymbol{n}} \cdot \boldsymbol{\Sigma}(\boldsymbol{r}; \boldsymbol{r}')) - \boldsymbol{t}^{(n)}(\boldsymbol{r}) \cdot \boldsymbol{G}(\boldsymbol{r}; \boldsymbol{r}') \Big] \, \mathrm{d}S$$

$$= \begin{cases} \boldsymbol{u}(\boldsymbol{r}'), & \boldsymbol{r}' \text{ outside } S \\ 0, & \boldsymbol{r}' \text{ inside } S \end{cases}$$

where $\boldsymbol{t}^{(n)} = \hat{\boldsymbol{n}} \cdot \boldsymbol{\sigma}$ is the traction vector on S. In the present case the surface S is composed of two disjunct parts, namely the surface of the half-space and a surface enclosing the crack. Thus "outside S" denotes the region between these two surfaces. The Green function is chosen as that of the half-space, it thus satisfies the traction-free boundary condition on the surface of the half-space. As the total wave $\boldsymbol{u}$ also satisfies the boundary condition on the surface of the half-space the integral, over this surface disappears and only the integral over the surface surrounding the crack remains. Let this surface approach the crack. As the normal to the surface has opposite directions on the two sides of the crack the crack-opening displacement $\boldsymbol{\Delta u}(y, z)$ appears and the integral representation becomes

$$\lim_{x' \to 0+} \frac{k_{\text{s}}}{\mu} \int_{-\infty}^{\infty} \int_{-a}^{a} \boldsymbol{\Delta u}(y, z) \cdot (\hat{\boldsymbol{n}} \cdot \boldsymbol{\Sigma}(0, y, z; x', y', z')) \, \mathrm{d}z \, \mathrm{d}y$$

$$= \boldsymbol{u}^{\text{sc}}(0, y', z')$$

The limit in front of the integrals can not be taken inside the integrals as the integrals become singular. Operate with the traction on this equation and use the boundary condition on the crack to obtain

$$\lim_{x' \to 0+} \frac{k_{\text{s}}}{\mu} \int_{-\infty}^{\infty} \int_{-a}^{a} \boldsymbol{\Delta u}(y, z) \cdot \boldsymbol{S}(0, y, z; x', y', z') \, \mathrm{d}z \, \mathrm{d}y = -\boldsymbol{t}^{\text{in}}(y', z') \tag{12.11}$$

where $\boldsymbol{t}^{\text{in}}$ is the traction of the incident wave on the crack. This is the desired hypersingular integral equation, which means that the limit can not be taken inside the integrals. The kernel $\boldsymbol{S}$ in the equation is obtained from the Green tensor by operating with the traction on both variables.

To express the Green tensor the plane vector wavefunctions are used, see Eq. (8.15). The expansion of the free space Green tensor is given in Eq. (8.18) and is

$$\boldsymbol{G}_0(\boldsymbol{r};\boldsymbol{r}') = 2\mathrm{i}\sum_j \int_{-\infty}^{\infty}\int_{-\infty}^{\infty} \boldsymbol{\phi}_j^-(-q,-p;\boldsymbol{r}_<)\boldsymbol{\phi}_j^+(q,p;\boldsymbol{r}_>)\,\frac{\mathrm{d}q\,\mathrm{d}p}{k_j h_j}$$

where $\boldsymbol{r}_> = \boldsymbol{r}$ if $z > z'$ and $\boldsymbol{r}_> = \boldsymbol{r}'$ if $z < z'$ and vice versa for $\boldsymbol{r}_<$. The index "0" is added to distinguish the free space Green tensor from that for the half-space. The half-space Green tensor is obtained from the free space one by adding a term to fulfil the boundary condition. The half-space Green tensor becomes

$$\boldsymbol{G}(\boldsymbol{r};\boldsymbol{r}') = \boldsymbol{G}_0(\boldsymbol{r};\boldsymbol{r}')$$
$$+2\mathrm{i}\sum_{j,j'} \int_{-\infty}^{\infty}\int_{-\infty}^{\infty} \boldsymbol{\phi}_j^+(-q,-p;\boldsymbol{r})R_{jj'}(q,p)\boldsymbol{\phi}_{j'}^+(q,p;\boldsymbol{r}')\,\frac{\mathrm{d}q\,\mathrm{d}p}{k_j h_j} \tag{12.12}$$

Here $R_{jj'}$ are the elements of the plane wave reflection matrix and these elements are given in Eq. (8.26). However, in the present case it is the reflection matrix from above that is needed and in Section 8.2.2 it is the reflection from below that is considered. This amounts to a change in the sign of h_s and h_p with the result that the only change needed is an overall sign change on R_{23} and R_{32}.

Above the free space Green tensor is given as a double Fourier transform in the xy-plane. This is needed when the half-space part of the Green tensor is calculated, but it is not well suited in the integral equation due to the awkward term $h_j|z-z'|$ that appears in the exponential. But the free space Green tensor has the same appearance in all coordinate systems, so it can be written as a double Fourier transform in the yz-plane instead. This gives the term $h_j|x-x'|$ instead, but as $x = x' = 0$ on the crack this causes no problem. Operating twice with the traction operator (once on each variable) the final form of the double traction half-space Green tensor becomes (written in tensor notation for clarity)

$$S_{ii'}(\boldsymbol{r};\boldsymbol{r}') = 2\mathrm{i}\mu^2 \sum_j \int_{-\infty}^{\infty}\int_{-\infty}^{\infty} F_{ji}F^{\star}_{ji'}\mathrm{e}^{\mathrm{i}(h_j|x-x'|+p(y-y')+q(z-z'))}\,\frac{\mathrm{d}q\,\mathrm{d}p}{k_j h_j}$$

$$+2\mathrm{i}\mu^2 \sum_{j,j'} \int_{-\infty}^{\infty}\int_{-\infty}^{\infty} E_{ji}R_{jj'}E^{+}_{j'i'}$$

$$\times \mathrm{e}^{\mathrm{i}(q(x-x')+p(y-y')+h_j z+h_{j'}z'+d(h_j+h_{j'}))}\,\frac{\mathrm{d}q\,\mathrm{d}p}{k_j h_j} \qquad (12.13)$$

For a fixed j, F_{ji} are vectors $\boldsymbol{F}_j$, $j = 1, 2, 3$, which become

$$\boldsymbol{F}_1 = \frac{1}{4\pi s}\left(-\hat{\boldsymbol{y}} q h_{\mathrm{s}} + \hat{\boldsymbol{z}} p h_{\mathrm{s}}\right)$$

$$\boldsymbol{F}_2 = \frac{\mathrm{i}}{4\pi k_{\mathrm{s}} s}\left(\hat{\boldsymbol{x}} 2s^2 h_{\mathrm{s}} + \hat{\boldsymbol{y}} p(2s^2 - k_{\mathrm{s}}^2) + \hat{\boldsymbol{z}} q(2s^2 - k_{\mathrm{s}}^2)\right)$$

$$\boldsymbol{F}_3 = \frac{1}{4\pi}\sqrt{\frac{k_{\mathrm{p}}}{k_{\mathrm{s}}^3}}\left(\hat{\boldsymbol{x}}(2s^2 - k_{\mathrm{s}}^2) - \hat{\boldsymbol{y}} 2p h_{\mathrm{p}} - \hat{\boldsymbol{z}} 2q h_p\right)$$

and $\boldsymbol{F}^{\star}_j$, $j = 1, 2, 3$, are the same except for a sign change in $\boldsymbol{F}^{\star}_2$ (a change of "i" to "−i"). Similarly for $\boldsymbol{E}_j$, $j = 1, 2, 3$,

$$\boldsymbol{E}_1 = \frac{1}{4\pi s}\left(-\hat{\boldsymbol{x}} 2qp + \hat{\boldsymbol{y}}(q^2 - p^2) + \hat{\boldsymbol{z}} p h_{\mathrm{s}}\right)$$

$$\boldsymbol{E}_2 = \frac{\mathrm{i}}{4\pi k_{\mathrm{s}} s}\left(-\hat{\boldsymbol{x}} 2q^2 h_{\mathrm{s}} - \hat{\boldsymbol{y}} 2qp h_{\mathrm{s}} - \hat{\boldsymbol{z}} q(2s^2 - k_{\mathrm{s}}^2)\right)$$

$$\boldsymbol{E}_3 = \frac{1}{4\pi}\sqrt{\frac{k_{\mathrm{p}}}{k_{\mathrm{s}}^3}}\left(\hat{\boldsymbol{x}} 2(k_{\mathrm{p}}^2 - q^2) - \hat{\boldsymbol{y}} qp - \hat{\boldsymbol{z}} 2q h_p\right)$$

and $\boldsymbol{E}^{+}_j$, $j = 1, 2, 3$, are the same except for changing "i" to "−i" and "h_j" to "$-h_j$". Here $s^2 = q^2 + p^2$, $h_{\mathrm{s}}^2 = (k_{\mathrm{s}}^2 - s^2)^{1/2}$, and $h_{\mathrm{p}}^2 = (k_{\mathrm{p}}^2 - s^2)^{1/2}$ with $\operatorname{Im} h_{\mathrm{s}} \geq 0$ and $\operatorname{Im} h_{\mathrm{p}} \geq 0$.

To solve the integral equation the COD is represented as a Fourier transform in y and a sum in Chebyshev functions in z

$$\Delta u_i(y,z) = \int_{-\infty}^{\infty} \sum_n \beta_{in}(p)\psi_n(z/a)\,\mathrm{e}^{\mathrm{i}py}\,\frac{\mathrm{d}p}{k_{\mathrm{s}}} \qquad (12.14)$$

As usual the Chebyshev functions ensure that the square root behaviour at the crack edges are correctly represented. Inserting this expansion and the expansion Eq. (12.13) of the double traction half-space Green tensor into the integral equation (12.11), projecting on the Chebyshev functions, and taking a Fourier transform in y', this finally gives [using the integral Eq. (10.4) and the completeness of the Fourier transform]

$$\sum_{i',n'} Q_{in,i'n'}(p)\beta_{i'n'}(p) = T_{in}(p) \tag{12.15}$$

where

$$\begin{aligned} Q_{in,i'n'}(p) &= 4\mathrm{i}\pi^3 k_\mathrm{s}(-1)^{n'} nn' \int_{-\infty}^{\infty} \mathrm{J}_n(qa)\mathrm{J}_{n'}(qa) \sum_j \frac{F_{jn}F^\star_{jn'}}{k_j q^2 h_j}\,\mathrm{d}q \\ &\quad -4\mathrm{i}\pi^3 k_\mathrm{s}(-1)^{n+n'} nn' \int_{-\infty}^{\infty} \sum_{j,j'} \mathrm{J}_n(h_j a)\mathrm{J}_{n'}(h_{j'} a) \\ &\quad \times \mathrm{e}^{\mathrm{i}d(h_j+h_{j'})} \frac{E_{jn}R_{jj'}E^+_{j'n'}}{k_j h_j^2 h_{j'}}\,\mathrm{d}q \end{aligned} \tag{12.16}$$

The first term can be divided into a symmetric and an antisymmetric part, so that the matrix contains the corresponding matrices from Section 10.3.2 (with a somewhat different normalization). The right-hand side is

$$T_{in}(p) = \frac{k_\mathrm{s}}{2\pi\mu} \int_{-a}^{a} \int_{\infty}^{\infty} t_i^\mathrm{in}(z,y)\psi_n(z/a)\mathrm{e}^{\mathrm{i}py}\,\mathrm{d}y\,\mathrm{d}z \tag{12.17}$$

The limit $\lim_{x'\to 0+}$ in front of the integral equation has been moved inside the integrals as these are now convergent in the ordinary sense. This is not at all obvious as the elements of F_{jn}, G_{jn}, and $R_{jj'}$ all behave as $O(q^2)$, so that the integrals in fact seem to be divergent. However, due to cancellations the following asymptotic estimates are valid

$$\begin{aligned} 16\pi^2 \sum_j \frac{F_{jn}F^\star_{jn'}}{k_j h_j} &= \delta_{nn'}\alpha_n|q| + O(1) \\ 16\pi^2 \sum_{j,j'} \frac{E_{jn}R_{jj'}E^+_{j'n'}}{k_j h_j} &= \delta_{nn'}\alpha_n|q| + O(1) \end{aligned} \tag{12.18}$$

where $\alpha_1 = \alpha_3 = 2\mathrm{i}(k_\mathrm{s}^2 - k_\mathrm{p}^2)$ and $\alpha_2 = \mathrm{i}$. The $O(1)$ terms only appear for $n \neq n'$, for $n = n'$ the next order terms are $O(q^{-1})$. The first integral in Eq. (12.16) thus behaves as $O(q^{-1})$ times two Bessel functions and is thus convergent, although slowly so [to increase the convergence the dominating term can be integrated analytically using Eq. (10.11)]. The second term in Eq. (12.16) is exponentially convergent when $d > a$ and in the limit $d \to a$ it is still convergent due to Eq. (12.18). When there is no damping the integrals contain cuts and poles (the Rayleigh pole) that are located on the real axis in the complex q-plane and the integration contour must therefore be deformed into the second and fourth quadrants (with damping the cuts and poles moves into the first and third quadrants). The expansion coefficients β_{in} of the COD are obtained by solving the system of equations (12.15) and the integral representation can then be used to determine the scattered wave if this is of interest.

12.4.2 Surface-breaking crack

It may seem that the solution for the interior crack in a half-space could be used also for a surface-breaking crack. There is, however, one problem with this, namely that the COD vanishes at the crack mouth as opposed to a finite value that is expected. This difficulty can be avoided by choosing a different COD expansion [16].

Consider a crack of width a (as opposed to $2a$ for the interior crack) in a half-space that is surface-breaking and normal to the surface of the half-space. Introduce the coordinate system (x, y, z) with the x-axis normal to the crack, the y-axis along the crack mouth, and the z-axis along the crack pointing into the half-space (the xy-plane is thus the surface of the half-space). The integral equation and the Green tensor are the same as in the previous subsection except that the different location of the origin means that the distance d (the distance between the origin and the half-space) must be put to zero (d appears in the reflection coefficients $R_{jj'}$). To expand the COD the Chebyshev functions are still used, but only the even ones are used on half the interval

$$\Delta u_i(y, z) = \int_{-\infty}^{\infty} \sum_{n=1}^{\infty} \beta_{in}(p)\psi_{2n-1}(z/a)\, \mathrm{e}^{\mathrm{i}py}\, \frac{\mathrm{d}p}{k_\mathrm{s}} \tag{12.19}$$

The used Chebyshev functions form a complete set on the crack, they have a finite value at $z = 0$, and they have the square root behaviour $(a - z)^{1/2}$

at the crack tip, exactly as the COD must behave. When the COD is inserted into the integral equation the following integral appears

$$I_n(\gamma) = \int_0^1 \cos\left((2n-1)\arcsin t\right) \mathrm{e}^{\mathrm{i}\gamma t}\, \mathrm{d}t \tag{12.20}$$

It seems impossible to calculate this integral analytically, but except for large γ it is straightforward to compute numerically. It is noted that the corresponding integral appearing for the interior crack essentially yields a Bessel function, exactly as the real part of $I_n(\gamma)$ gives a Bessel function. Proceeding as in the previous subsection (the only change being a different set of Chebyshev functions) leads to the following set of equations for the unknown COD coefficients

$$\sum_{i',n'} Q_{in,i'n'}(p)\beta_{i'n'}(p) = T_{in}(p) \tag{12.21}$$

where

$$Q_{in,i'n'}(p) = -4\mathrm{i}\pi k_{\mathrm{s}} a^2 \int_{-\infty}^{\infty} \left(\sum_j \frac{F_{jn}F^{\star}_{jn'}}{k_j h_j} I_n(qa) I_{n'}(-qa) \right.$$
$$\left. + \sum_{j,j'} \frac{E_{jn}R_{jj'}E^{+}_{j'n'}}{k_j h_j} I_n(h_j a) I_{n'}(-h_{j'}a) \right) \mathrm{d}q \tag{12.22}$$

The right-hand side is

$$T_{in}(p) = \frac{k_{\mathrm{s}}}{2\pi\mu} \int_0^a \int_{\infty}^{\infty} t_i^{\mathrm{in}}(y,z)\psi_{2n-1}(z/a)\, \mathrm{e}^{\mathrm{i}py}\, \mathrm{d}y\, \mathrm{d}z \tag{12.23}$$

With this the problem is in principle solved. However, it is not at all obvious that the integral in Eq. (12.22) is convergent. To show this the asymptotic behaviour for large arguments of the function $I_n(\gamma)$ is needed. For a purely imaginary argument [which is relevant for the second term in Eq. (12.22)] $\gamma = \mathrm{i}s$, s real, the definition becomes

$$I_n(\mathrm{i}s) = \int_0^1 \cos\left((2n-1)\arcsin t\right) \mathrm{e}^{-st}\, \mathrm{d}t$$

A repeated integration by parts (Olver [110, Chapter 3]), note that only the lower limit contributes) gives

$$I_n(\mathrm{i}s) = \frac{1}{s} + O(s^{-3})$$

For a purely real argument γ the situation is a little more complicated. Start with a partial integration and a change of variable to $x = 1 - t$ to obtain

$$I_n(\gamma) = \frac{\mathrm{i}}{\gamma} - \frac{\mathrm{i}}{\gamma}\mathrm{e}^{\mathrm{i}\gamma}\int_0^1 \mathrm{e}^{-\mathrm{i}\gamma x}\sin\left((2n-1)\arcsin\left(1-x\right)\right)\frac{2n-1}{\sqrt{2x-x^2}}\,\mathrm{d}x$$

The dominating contribution here comes from the lower limit. Using a theorem by Olver [110, Chapter 3] gives

$$I_n(\gamma) = \frac{\mathrm{i}}{\gamma} + \frac{1}{\gamma^{3/2}}\sqrt{\frac{\pi}{2}}(-1)^n(2n-1)\mathrm{e}^{\mathrm{i}(\gamma+\pi/4)} + O(\gamma^{-5/2})$$

With these estimates of $I_n(\gamma)$ and the behaviour of the rest of the integrands given in Eq. (12.18) the two terms in Eq. (12.22) actually each diverges as the two integrands behave as $O(q^{-1})$. However, these leading order terms cancel because the same factor α_n appears in both terms in Eq. (12.18). It is thus essential to keep both terms under the same integral as they would diverge separately. The remaining leading order terms then behave as $O(q^{-3/2})$ and have to be treated by care in a numerical integration.

12.5 Cracks in anisotropic solids

The methods used above for cracks in isotropic structures can in most cases be extended to cracks in anisotropic structures, although not much is to be found in the literature on this. For 2D SH waves multiple cracks in a half-space are considered by Ang *et al.* [6] and a crack in a layered structure is investigated by Grahn [55]. For 2D P-SV waves Kuo [82, 83] studies the scattering by an interface crack between two anisotropic half-spaces, and in particular investigates the singularity at the crack tips. Shen *et al.* [122] investigate the same problem in a layered structure. In 3D Boström *et al.* [24] consider a rectangular crack in an anisotropic half-space.

13 Applications: nondestructive evaluation and material characterization

There are many applications of the scattering of elastic waves in solids. One area of interest is waves that are propagating in the earth due to seismic events, explosions, or traffic vibrations. Another area is ultrasound with applications in nondestructive testing, material characterization, and medical diagnostics. In this chapter two applications of ultrasound are briefly taken up, namely modelling of ultrasonic testing and material characterization.

13.1 Nondestructive evaluation

Ultrasonic nondestructive testing is used extensively in industrial applications, particularly in the aerospace and nuclear power industries. It is very useful to have a mathematical model of the whole testing situation for several reasons. It can give a physical feeling for the process and it is easy to perform parametric studies (which may not be the case when performing experiments). It is thus a useful tool for the development of testing procedures and signal processing and it can be part of the qualification of procedures and personnel. In this connection it is customary to use the designation "nondestructive evaluation".

In nondestructive testing ultrasound is usually generated by a transducer, it then propagates through the component being tested and is scattered by an obstacle (usually called "defect" in this context), and finally propagates to a receiver, where it is converted to an electrical signal. All these steps are to be modelled, although the scattering (as taken up in previous chapters) is the most vital one. However, the action of a transducer in both transmission and reception is important also. In this section the modelling of transducers in both transmission and reception is taken up and it is shown how this can be combined with the scattering by a defect in some situations. Some work on modelling the whole testing situation is described by Chapman [36] and Schmerr [119].

13.1.1 Modelling of transmitting ultrasonic transducers

To generate ultrasound in a component some sort of transducer is usually used, although there are alternatives, for instance laser generated ultrasound. Most transducers use piezoelectric materials to convert electric energy into mechanical energy, or vice versa for a receiver. Examples of modelling of ultrasonic transducers include Harris [61], Bresse and Hutchins [33], Schmerr and Sedov [120], McNab *et al.* [97], and Baboux and Kazys [10], see also the book by Schmerr [119, Chapter 8], where both contact transducers and immersion testing are investigated. The development in this section follows that of Boström and Wirdelius [30].

Consider an ultrasonic contact transducer scanning over the plane surface of an isotropic elastic half-space. This model should work well also in cases where the scanning surface is locally plane; the important point is that there are no disturbing surfaces close to the transducer to ensure that the radiation from the transducer is correctly modelled. Once a wave is transmitted its interaction with obstacles or other disturbances has to be dealt with in some way. This division into the transmission from the transducer and the scattering by an obstacle is more clearly envisaged in the time domain.

The action of the transducer as a transmitter can most simply be modelled by a traction boundary condition on the surface of the half-space where the transducer is situated. Usually only the normal component of the traction (the pressure) is needed as the transducer is coupled to the tested component with some type of (viscous) fluid, but to begin with all three components are retained. This model may seem very simplistic as it disregards the internal structure of the transducer. For a P wave transducer radiating normal to the surface this seems not very severe, but for an angled transducer with a plastic wedge this may seem more drastic. However, results of this type of modelling have been compared to experimental results with good or reasonable correspondence in most cases.

Let the elastic half-space occupy $z < 0$ with the origin at the centre of the transducer and the xy-plane as the surface of the half-space. The Lamé constants of the half-space are λ and μ and the wavenumbers are k_{p} and k_{s}.

As always time harmonic conditions are assumed. The boundary condition on the half-space is

$$\boldsymbol{t}^{(z)}(x,y,0)=\begin{cases}\boldsymbol{t}^{\mathrm{tr}}(x,y), & \boldsymbol{r} \text{ on } S\\ \boldsymbol{0}, & \boldsymbol{r} \text{ not on } S\end{cases} \tag{13.1}$$

where $\boldsymbol{t}^{\mathrm{tr}}(x,y)$ is the applied traction from the transducer and S is the (effective) area of the transducer. To solve this boundary value problem it is natural to employ Fourier transforms in the x- and y-directions (using a Green function is in principle equivalent). This is equivalent to using the vector plane waves discussed in Section 8.2 and the radiated wave can thus be assumed as a superposition of down-going plane waves

$$\boldsymbol{u}(\boldsymbol{r})=\sum_j\int_{-\infty}^{\infty}\int_{-\infty}^{\infty}\xi_j(q,p)\boldsymbol{\phi}_j^-(q,p;\boldsymbol{r})\,\frac{\mathrm{d}q\,\mathrm{d}p}{k_jh_j} \tag{13.2}$$

where the sum is over $j=1,2,3$ and $\xi_j(q,p)$, $j=1,2,3$, are coefficients to be determined. Applying the boundary condition, inverting the Fourier transforms, and solving gives

$$\begin{aligned}\xi_1(q,p)&=\frac{1}{s}\left(pT_x-qT_y\right)\\ \xi_2(q,p)&=\frac{\mathrm{i}k_\mathrm{s}h_\mathrm{s}}{sR}\left[\left(k_\mathrm{s}^2-2s^2\right)\left(qT_x+pT_y\right)+2s^2h_\mathrm{p}T_z\right]\\ \xi_3(q,p)&=\frac{\sqrt{k_\mathrm{p}k_\mathrm{s}}\,h_\mathrm{p}}{R}\left[2h_\mathrm{s}\left(qT_x+pT_y\right)-\left(k_\mathrm{s}^2-2s^2\right)T_z\right]\end{aligned} \tag{13.3}$$

Here $s=(q^2+p^2)^{1/2}$, $h_n=(k_n^2-s^2)^{1/2}$ with $\operatorname{Im}h_n\geq 0$, $n=\mathrm{s},\mathrm{p}$, and the Rayleigh function is

$$R=4s^2h_\mathrm{s}h_\mathrm{p}+\left(k_\mathrm{s}^2-2s^2\right)^2$$

The normalized Fourier transform of the applied traction vector is

$$\boldsymbol{T}(q,p)=\frac{k_\mathrm{s}}{\pi\mu}\iint_S\boldsymbol{t}^{\mathrm{tr}}(x,y)\,\mathrm{e}^{-\mathrm{i}(qx+py)}\,\mathrm{d}x\,\mathrm{d}y \tag{13.4}$$

where the integration is over the transducer surface and $\boldsymbol{t}^{\mathrm{tr}}(x,y)$ is the applied traction from Eq. (13.1). If there is only a normal component of the applied pressure $\xi_1(q,p) = 0$, so there is no SH wave excited in this case. The radiation from the transducer is, in principle, now determined, but the solution is not very useful as it appears as a quadruple integral (fivefold if a Fourier transform in frequency is also performed). However, for many applied tractions of interest the double Fourier transform in Eq. (13.4) can be calculated analytically. In addition the integrals in Eq. (13.1) can be integrated approximately in some cases.

The applied traction can be of different type depending on the transducer. Particularly simple to model are unfocussed transducers which can be unangled or angled. The traction can then be taken to be constant except for a phase factor determining the angle. If the tangential traction components can be neglected the applied traction can the be taken as

$$\boldsymbol{t}^{\mathrm{tr}}(x,y) = \hat{\boldsymbol{z}} T_0 \,\mathrm{e}^{-\mathrm{i}k_n x \sin\gamma} \tag{13.5}$$

where T_0 is an amplitude factor and γ is the angle to the normal of the half-space of the beam from the transducer in the third quadrant of the xz-plane. The wavenumber k_n is equal to k_{s} for an S wave transducer and to k_{p} for a P wave transducer. However, it must be remembered that both S and P waves are generated in general and that, furthermore, Rayleigh surface waves are generated along the surface of the half-space. For an unangled transduced mostly P waves propagating normal to the half-space are generated and for an angled S wave transducer above the critical angle mainly S waves are generated.

With the chosen form of the applied traction in Eq. (13.5) the integrals in Eq. (13.4) can be integrated analytically for rectangular and elliptic transducers. For a rectangular transducer $|x| < a$, $|y| < b$ the result is

$$\boldsymbol{T} = \hat{\boldsymbol{z}}\, \frac{4T_0 k_{\mathrm{s}}}{\pi\mu Q p} \sin Qa \, \sin pb \tag{13.6}$$

and for an elliptic transducer $x^2/a^2 + y^2/b^2 < 1$ the result is

$$\boldsymbol{T} = \hat{\boldsymbol{z}}\, \frac{4T_0 k_{\mathrm{s}} ab}{\mu\sqrt{Q^2a^2 + p^2b^2}} \,\mathrm{J}_1(\sqrt{Q^2a^2 + p^2b^2}) \tag{13.7}$$

where $Q = q + k_n \sin\gamma$.

A phased array transducer is composed of a number of small piezoelectric elements, often in a rectangular array, that can be controlled individually, and as a result such a transducer can be used to generate waves in any direction (and the wave can also be focussed). If each element is rectangular Eq. (13.6) with an appropriate phase factor (corresponding to the position and a time delay) can be used for each element, which are then simply summed to give the overall Fourier transform of the traction.

To proceed with an approximate far field calculation of Eq. (13.1) it is best to write the integral with the spherical angles of propagation as

$$\boldsymbol{u}(\boldsymbol{r}) = \sum_j \int_0^{2\pi} \int_{C_-} \xi_j(q,p)\boldsymbol{\phi}_j(\alpha,\beta;\boldsymbol{r}) \sin\alpha \, \mathrm{d}\alpha \, \mathrm{d}\beta \tag{13.8}$$

where thus $q = k_j \sin\alpha\cos\beta$ and $p = k_j \sin\alpha\sin\beta$. The evanescent waves are now discarded and the integral is then over the lower half of the unit sphere. Using the 2D version of the stationary-phase approximation [Eq. (5.18)] then gives (where only the first term contributes as the integration is over half of the unit sphere)

$$\boldsymbol{u}(\boldsymbol{r}) \simeq \frac{1}{2}\left[\left(-\hat{\boldsymbol{\varphi}}\xi_1(\hat{\boldsymbol{r}}) + \mathrm{i}\hat{\boldsymbol{\theta}}\xi_2(\hat{\boldsymbol{r}})\right)\frac{\mathrm{e}^{\mathrm{i}k_\mathrm{s}r}}{k_\mathrm{s}r} + \hat{\boldsymbol{r}}\xi_3(\hat{\boldsymbol{r}})\frac{\mathrm{e}^{\mathrm{i}k_\mathrm{p}r}}{k_\mathrm{p}r}\left(\frac{k_\mathrm{p}}{k_\mathrm{s}}\right)^{3/2}\right] \tag{13.9}$$

where the argument of ξ_j means that the spherical angles of the observation point $\boldsymbol{r}$ are to be used when calculating q and p. Once the expansion coefficients ξ_j are known this result is very simple.

For the stationary-phase approximation to be valid three conditions must be satisfied involving the distance to the observation point r, the wavenumber k (k_s or k_p), and the radius b of the transducer. These conditions require that the distance to the observation point must be large compared to the wavelength, the transducer radius, and the near field length of the transducer (which is defined as the transducer radius squared divided by the wavelength; in electromagnetics this is known as the Frauenhofer distance). This means that the following requirements all have to be satisfied:

$$kr \gg 1$$

$$kr \gg kb$$

$$kr \gg (kb)^2$$

Schmerr [119, Chapter 8] investigates transducer fields at various distances and shows the meaning of the near field length. If the transducer radius b is about a wavelength or smaller, then only the first condition is of importance; on the other hand, when b is larger than the wavelength only the last condition have to be satisfied. (It thus seems that the middle condition is unnecessary, but usually all three conditions are stated.)

In practice many transducers are larger than the wavelength and are operating at a distance that is not large compared to the near field length and the stationary-phase approximation is thus not applicable. However, the size of the elements in a phased array transducer are comparable to the wavelength or smaller, and for such transducers the stationary-phase approximation is often valid for each element. So for a phased array transducer with rectangular elements both the Fourier transform of the applied traction and the field calculation can be performed analytically (at the cost of a sum over all elements twice). The same approach can of course be used for ordinary transducers by dividing them into a number of elements.

13.1.2 Calibration by side-drilled hole

A side-drilled hole (sdh for short) is often used for calibration in ultrasonic nondestructive testing. The sdh is well defined and easily reproducible and is therefore well suited for this purpose. The scattering by an sdh is considered by, for example, Krautkrämer and Krautkrämer [76], Chapman [36], and Schmerr [119] using high frequency approximations (which work for surprisingly low frequencies). An sdh is of course a circular cylindrical cavity and the scattering by the sdh is thus described in Section 2.3 using the exact solution in terms of cylindrical vector wavefunctions. The task here is to couple the scattering by the sdh to the transducer model from the previous subsection and to model a transducer also as a receiver. The approach in this subsection follows that of Boström and Bövik [21].

The geometry of the problem is depicted in Figure 13.1 which shows the transmitting and receiving transducers (which are often the same) scanning on a plane surface of a half-space above the sdh, which has its axis parallel to the plane. It is assumed that the transducers emit beams in the plane normal to the sdh axis. Each transducer has an attached coordinate system, (x_t, y_t, z_t) for the transmitter and (x_r, y_r, z_r) for the receiver. The z_t- and z_r-axes are normal to the scanning surface, the x_t- and x_r-axes are normal to the sdh axis, and the y_t- and y_r-axes are parallel to the sdh axis. The two coordinate systems are thus collinear. The (x, y, z) coordinate system is attached to the sdh with the origin in the $x_t z_t$-plane, the z-axis as the sdh

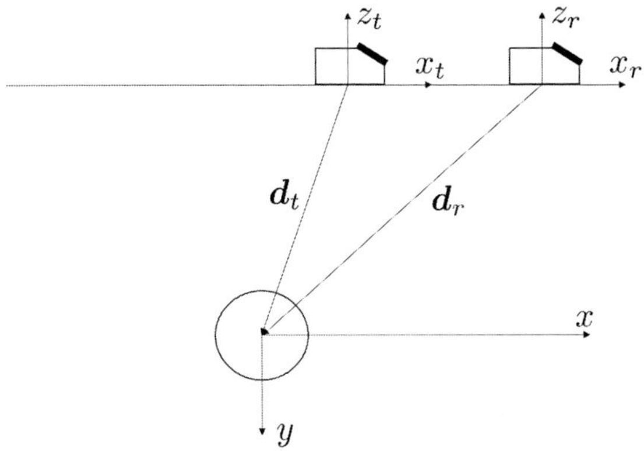

Figure 13.1: A side-drilled hole in a half-space with transmitting and receiving transducers.

axis, the x-axis parallel with the x_t- and x_r-axes, and the y-axis antiparallel with the z_t- and z_r-axes. The vectors from the transducers to the sdh are

$$\begin{aligned} \boldsymbol{d}_t &= \hat{\boldsymbol{x}}_t d_{tx} - \hat{\boldsymbol{z}}_t d_z \\ \boldsymbol{d}_r &= \hat{\boldsymbol{x}}_r d_{rx} + \hat{\boldsymbol{y}}_r d_y - \hat{\boldsymbol{z}}_r d_z \end{aligned}$$

where d_y is the distance between the two transmitters in the axial direction of the sdh. Usually $d_y = 0$ and often the same transducer is used both as transmitter and receiver (pulse-echo testing).

The scattering by the sdh (a circular cylinder) is described in Section 2.3. The incident wave from the transmitting receiver must thus be expanded in the regular cylindrical wavefunctions [Eq. (2.43)]

$$\boldsymbol{u}^{\mathrm{in}}(\boldsymbol{r}) = \sum_{\tau,\sigma,m} \int_{-\infty}^{\infty} b_{\tau\sigma m}(h) \boldsymbol{\chi}^0_{\tau\sigma m}(h, \boldsymbol{r})\,\mathrm{d}h$$

where the coefficients $b_{\tau\sigma m}(h)$ are to be determined by the wave from the transmitter. The wave scattered by the sdh is expanded in the outgoing cylindrical vector wavefunction [Eq. (2.44)]

$$\boldsymbol{u}^{\mathrm{sc}}(\boldsymbol{r}) = \sum_{\tau,\sigma,m} \int_{-\infty}^{\infty} f_{\tau\sigma m}(h) \boldsymbol{\chi}^+_{\tau\sigma m}(h, \boldsymbol{r})\,\mathrm{d}h$$

where the coefficients $f_{\tau\sigma m}(h)$ are given by [Eq. (2.45)]

$$f_{\tau\sigma m}(h) = \sum_{\tau',\sigma',m'} T_{\tau\sigma m,\tau'\sigma' m'}(h) b_{\tau'\sigma' m'}(h) \tag{13.10}$$

The T-matrix is diagonal in the azimuthal m-index and can thus be given as $T^{m}_{\tau\sigma,\tau'\sigma'}(h)$ and there is, furthermore, a decoupling into two groups for the τ- and σ-indices, with one group $\tau\sigma = 1\text{o}, 2\text{e}, 3\text{e}$ and the other $\tau\sigma = 1\text{e}, 2\text{o}, 3\text{o}$. The T-matrix elements coupling the two groups are vanishing and the elements of the two groups are the same except for some sign changes as described in Section 2.3.2.

From the previous subsection the wave from the transmitting transducer can be written as

$$\boldsymbol{u}(\boldsymbol{r}_t) = \sum_j \int_0^{2\pi} \int_{\pi/2}^{\pi} \xi_j(q,p) \boldsymbol{\phi}_j(\alpha,\beta;\boldsymbol{r}_t) \sin\alpha \,\mathrm{d}\alpha \,\mathrm{d}\beta$$

where the evanescent waves are neglected. The transformation of this into an expansion in terms of the regular cylindrical vector wavefunctions is performed in three steps. A translation of the plane waves (a trivial step) is followed by a rotation to give an expansion in plane waves in the (x, y, z) system, and finally the plane waves are transformed to cylindrical waves. A translation and a rotation by $-90°$ around the x-axis give

$$\boldsymbol{u}(\boldsymbol{r}) = \sum_{j,j'} \int_0^{\pi} \int_0^{\pi} \xi_j(q,p) \exp\left(\mathrm{i}k_j \hat{\boldsymbol{\gamma}} \cdot \boldsymbol{d}_t\right) R_{jj'} \boldsymbol{\phi}_{j'}(\alpha',\beta';\boldsymbol{r}) \,\sin\alpha' \,\mathrm{d}\alpha' \,\mathrm{d}\beta'$$

where the plane waves are now with respect to the (x, y, z) system. The wavenumbers are $k_1 = k_2 = k_{\mathrm{s}}$ and $k_3 = k_{\mathrm{p}}$. The relations between the spherical angles in the $-90°$ rotation are

$$\sin\alpha\cos\beta = \sin\alpha'\cos\beta'$$

$$\sin\alpha\sin\beta = \cos\alpha'$$

$$\cos\alpha = -\sin\alpha'\sin\beta'$$

and the elements of the rotation matrix are [27]

$$
\begin{aligned}
R_{11} &= R_{22} = \sin\beta \sin\beta' \\
R_{12} &= R_{21} = -\mathrm{i}\,\frac{\cos\beta'}{\sin\alpha} \\
R_{33} &= 1 \\
R_{13} &= R_{31} = R_{23} = R_{32} = 0
\end{aligned}
$$

The structure of the rotation matrix means that SH and SV waves ($j = 1, 2$) are mixed by the rotation whereas the P wave ($j = 3$) is left unaffected.

The final step is the transformation to the cylindrical vector wavefunctions. As the plane and cylindrical vector wavefunctions are defined in the same manner from the scalar ones and have the exponential factor in the z-direction in common, the transformation is the same as in the scalar case or even in the 2D scalar case as given in Eq. (8.1) and can be written

$$
\boldsymbol{\phi}_\tau(\alpha', \beta'; \boldsymbol{r}) = \sum_{\sigma,m} D^\dagger_{\tau\sigma m}(\beta')\boldsymbol{\chi}^0_{\tau\sigma m}(k_\tau \cos\alpha', \boldsymbol{r}) \tag{13.11}
$$

where the transformation function is (the dagger may seem strange but follows the notation of Boström *et al.* [27])

$$
D^\dagger_{\tau\sigma m}(\beta') = \mathrm{i}^m \sqrt{\frac{\epsilon_m}{2\pi}} \left\{ \begin{array}{c} \cos m\beta' \\ \sin m\beta' \end{array} \right\} \tag{13.12}
$$

where the upper row is for $\sigma = \mathrm{e}$ and the lower for $\sigma = \mathrm{o}$. Use of this transformation gives an expansion in the cylindrical vector wavefunctions with expansion coefficients

$$
b_{\tau\sigma m}(h) = \sum_j \int_0^\pi \xi_j(q, p) \exp(\mathrm{i}k_j \hat{\boldsymbol{\gamma}} \cdot \boldsymbol{d}_t) R_{j\tau} D^\dagger_{\tau\sigma m}(\beta')\,\mathrm{d}\beta' \tag{13.13}
$$

As the evanescent waves are neglected this equation is only valid for $|h| \leq k_\tau$. For $|h| > k_\tau$, $b_{\tau\sigma m}(h) = 0$.

So far the action of the transmitting transducer and the scattering by the sdh are treated. Next the action of the receiving transducer should be modelled and a good way to do this is by the electromechanical reciprocity relation due to Auld [8]. This result relates two different time harmonic elastodynamic states 1 and 2 generated by two transducers, which as a special case may be same. State 1 with displacement field $\boldsymbol{u}_1$ is due to an incident power P to the transmitting transducer in the *presence* of the obstacle (the sdh in this case) and state 2 with displacement field $\boldsymbol{u}_2$ is due to an incident power P to the receiving transducer in the *absence* of the obstacle. The result of Auld states that the change in the electric reflection coefficient $\delta\Gamma$ of the receiving transducer due to the presence of the obstacle, i.e. more or less the quantity measured in practice, is

$$\delta\Gamma = -\frac{\mathrm{i}\omega}{4P}\int_S (\boldsymbol{u}_2 \cdot \boldsymbol{t}_1 - \boldsymbol{u}_1 \cdot \boldsymbol{t}_2)\,\mathrm{d}S \tag{13.14}$$

The surface S is any control surface enclosing the obstacle and $\boldsymbol{t}_1$ and $\boldsymbol{t}_2$ are the tractions on S corresponding to states 1 and 2, respectively.

Let state 1 be the actual situation of interest with the transmitting transducer and the sdh. For the auxiliary state 2 the receiving transducer is acting as a transmitter (but all the characteristics of this transducer is otherwise unchanged) and there is no obstacle. The surface S can be the surface of the sdh or any other surface enclosing the sdh but not touching the surface where the transducers are located. State 1 can be expanded as

$$\boldsymbol{u}_1(\boldsymbol{r}) = \sum_{\tau,\sigma,m}\int_{-\infty}^{\infty}\left[b^{(t)}_{\tau\sigma m}(h)\boldsymbol{\chi}^0_{\tau\sigma m}(h,\boldsymbol{r}) + f^{(t)}_{\tau\sigma m}(h)\boldsymbol{\chi}^+_{\tau\sigma m}(h,\boldsymbol{r})\right]\mathrm{d}h \tag{13.15}$$

which is just the sum of the incident and scattered waves. Likewise state 2 can be expanded as

$$\boldsymbol{u}_2(\boldsymbol{r}) = \sum_{\tau,\sigma,m}\int_{-\infty}^{\infty} b^{(r)}_{\tau\sigma m}(-h)\boldsymbol{\chi}^0_{\tau\sigma m}(-h,\boldsymbol{r})\,\mathrm{d}h \tag{13.16}$$

where $b^{(r)}_{\tau\sigma m}(-h)$ is the expansion coefficient for the receiver acting as transmitter. For a reason soon to be apparent the latter expansion is with wavefunctions with argument $-h$ instead of h.

When the expansions Eqs. (13.15) and (13.16) are inserted into Eq. (13.14) the resulting integrals over S can be calculated by orthogonality identities for cylindrical wavefunctions. These are derived in the same way as the spherical counterparts in Section 3.1.4 and they are

$$\int_S \Big[\boldsymbol{t}^{(n)} \left(\boldsymbol{\chi}^0_{\tau\sigma m}(h;\boldsymbol{r})\right) \cdot \boldsymbol{\chi}^0_{\tau'\sigma' m'}(-h';\boldsymbol{r})$$

$$- \boldsymbol{t}^{(n)} \left(\boldsymbol{\chi}^0_{\tau'\sigma' m'}(-h';\boldsymbol{r})\right) \cdot \boldsymbol{\chi}^0_{\tau\sigma m}(h;\boldsymbol{r}) \Big] \mathrm{d}S = 0$$

$$\int_S \Big[\boldsymbol{t}^{(n)} \left(\boldsymbol{\chi}^+_{\tau\sigma m}(h;\boldsymbol{r})\right) \cdot \boldsymbol{\chi}^+_{\tau'\sigma' m'}(-h';\boldsymbol{r})$$

$$- \boldsymbol{t}^{(n)} \left(\boldsymbol{\chi}^+_{\tau'\sigma' m'}(-h';\boldsymbol{r})\right) \cdot \boldsymbol{\chi}^+_{\tau\sigma m}(h;\boldsymbol{r}) \Big] \mathrm{d}S = 0$$

$$\int_S \Big[\boldsymbol{t}^{(n)} \left(\boldsymbol{\chi}^+_{\tau\sigma m}(h;\boldsymbol{r})\right) \cdot \boldsymbol{\chi}^0_{\tau'\sigma' m'}(-h';\boldsymbol{r})$$

$$- \boldsymbol{t}^{(n)} \left(\boldsymbol{\chi}^0_{\tau'\sigma' m'}(-h';\boldsymbol{r})\right) \cdot \boldsymbol{\chi}^+_{\tau\sigma m}(h;\boldsymbol{r}) \Big] \mathrm{d}S$$

$$= \frac{\mathrm{i}\mu k_\tau}{k_\mathrm{s}} \delta_{\tau\tau'} \delta_{\sigma\sigma'} \delta_{mm'} \delta(h-h') \tag{13.17}$$

where $\boldsymbol{t}^{(n)}$ is the traction operator on the surface S with normal $\hat{\boldsymbol{n}}$ and S is a cylinder-like surface (an infinite surface situated between two cylinders). Using the orthogonality the signal response becomes

$$\delta\Gamma = \frac{\mu c_\mathrm{s}}{4P} \sum_{\sigma,m} \sum_{\tau,\tau'} \int_{-k_{\tau,\tau'}}^{k_{\tau\tau'}} b^{(r)}_{\tau\sigma m}(-h) T^m_{\tau\sigma,\tau'\sigma'}(h) b^{(t)}_{\tau' sigma' m}(h) k_\tau \, \mathrm{d}h \tag{13.18}$$

where the integration limit is $k_{\tau\tau'} = \min(k_\tau, k_{\tau'})$. The determination of σ' is described below Eq. (13.10). In Eq. (13.18) the effect of the transmitter, the sdh, and the receiver are separated in a very nice fashion.

If the sdh is in the far field of the transducer(s) the stationary-phase approximation can be used in two steps, first to approximate the β'-integral in Eq. (13.13) and then to approximate the h-integral in Eq. (13.18), see Boström and Bövik [21] for details and examples.

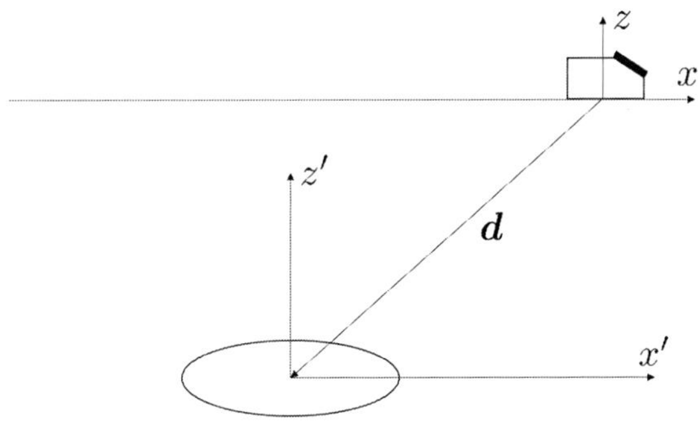

Figure 13.2: A bounded defect in a half-space and the transmitting transducer.

13.1.3 Scattering by internal obstacle

For an internal bounded obstacle (usually termed "defect" in ultrasonic testing) a similar approach as in the previous subsection is possible. It is thus assumed that the obstacle is not lying close to other scattering surfaces that contribute to the scattering. The approach here follows that of Boström and Wirdelius [30].

Figure 13.2 shows the geometry with transmitting transducer and the obstacle. A coordinate system (x, y, z) is attached to the transmitting transducer and a collinear system (x', y', z') is attached to the obstacle. If convenient a third system (x'', y'', z'') is also attached to the obstacle; this system being rotated relative to the (x', y', z') system. This is useful if the scattering by the obstacle is better treated in the (x'', y'', z'') system, for example if the obstacle is a circular crack. The vector from the transmitter to the obstacle is denoted $\boldsymbol{d}$.

To describe the scattering by the obstacle the T-matrix is employed, this matrix being calculated for a number of simple obstacles in previous chapters. In the coordinate system of the obstacle the incident and scattered waves are thus expanded in the spherical vector wavefunctions

$$\boldsymbol{u}^{\text{in}}(\boldsymbol{r}'') = \sum_n b_n \boldsymbol{\psi}_n^0(\boldsymbol{r}'') \tag{13.19}$$

$$\boldsymbol{u}^{\text{sc}}(\boldsymbol{r}'') = \sum_n f_n \boldsymbol{\psi}_n^+(\boldsymbol{r}'') \tag{13.20}$$

where the multiple index $n = \tau\sigma ml$ is used and the summation is thus over all these indices. The first expansion is valid inside a sphere centered at the obstacle out to the closest sources of the incident wave (or a surface where the incident wave is scattered) and the second expansion is valid outside the circumscribed sphere to the obstacle. The relationship between the expansion coefficients of the incident and scattered wave defines the T-matrix

$$f_n = \sum_{n'} T_{nn'} b_{n'} \tag{13.21}$$

The first task is to take the expansion of the incident wave in down-going plane waves from Eq. (13.1) and transform it to an expansion in terms of the regular spherical wavefunctions centered at the obstacle. This is done in three steps. First the plane waves are translated from the transducer system to the obstacle and then a transformation to the spherical vector waves is performed. If convenient the spherical waves are rotated to the final obstacle system. The translation is trivial so the incident wave on the obstacle can be written (where the form of the incident wave given in Eq. (13.8) is used)

$$\boldsymbol{u}^{\mathrm{in}}(\boldsymbol{r}) = \sum_j \int_0^{2\pi} \int_{C_-} \xi_j(q,p) \exp(\mathrm{i}k_j \hat{\boldsymbol{\gamma}} \cdot \boldsymbol{d}) \boldsymbol{\phi}_j(\alpha, \beta; \boldsymbol{r}) \sin\alpha \, \mathrm{d}\alpha \, \mathrm{d}\beta$$

The transformation to the spherical waves is performed in Eq. (8.22) and gives

$$\begin{aligned} \boldsymbol{u}^{\mathrm{in}}(\boldsymbol{r}) = \sum_j \int_0^{2\pi} \int_{C_-} & \xi_j(q,p) \exp(\mathrm{i}k_j \hat{\boldsymbol{\gamma}} \cdot \boldsymbol{d}) \\ & \times \sum_n B_{nj}^{\dagger}(\alpha, \beta) \boldsymbol{\psi}_n^0(\boldsymbol{r}') \sin\alpha \, \mathrm{d}\alpha \, \mathrm{d}\beta \end{aligned}$$

A final rotation finally gives an expansion in spherical vector wavefunctions as in Eq. (13.19) with expansion coefficients

$$b_n = \sum_{n'} D_{n'n} \sum_j \int_0^{2\pi} \int_{C_-} \xi_j(q,p) \exp(\mathrm{i}k_j \hat{\boldsymbol{\gamma}} \cdot \boldsymbol{d}) B_{nj}^{\dagger}(\alpha, \beta) \sin\alpha \, \mathrm{d}\alpha \, \mathrm{d}\beta \tag{13.22}$$

The rotation matrix $D_{n'n}$ is parametrized by the Euler angles of rotation and it is diagonal in the l and l' indices, see Boström *et al.* [27] for explicit expressions and Edmonds [46] for more details.

The integrals in Eq. (13.22) can be estimated by the stationary-phase method in the same way as the wave from the transducer is calculated in Eq. (13.9) and thus

$$b_n \simeq -2\pi \mathrm{i} \sum_{n'} D_{n'n} \sum_j \frac{\mathrm{e}^{\mathrm{i}k_j d}}{k_j d} \xi_j(\hat{\boldsymbol{d}}) B^{\dagger}_{nj}(\hat{\boldsymbol{d}}) \tag{13.23}$$

where the $\hat{\boldsymbol{d}}$ argument denotes the spherical angles [in the (x, y, z) system]. The range of validity of the approximation in Eq. (13.23) is the same as in Eq. (13.9).

As in the previous subsection the signal response can be obtained by employing the electromechanical reciprocity relation Eq. (13.14). State 1 is the actual state of interest with the transmitting transducer and the obstacle and state 2 is an auxiliary state with the receiving transducer acting as transmitter and with no obstacle. Thus

$$\boldsymbol{u}_1(\boldsymbol{r}'') = \sum_n \left[b_n^{(t)} \boldsymbol{\psi}_n^0(\boldsymbol{r}'') + f_n^{(t)} \boldsymbol{\psi}_n^+(\boldsymbol{r}'') \right] \tag{13.24}$$

$$\boldsymbol{u}_2(\boldsymbol{r}'') = \sum_n b_n^{(r)} \boldsymbol{\psi}_n^0(\boldsymbol{r}'') \tag{13.25}$$

Using the orthogonality of the spherical vector wavefunctions in Eq. (3.32) this leads to the signal response in the form

$$\delta\Gamma = \frac{\mu c_{\mathrm{s}}}{4P} \sum_{n,n'} b_n^{(r)} T_{nn'} b_{n'}^{(t)} \tag{13.26}$$

The effects of the transmitting transducer (given by $b_n^{(t)}$), the obstacle (given by $T_{nn'}$), and the receiving transducer (given by $b_n^{(r)}$) are clearly separated in this equation. As the T-matrix is symmetric an exchange of transmitter and receiver leads to no change in the signal response.

It is remarked that the signal response in Eq. (13.26) is in the frequency domain. In practice interest is often on the time domain signal (where time of flight is apparent) and to obtain this the signal must be multiplied by the frequency content of the exciting current and the frequency transfer function of the transducers and then Fourier transformed to the time domain.

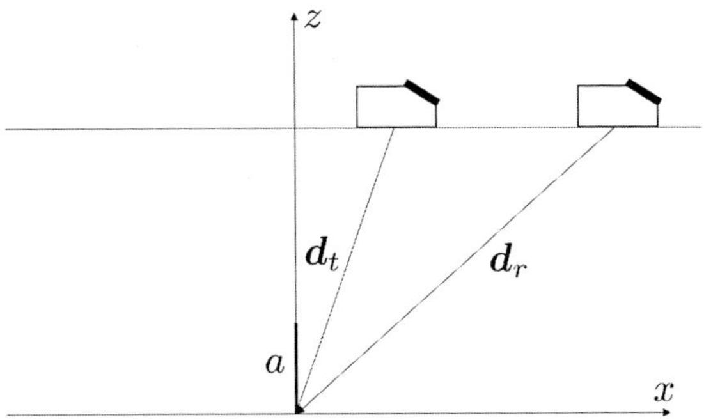

Figure 13.3: A surface-breaking crack with transmitting and receiving transducers.

13.1.4 Scattering by strip-like crack in a half-space

In the previous two subsections the modelling of ultrasonic testing by a side-drilled hole and a finite obstacle treated by its T-matrix is taken up. For the circular crack the methodology with computing the T-matrix works well, but for other cracks in 3D it is awkward to compute the T-matrix. A more direct approach, more similar to the side-drilled hole, is preferred. As an illustration a strip-like surface-breaking crack is taken up in this subsection, this being a very important defect in practice. The scattering by this crack is investigated in Section 12.4, but here this solution is coupled to transmitting and receiving transducers. For simplicity the crack is normal to the plane surface (henceforth called the "back surface") and this surface is in turn parallel to the plane surface where the transducers are scanning. The more general case with tilt of both the crack and the back surface is described by Boström and Bövik [20] for an interior crack close to a plane surface (and the limit with the crack surface-breaking is also taken although this is questionable, especially for low frequencies, see also Boström [16]).

The scattering geometry is depicted in Figure 13.3 which shows the crack situated at $x = 0$, $-\infty < y < \infty$, $0 < z < a$. The vectors from the transducers to the crack mouth (the origin) are

$$\boldsymbol{d}_t = \hat{\boldsymbol{x}} d_{xt} - \hat{\boldsymbol{z}} d_z$$

$$\boldsymbol{d}_r = \hat{\boldsymbol{x}} d_{xr} + \boldsymbol{y} d_{yr} - \hat{\boldsymbol{z}} d_z$$

The transmitter emits ultrasound which is scattered by the crack and the back surface and is then picked up by the receiving transducer. All multiple scattering between the scanning surface and the crack or back surface is neglected and this is reasonable if the distance between the scanning and back surfaces is at least a couple of wavelengths.

The incident wave from the transmitting transducer is given in Eq. (13.2) which can be written

$$u_i(\boldsymbol{r}) = \sum_j \int_{-\infty}^{\infty} \int_{-\infty}^{\infty} \xi_j^t(q,p)\, \mathrm{e}^{\mathrm{i}(qd_{xt}+h_j d_z)} g_{j,i}^-(q,p)\, \mathrm{e}^{\mathrm{i}(qx+py-h_j z)} \frac{\mathrm{d}q\,\mathrm{d}p}{k_j h_j} \tag{13.27}$$

given in the (x,y,z) system. Here $\xi_j^t(q,p)$ (with "t" for transmitter) is determined in the first subsection and $g_{j,i}^-$ is a vector $\boldsymbol{g}_j^-$, $j = 1,2,3$, with components

$$\begin{aligned}
\boldsymbol{g}_1^-(q,p) &= \frac{\mathrm{i}}{4\pi s}\,(\hat{\boldsymbol{x}}p - \hat{\boldsymbol{y}}q) \\
\boldsymbol{g}_2^-(q,p) &= \frac{1}{4\pi k_\mathrm{s} s}\left(\hat{\boldsymbol{x}}qh_\mathrm{s} + \hat{\boldsymbol{y}}ph_\mathrm{s} + \hat{\boldsymbol{z}}s^2\right) \\
\boldsymbol{g}_3^-(q,p) &= \frac{\mathrm{i}}{4\pi}\sqrt{\frac{k_\mathrm{p}}{k_\mathrm{s}^3}}\,(\hat{\boldsymbol{x}}q + \hat{\boldsymbol{y}}p - \hat{\boldsymbol{z}}h_\mathrm{p})
\end{aligned} \tag{13.28}$$

The solution to the scattering by the surface-breaking crack in Section 12.4 demands that the incident wave satisfies the boundary condition on the back surface. This is accomplished by adding reflected plane waves to the direct wave from the transducer so that the total incident wave becomes

$$\begin{aligned}
u_i^{\mathrm{in}}(\boldsymbol{r}) = \sum_j \int_{-\infty}^{\infty} \int_{-\infty}^{\infty} \xi_j^t(q,p)\, \mathrm{e}^{\mathrm{i}(qd_{xt}+h_j d_z)} \Big[g_{j,i}^-(q,p)\, \mathrm{e}^{\mathrm{i}(qx+py-h_j z)} \\
+ \sum_{j'} R_{j'j} g_{j',i}^+(q,p)\, \mathrm{e}^{\mathrm{i}(qx+py+h_{j'} z)} \Big] \frac{\mathrm{d}q\,\mathrm{d}p}{k_j h_j}
\end{aligned} \tag{13.29}$$

The plane wave reflection coefficients are given in Eq. (8.27), however, two changes are necessary. In Eq. (8.27) it is required to put $d = 0$ (as the reflecting surface is at $z = 0$) and the sign of all h_j must be changed (because the reflection matrix from above is needed, not from below) and the only change this gives is a sign change on R_{23} and R_{32}. The vectors $\boldsymbol{g}_j^+$ are obtained from $\boldsymbol{g}_j^-$ by changing all h_j to $-h_j$. The traction of the incident wave on the crack is needed in the solution to the integral equation given in Eqs. (12.21–12.23) and it becomes

$$t_i^{\mathrm{in}}(\boldsymbol{r}) = \mu \sum_j \int_{-\infty}^{\infty}\int_{-\infty}^{\infty} \xi_j^t(q,p)\,\mathrm{e}^{\mathrm{i}(qd_{xt}+h_j d_z)} \Bigg[f_{j,i}^-(q,p)\,\mathrm{e}^{\mathrm{i}(qx+py-h_j z)} + \sum_{j'} R_{j'j} f_{j',i}^+(q,p)\,\mathrm{e}^{\mathrm{i}(qx+py+h_{j'} z)} \Bigg] \frac{\mathrm{d}q\,\mathrm{d}p}{k_j h_j} \tag{13.30}$$

where

$$\begin{aligned} \boldsymbol{f}_1^-(q,p) &= \frac{1}{4\pi s}\left(-\hat{\boldsymbol{x}} p h_{\mathrm{s}} + \hat{\boldsymbol{y}} q h_{\mathrm{s}}\right) \\ \boldsymbol{f}_2^-(q,p) &= \frac{\mathrm{i}}{4\pi k_{\mathrm{s}} s}\left(\hat{\boldsymbol{x}} q(2s^2-k_{\mathrm{s}}^2) + \hat{\boldsymbol{y}} p(2s^2-k_{\mathrm{s}}^2) - \hat{\boldsymbol{z}}\, 2s^2 h_{\mathrm{s}}\right) \\ \boldsymbol{f}_3^-(q,p) &= \frac{1}{4\pi}\sqrt{\frac{k_{\mathrm{p}}}{k_{\mathrm{s}}^3}}\left(\hat{\boldsymbol{x}}\, 2qh_{\mathrm{p}} + \hat{\boldsymbol{y}}\, 2ph_{\mathrm{p}} - \hat{\boldsymbol{z}}(2s^2-k_{\mathrm{s}}^2)\right) \end{aligned} \tag{13.31}$$

and $\boldsymbol{f}_j^+$ is obtained from $\boldsymbol{f}_j^-$ by changing sign on all h_j. Insertion of the traction in Eq. (13.30) into Eq. (12.23) gives

$$T_{in}^t(p) = k_{\mathrm{s}} a \int_{-\infty}^{\infty} \sum_j \xi_j^t(q,-p)\,\mathrm{e}^{\mathrm{i}(qd_{xt}+h_j d_z)} \times\left(f_{j,i}^-(q,-p) I_n(-h_j a) + \sum_{j'} R_{j'j} f_{j',i}^+(q,-p) I_n(h_{j'} a)\right) \frac{\mathrm{d}q}{k_j h_j} \tag{13.32}$$

where $I_n(\gamma)$ is defined in Eq. (12.19).

To obtain the response of the receiving transducer the electromechanical reciprocity relation stated in Eq. (13.14) is again used. State 1 is the situation of interest with the incident wave as above in the presence of the crack whereas in state 2 the receiver is acting as transmitter and the crack is absent (but the back surface is present). In this way the wave scattered by the crack, including all multiple scattering between the crack and back surface, is included, but the wave directly reflected by the back surface is not included. Letting the control surface S in Eq. (13.14) be the crack (both sides) the reciprocity relation is reduced to

$$\delta\Gamma = \frac{\mathrm{i}\omega}{4P}\int_{-\infty}^{\infty}\int_{0}^{a}\boldsymbol{\Delta u}_1\cdot\boldsymbol{t}_2\,\mathrm{d}z\,\mathrm{d}y \tag{13.33}$$

Here $\boldsymbol{\Delta u}_1$ is the solution to Eq. (12.21) with the right-hand side given in Eq. (13.32) and $\boldsymbol{t}_2$ is given in Eq. (13.30) with the properties of the receiving transducer which means that the factor $\xi_j^t(q,-p)\,\mathrm{e}^{\mathrm{i}(qd_{xt}+h_jd_z)}$ is changed to $\xi_j^r(q,-p)\,\mathrm{e}^{\mathrm{i}(qd_{xr}+pd_{yr}+h_jd_z)}$. Note in particular the term pd_{yr} in the exponential which gives the location of the receiver in the y-direction relative to the transmitter. In practice usually $d_{yr}=0$. Inserting the expansion of the COD in Eq. (12.19) into Eq. (13.33) and performing some of the integrations give

$$\begin{aligned}\delta\Gamma = \frac{\mathrm{i}\pi c_{\mathrm{S}}a}{2P}\sum_{i,n}\sum_{j}\int_{-\infty}^{\infty}\int_{-\infty}^{\infty}\beta_{in}(-p)\,\xi_j^r(q,p)\,\mathrm{e}^{\mathrm{i}(qx+qd_{xr}+pd_{yr}+h_jd_z)}\\ \times\left(f_{j,i}^-(q,p)I_n(-h_ja)+\sum_{j'}R_{j'j}f_{j',i}^+(q,p)I_n(h_{j'}a)\right)\frac{\mathrm{d}q\,\mathrm{d}p}{k_jh_j}\end{aligned}$$

This can also be written

$$\delta\Gamma = \frac{\mathrm{i}\pi\mu c_{\mathrm{S}}}{2P}\sum_{in,i'n'}\int_{-\infty}^{\infty}T_{in}^t(-p)\left(Q^{-1}\right)_{in,i'n'}(-p)T_{i'n'}^r(p)\,\mathrm{d}p \tag{13.34}$$

where

$$\begin{aligned}T_{in}^r(p) = k_{\mathrm{S}}a\int_{-\infty}^{\infty}\sum_j\xi_j^r(q,p)\,\mathrm{e}^{\mathrm{i}(qd_{xt}+pd_{yr}+h_jd_z)}\\ \times\left(f_{j,i}^-(q,p)I_n(-h_ja)+\sum_{j'}R_{j'j}f_{j',i}^+(q,p)I_n(h_{j'}a)\right)\frac{\mathrm{d}q}{k_jh_j}\end{aligned} \tag{13.35}$$

As in previous subsections this is clearly divided into parts due to the transmitting transducer, the scattering by the crack, and the receiving transducer.

To get simpler expressions it is possible to perform a stationary-phase approximation of the p-integral in Eq. (13.34) and the q-integrals in Eqs.

(13.32) and (13.35). If $d_y = 0$, which is often the case, the stationary point to the p-integral in Eq. (13.34) appears at $p = 0$. For more details and examples, see Boström and Bövik [21]. For the stationary-phase approximations to be valid the restrictions are the same as in Section 13.1.1 when approximating the wave from a transducer.

13.2 Material characterization in polycrystalline materials

Ultrasound can be used for material characterization in various ways. A simple application is the estimation of the wave speeds and this can lead to a determination of the stiffness constants of the material. This is, however, often complicated by the fact that many materials are anisotropic and lossy and may also be inhomogeneous. Furthermore, the material may consist of more than one phase as in composites. Metals are usually grainy, where each grain is anisotropic with orientation that may be random or not. If the particles (or fibers) in a composite or the grains in a metal are small compared to relevant wavelengths it is useful to determine "effective" wave speeds and attenuations of the material, and a way to do this is discussed in the present section.

Wave propagation in a material consisting of a matrix with a distribution of obstacles is a classical subject in several branches of physics. Several methods are used to investigate such problems, but these are not reviewed here. An introduction to the subject with further references is given by Martin [88]. Anson and Chivers [7] compare twelve different theories with experiments for the case with solid spheres in a solid matrix and concludes "that they give widely disparate results when tested on a range of practically realizable systems" and that none is satisfactory in all cases.

The well-known paper by Foldy [51] treats the scattering by a random distribution of identical isotropic point scatterers in the scalar case and gives the following formula for the effective wavenumber

$$k_{\text{eff}}^2 = k^2 - 4\pi \mathrm{i} N g / k \tag{13.36}$$

Here k is the wavenumber of the matrix, N is the number density of scatterers, and g is the "scattering coefficient" which determines the strength of the scattering by a single scatterer. This expression neglects multiple scattering and is in general only valid for small concentration of scatterers. It is generalized in various ways by different authors. For waves in an elastic solid with equal obstacles with random orientation Gubernatis and Domany [58] give an equation for the effective wavenumber

$$K^2 = k^2 + 4\pi N\overline{F}(K)/K \tag{13.37}$$

Here K is the effective P or S wavenumber, k is the P or S wavenumber of the matrix, and $\overline{F}(K)$ is the average P or S forward scattering amplitude of a single obstacle. Note that this is an equation that is to be solved for K as the forward scattering amplitude is calculated as if the obstacle is located in the effective material. As for the formula of Foldy multiple scattering is neglected, so the equation is only valid for small concentration of obstacles. A common approximation is to calculate the forward scattering amplitude in the matrix material instead of in the effective material and this gives the explicit form

$$K^2 = k^2 + 4\pi N\overline{F}(k)/k \tag{13.38}$$

The validity of this expression is of course even more restricted. Earlier similar work is performed by Mal and Knopoff [87] and Datta [40].

It is useful to express the equation for the effective wavenumber with the help of the T-matrix [28]. For an incident plane P wave the P wave scattered far field amplitude becomes [Eqs. (3.46) and (5.11)]

$$F_{\mathrm{pp}}(\hat{\boldsymbol{k}}) = 4\pi \sum_{n,n'} \mathrm{i}^{l-1-l'} T_{3n,3n'} \mathrm{Y}_n(\hat{\boldsymbol{k}}) \mathrm{Y}_{n'}(\hat{\boldsymbol{k}})$$

Here the multi-index $n = \sigma ml$ is used for convenience and the sums are thus over these three indices. If the effective material is assumed to be isotropic, a mean over all directions of incidence (which is the same as a mean over all orientations of the obstacle) gives

$$\overline{F}_{\mathrm{pp}} = -\mathrm{i} \sum_n T_{3n,3n}$$

This is the trace of the P part of the T-matrix. The effective P wavenumber becomes

$$(K_{\mathrm{p}}/k_{\mathrm{p}})^2 = 1 - \frac{4\pi \mathrm{i} c}{V k_{\mathrm{p}}^3} \sum_n T_{3n,3n} \tag{13.39}$$

Here V is the volume of one obstacle and c is the volume concentration of obstacles.

For an incident plane S wave the S wave scattered far field amplitude becomes [Eqs. (3.45) and (5.13)]

$$\hat{\boldsymbol{e}} \cdot F_{\mathrm{ss}}(\hat{\boldsymbol{k}}) = 4\pi \sum_{n,n'} \sum_{\tau=1}^{2} \sum_{\tau'=1}^{2} \mathrm{i}^{l-1-l'} T_{\tau' n', \tau n}$$

$$\times\, \hat{\boldsymbol{e}} \cdot \left[-\mathrm{i}\delta_{\tau' 1} \boldsymbol{A}_{1n'}(\hat{\boldsymbol{k}}) + \delta_{\tau' 2} \boldsymbol{A}_{2n'}(\hat{\boldsymbol{k}}) \right] \hat{\boldsymbol{e}} \cdot \left[\mathrm{i}\delta_{\tau 1} \boldsymbol{A}_{1n}(\hat{\boldsymbol{k}}) + \delta_{\tau 2} \boldsymbol{A}_{2n}(\hat{\boldsymbol{k}}) \right]$$

where $\hat{\boldsymbol{e}}$ is the polarization of the incident wave, with $\hat{\boldsymbol{e}} \cdot \hat{\boldsymbol{k}} = 0$. When taking the mean over all directions of incidence and all polarizations it is first noted that the mean over all polarizations gives (in dyadic notation)

$$\overline{\hat{\boldsymbol{e}}\hat{\boldsymbol{e}}} = \frac{1}{2} \left(I - \hat{\boldsymbol{k}}\hat{\boldsymbol{k}} \right)$$

i.e. half the transverse part of the unit dyadic I. The mean over all angles is performed using the orthonormality of the vector spherical harmonics and gives

$$(K_{\mathrm{s}}/k_{\mathrm{s}})^2 = 1 - \frac{2\pi \mathrm{i} c}{V k_{\mathrm{s}}^3} \sum_{n} \sum_{\tau=1}^{2} T_{\tau n, \tau n} \tag{13.40}$$

As written the two equations for the effective wavenumber use the further approximation that the right-hand side is calculated for the scattering in the matrix, but the equations are also valid if the right-hand side is calculated for the scattering in the effective material. With the effective wavenumber calculated the attenuation α and the phase velocity c can be obtained

$$\alpha = \operatorname{Im} K$$

$$c = \omega / \operatorname{Re} K$$

Equations (13.39) and (13.40) are derived for identical obstacles that are randomly oriented. As multiple scattering is neglected, the equations are just a sum of the scattering from each particle. Therefore, it is possible to generalize the equations when there are different types of obstacles by just adding one term for each type with its concentration and material properties. This is also useful if it is necessary to consider a size distribution of obstacles.

Often the effective properties are investigated for small obstacles (low frequencies) as the attenuation in many cases becomes too large for effective properties to be meaningful otherwise. It is then possible to use the low frequency expression for the T-matrix elements as calculated for some simple obstacles in previous chapters, in particular spheres (also anisotropic ones) and circular cracks, and this leads to explicit expressions for the attenuation and phase velocity. However, it is noted that all T-matrix elements are purely imaginary to leading order, so that the effective wavenumber becomes real and the attenuation is not determined. This issue can easily be resolved by using the "hermiticity" property of the T-matrix [Eq. (5.43)]

$$\sum_{\tau'',\sigma'',m'',l''} T^*_{\tau''\sigma''m''l'',\tau\sigma ml} T_{\tau''\sigma''m''l'',\tau'\sigma'm'l'} = -\mathrm{Re}\, T_{\tau\sigma ml,\tau'\sigma'm'l'}$$

This gives the real part of the T-matrix and as the imaginary part goes as $(ka)^3$ for low frequencies in all cases from previous chapters, the real part goes as $(ka)^6$. The attenuation then goes as $(ka)^3$ and the phase velocity is constant for low frequencies.

For a composite the contrast between the particles (fibers) and the matrix is usually large and Eqs. (13.39) and (13.40) are then valid for low concentration of particles, say $c < 0.05$. A quite different application of these equations is to the attenuation and phase velocity of polycrystalline materials, typically metals. A grain must then be supposed to scatter as if it is situated in the effective material due to all the other grains. As the grains fill the whole volume the concentration is put to $c = 1$. This is certainly not a low concentration, but the argument is that each grain scatters very little so that multiple scattering may still be neglected which makes it reasonable to apply Eqs. (13.39) and (13.40). Another issue is the shape of the grains, which are largely unknown in practice. A simple approach is to take them spherical, as it is then possible to employ the explicit low frequency T-matrix elements from Chapter 6. This is motivated by the fact that for low frequencies the scattering is mainly a volume effect, see e.g. [144, 118] [it is also apparent in the Born approximation, Eq. (4.29)]. Jafarzadeh *et al.* [70] use this approach for polycrystalline metals with transversely isotropic grains.

The scattering in grainy materials is a subject that has been investigated for a long time. Stanke and Kino [123] summarize earlier work and use the Keller approximation to obtain a unified theory valid for all frequencies. Weaver [141] uses the first-order smoothing approximation in the Dyson equation and obtains an equivalent theory. The original work by Stanke and Kino [123] and Weaver [141] is extended in various ways, all the way to generally anisotropic grains, see Sha *et al.* [121], where also many references to earlier work is given. These models [collectively called SOA (second order approximation) in the following] replace the actual polycrystalline material with a continuous elastic material with random fluctuations of the elastic stiffness tensor with statistics similar to the polycrystalline material. The wave propagation in this model is then solved with an approximate perturbation approach and often the Born approximation is also used. The mean material stiffnesses needed are usually taken as the Voigt average [123, 121], although this may be inadequate for higher degree of anisotropy. As SOA is a perturbative approach in the degree of anisotropy the errors are believed to decrease with decrease in material anisotropy [121].

Another way to investigate the scattering in polycrystalline materials is to employ a wholly numerical approach. With FEM this is pursued by Huang *et al.* [66, 67], where also many references are given. This has the advantage that arbitrary material properties and geometry can be accommodated, but there are issues with the size of the model, the generation of the deterministic grain structure, the boundary conditions, and the computational effort [67]. Low frequencies are also problematic because of the great size of the model that is needed (but static computations are possible). In general it is believed that FEM results for attenuation and phase velocity that are more or less accurate can be obtained. Comparisons with SOA and the method of Jafarzadeh *et al.* [70] give a good or very good correspondence in most cases, in particular for materials with weak anisotropy.

Bibliography

[1] Achenbach, J. D. 1973. *Wave Propagation in Elastic Solids*. Amsterdam: North-Holland.

[2] Achenbach, J. D., and Ch. Zhang. 1990. "Reflection and transmission of ultrasound by a region of damaged material." *Journal of Nondestructive Evaluation* 9: 22–32. https://doi.org/10.1007/BF00566385.

[3] Achenbach, J. D. 2003. *Reciprocity in Elastodynamics*. Cambridge: Cambridge University Press.

[4] Achenbach, J. D., and Y. Xu. 1999. "Wave motion in an isotropic elastic layer generated by a time-harmonic point load of arbitrary direction." *Journal of the Acoustical Society of America* 106: 83–90. https://doi.org/10.1121/1.427037.

[5] Aki, K., and P. G. Richards. 1980. *Quantitative Seismology*, volume 1. New York: W. H. Freeman and Company.

[6] Ang, W. T., D. L. Clements, and M. Dehghan. 1993. "Scattering and diffraction of SH-waves by multiple planar cracks in an anisotropic half-space: A hypersingular integral formulation." *International Journal of Solids and Structures* 30: 1301–1312. https://doi.org/10.1016/0020-7683(93)90213-Q.

[7] Anson, L. W., and R. C. Chivers. 1993. "Ultrasonic velocity in suspensions of solids in solids—a comparison of theory and experiment." *Journal of Physics D: Applied Physics* 26: 1566–1575. https://doi.org/10.1088/0022-3727/26/10/005.

[8] Auld, B. A. 1979. "General electromechanical reciprocity relations applied to the calculation of elastic wave coefficients." *Wave Motion* 1: 3–10. https://doi.org/10.1016/0165-2125(79)90020-9.

[9] Auld, B. A. 1990. *Acoustic Waves and Fields in Solids*, volume 1, second edition. Malabar, Florida: Krieger.

[10] Baboux, J. C., and R. Kazys. 1992. "Analysis of the transient ultrasonic fields radiated in solids by circular and annular sources." *Journal of the Acoustical Society of America* 92: 2942–2949. https://doi.org/10.1121/1.405292.

[11] Baik, J. M., and R. B. Thompson. 1984. "Ultrasonic scattering from imperfect interfaces: a quasistatic model." *Journal of Nondestructive Evaluation* 4: 177–196. https://doi.org/10.1007/BF00566223.

[12] Boström, A. 1980. "Multiple scattering of elastic waves by bounded obstacles." *Journal of the Acoustical Society of America* 67: 399–413. https://doi.org/10.1121/1.383925.

[13] Boström, A. 1980. "Scattering by a smooth elastic obstacle." *Journal of the Acoustical Society of America* 67: 1904–1913. https://doi.org/10.1121/1.384455.

[14] Boström, A. 1980. "Transmission and reflection of acoustic waves by an obstacle in a waveguide." *Wave Motion* 2: 167–184. https://doi.org/10.1016/0165-2125(80)90026-8.

[15] Boström, A. 1987. "Elastic wave scattering from an interface crack: antiplane strain." *Journal of Applied Mechanics* 54: 503–508. https://doi.org/10.1115/1.3173060.

[16] Boström, A. 1999. "On the importance of edge conditions in an elastodynamic crack problem." In *Festskrift till Staffan Ström*, edited by A. Karlsson and G. Kristensson, 19–27. Lund, Sweden.

[17] Boström, A. 2003. "Review of integral equation method for crack scattering and application to modelling of ultrasonic nondestructive evaluation." *Applied Mechanics Reviews* 56: 383–405. https://doi.org/10.1115/1.1574522.

[18] Boström, A. 2015. "Scattering by an anisotropic circle." *Wave Motion* 57: 239–244. https://doi.org/10.1016/j.wavemoti.2015.04.007.

[19] Boström, A. 2018. "Scattering of in-plane elastic waves by an anisotropic circle." *Quarterly Journal of Mechanics and Applied Mathematics* 71: 139–155. https://doi.org/10.1093/qjmam/hbx029.

[20] Boström, A., and P. Bövik. 1997. "A model of ultrasonic nondestructive testing for internal and subsurface cracks." *Journal of the Acoustical Society of America* 102: 2723–2733. https://doi.org/10.1121/1.420326.

[21] Boström, A., and P. Bövik. 2003. "Ultrasonic scattering by a side-drilled hole." *International Journal of Solids and Structures* 40: 3493–3505. https://doi.org/10.1016/S0020-7683(03)00150-1.

[22] Boström, A., P. Bövik, and P. Olsson. 1992. "A comparison of exact first order and spring boundary conditions for scattering by thin layers." *Journal of Nondestructive Evaluation* 11: 175–184. https://doi.org/10.1007/BF00566408.

[23] Boström, A., and A. S. Eriksson. 1993. "Scattering by two penny-shaped cracks with spring boundary conditions." *Proceedings of the Royal Society of London* A 443: 183–201. https://doi.org/10.1098/rspa.1993.0139.

[24] Boström, A., T. Grahn, and A. J. Niklasson. 2003. "Scattering of elastic waves by a rectangular crack in an anisotropic half-space." *Wave Motion* 38: 91–107. https://doi.org/10.1016/S0165-2125(03)00025-8.

[25] Boström, A., and A. Karlsson. 1987. "Broad-band synthetic seismograms for a spherical inhomogeneity in a many-layered elastic half-space." *Geophysical Journal of the Royal Astronomical Society* 89: 527–547. https://doi.org/10.1111/j.1365-246X.1987.tb05182.x.

[26] Boström, A., and Kristensson, G. 1980. "Elastic wave scattering by a three-dimensional inhomogeneity in an elastic half-space." *Wave Motion*, 2: 335–353. https://doi.org/10.1016/0165-2125(80)90013-X.

[27] Boström, A., G. Kristensson, and S. Ström. 1991. "Transformation properties of plane, spherical and cylindrical scalar and vector wave functions." In *Field Representations and Introduction to Scattering*, edited by V. V. Varadan, A. Lakhtakia, and V. K. Varadan, 165–210. Amsterdam: North-Holland.

[28] Boström, A., P. Olsson, and S. K. Datta. 1992. "Effective plane wave propagation through a medium with spheroidal inclusions surrounded by thin interface layers." *Mechanics of Materials* 14: 59–66. https://doi.org/10.1016/0167-6636(92)90018-9.

[29] Boström, A., and G. R. Wickham. 1991. "On the boundary conditions for ultrasonic transmission by partially closed cracks." *Journal of Nondestructive Evaluation* 10: 139–149. https://doi.org/10.1007/BF00567096.

[30] Boström, A., and H. Wirdelius. 1995. "Ultrasonic probe modeling and nondestructive crack detection." *Journal of the Acoustical Society of America* 97: 2836–2848. https://doi.org/10.1121/1.411850.

[31] Bövik, P. 1993. "On the modelling of thin interface layers in elastic and acoustic scattering problems." *Quarterly Journal of Mechanics and Applied Mathematics* 47: 17–42. https://doi.org/10.1093/qjmam/47.1.17.

[32] Bövik, P., and P. Olsson. 1992. "Effective boundary conditions for the scattering of two-dimensional SH waves from a curved thin elastic layer." *Proceedings of the Royal Society of London* A 439: 257–269. https://doi.org/10.1098/rspa.1992.0148.

[33] Bresse, L. F., and D. A. Hutchins. 1989. "Transient generation of elastic waves in solids by a disk-shaped normal force model." *Journal of the Acoustical Society of America* 86: 810–817. https://doi.org/10.1121/1.398204.

[34] Buchwald, V. T. 1961. "Rayleigh waves in transversely isotropic media." *Quarterly Journal of Mechanics and Applied Mathematics* 14: 293–317. https://doi.org/10.1093/qjmam/14.3.293.

[35] Carcione, J. M. 2022. *Wave Fields in Real Media: Wave Propagation in Anisotropic, Anelastic, Porous and Electromagnetic Media*, fourth edition. Amsterdam: Elsevier.

[36] Chapman, R. K. 1990. “A system approach for the ultrasonic inspection of smooth planar cracks.” *Journal of Nondestructive Evaluation* 9: 197–211. https://doi.org/10.1007/BF00566394.

[37] Chevalier, Y., M. Louzar, and G. A. Maugin. “Influence of interface slits with elastic slip on the propagation of SH waves.” *Journal of the Acoustical Society of America* 98: 445–453. https://doi.org/10.1121/1.413701.

[38] Cruzan, O. R. 1962. “Translational addition theorems for spherical vector wave functions.” *Quarterly of Applied Mathematics* 20: 33–40. https://doi.org/10.21236/AD0260822.

[39] Dassios, G., and R. Kleinman. 2000. *Low Frequency Scattering*. Oxford: Oxford University Press.

[40] Datta, S. K. 1977. “A self-consistent approach to multiple scattering by elastic ellipsoidal inclusions”. *Journal of Applied Mechanics* 44: 657–662. https://doi.org/10.1115/1.3424153.

[41] de Hoop, A. T. 1995. *Handbook of Radiation and Scattering of Waves*. New York: Academic Press.

[42] Dhawan, G. K. 1982. “Interaction of elastic waves by a Griffith crack in an infinite transversely-isotropic medium.” *International Journal of Fracture* 19: 29–37. https://doi.org/10.1007/BF00012490.

[43] Ding, H., W. Chen, and L. Zhang. 2006. *Elasticity of Transversely Isotropic Materials*. Dordrecht, The Netherlands: Springer.

[44] Diligent, O., T. Grahn, A. Boström, P. Cawley, and M. J. S. Lowe. 2002. “The low-frequency reflection and scattering of the S_0 Lamb mode from a circular through-thickness hole in a plate: Finite Element, analytical and experimental studies.” *Journal of the Acoustical Society of America* 112: 2589–2601. https://doi.org/10.1121/1.1512292.

[45] Dundurs, J. 1969. “Discussion.” *Journal of Applied Mechanics* 36: 650–652. https://doi.org/10.1115/1.3564739.

[46] Edmonds, A. R. 1974. *Angular Momentum in Quantum Mechanics*, second edition. Princeton: Princeton University Press.

[47] Eriksson, A. S., J. Mattsson, and A. J. Niklasson. 2000. "Modeling of ultrasonic crack detection in anisotropic materials." *NDT & E International* 33: 441–451. https://doi.org/10.1016/S0963-8695(00)00016-5.

[48] Eringen, A. C., and E. S. Suhubi. 1973. *Elastodynamics, Volume II: Linear Theory*. New York: Academic Press.

[49] Fan, Y., A. N. Sinclair, and F. Honorvar. 1999. "Scattering of a plane acoustic wave from a transversely isotropic cylinder enclosed in a solid elastic medium." *Journal of the Acoustical Society of America* 106: 1229–1236. https://doi.org/10.1121/1.427174.

[50] Felsen, L. B., and N. Marcuvitz. 1994. *Radiation and Scattering of Waves*. Piscataway, New Jersy: IEEE Press.

[51] Foldy, L. L. 1945. "The multiple scattering of waves. I. General theory of isotropic scattering by randomly distributed scatterers." *Physical Review* 67: 107–119. https://doi.org/10.1103/PhysRev.67.107.

[52] Golub, M. V., O. V. Doroshenko, and A. Boström. 2016. "Effective spring boundary conditions for a damaged interface between dissimilar media in three-dimensional case." *International Journal of Solids and Structures* 81: 141–150. https://doi.org/10.1016/j.ijsolstr.2015.11.021.

[53] Gracewski, S. M., and D. B. Bogy. 1986. "Elastic wave scattering from an interface crack in a layered half space submerged in water. Part I: Applied tractions on the liquid-solid interface." *Journal of Applied Mechanics* 53: 326–332. https://doi.org/10.1115/1.3171760.

[54] Gracewski, S. M., and D. B. Bogy. 1986. "Elastic wave scattering from an interface crack in a layered half space submerged in water. Part II: Incident plane waves and bounded beams." *Journal of Applied Mechanics* 53: 333–338. https://doi.org/10.1115/1.3171761.

[55] Grahn, T. 2001. "Ultrasonic 2D SH crack detection in a layered anisotropic plate." *Journal of Nondestructive Evaluation* 20: 17–28. https://doi.org/10.1023/A:1010649713130.

[56] Gross, D., and C. Zhang. 1988. "Diffraction of SH waves by a system of cracks: solution by an integral equation method." *International Journal of Solids and Structures* 24: 41–49. https://doi.org/10.1016/0020-7683(88)90097-2.

[57] Guan, L., and A. N. Norris. "Elastic wave scattering by rectangular cracks." *International Journal of Solids and Strutures* 29: 1549–1565. https://doi.org/10.1016/0020-7683(92)90133-E.

[58] Gubernatis, J. E., and E. Domany. 1984. "Effects of microstructure on the speed and attenuation of elastic waves in porous materials." *Wave Motion* 6: 579–589. https://doi.org/10.1016/0165-2125(84)90048-9.

[59] Gubernatis, J. E., E. Domany, J. A. Krumhansl, and M. Huberman. 1977. "The Born approximation in the theory of the scattering of elastic waves by flaws." *Journal of Applied Physics* 48: 2812–2819. https://doi.org/10.1063/1.324142.

[60] Gurtin, M. E., J. Weissmuller, and F. Larche. 1998. "A general theory of curved deformable interfaces in solids in equilibrium." *Philosophical Magazine* 78: 1093–1109. https://doi.org/10.1080/01418619808239977.

[61] Harris, J. G. 1988. "The wavefield radiated into an elastic halfspace by a transducer of large aperture." *Journal of Applied Mechanics* 55, 398–404. https://doi.org/10.1115/1.3173689.

[62] "Hashheminejad, S. M., and M. Maleki. 2008. "Acoustic resonance scattering from a submerged anisotropic sphere." *Acoustical Physics* 54: 168–179. https://doi.org/10.1134/S1063771008020048.

[63] Helsing, J., A. Jonsson, and G. Peters. 2001. "Evaluation of the mode I stress intensity factor for a square crack in 3D." *Engineering Fracture Mechanics.* 68: 605–612. https://doi.org/10.1016/S0013-7944(00)00115-6.

[64] Honarvar, F., and A. N. Sinclair. 1996. "Acoustic wave scattering from transversely isotropic cylinders." *Journal of the Acoustical Society of America* 100: 57–63. https://doi.org/10.1121/1.415868.

[65] Huang, M., P. Huthwaite, S. I. Rokhlin, and M. J. S. Lowe. 2022. "Finite-element and semi-analytical study of elastic wave propagation in strongly scattering polycrystals." *Proceedings of the Royal Society* A 478: 20210850. https://doi.org/10.1098/rspa.2021.0850.

[66] Huang, M., S. I. Rokhlin, and M. J. S. Lowe. 2022. "Appraising scattering theories for polycrystals of any symmetry using finite elements." *Philosophical Transactions of the Royal Society A* 380: 20210382. https://doi.org/10.1098/rsta.2021.0382.

[67] Huang, M., G. Sha, P. Huthwaite, S. I. Rokhlin, and M. J. S. Lowe. 2020. "Maximizing the accuracy of finite element simulation of elastic wave propagation in polycrystals." *Journal of the Acoustical Society of America* 148: 1890–1910. https://doi.org/10.1121/10.0002102.

[68] Itou, S. 1980. "Dynamic stress concentration around a rectangular crack in an infinite elastic medium." *Zeitschrift für angewandte Mathematic und Mechanik* 60: 317–322. https://doi.org/10.1002/zamm.19800600806.

[69] Itou, S. 1986. "Transient stresses around an interface crack." *Engineering Fracture Mechanics* 25: 475–482. https://doi.org/10.1016/0013-7944(86)90261-4.

[70] Jafarzadeh, A., P. D. Folkow, and A. Boström. 2022. "Scattering of elastic waves by a transversely isotropic sphere and ultrasonic attenuation in hexagonal polycrystalline materials." *Wave Motion* 112: 102963. https://doi.org/10.1016/j.wavemoti.2022.102963.

[71] Jafarzadeh, A., P. D. Folkow, and A. Boström. 2023. "Scattering of elastic waves by a sphere with cubic anisotropy with application to attenuation in polycrystalline materials." *Proceedings of the Royal Society* A 479: 20220476. https://doi.org/10.1098/rspa.2022.0476.

[72] Jafarzadeh, A., P. D. Folkow, and A. Boström. 2024. "Scattering of elastic waves by a sphere with orthorhombic anisotropy and application to polycrystalline material characterization." *Ultrasonics* 138: 107199. https://doi.org/10.1016/j.ultras.2023.107199.

[73] Karlsson, A. 1982. "Scattering from inhomogeneities in layered structures." *Journal of the Acoustical Society of America* 71: 1083–1092. https://doi.org/10.1121/1.387779.

[74] Karlsson, A. 1984. "Scattering of Rayleigh-Lamb waves from cavities in an elastic plate." *Wave Motion* 6: 205–222. https://doi.org/10.1016/0165-2125(84)90016-7.

[75] Kausel, E. 2006. *Fundamental Solutions in Elastodynamics*. Cambridge: Cambridge University Press.

[76] Krautkrämer, J., and H. Krautkrämer. 1990. *Ultrasonic Testing of Materials*, fourth edition. Berlin: Springer.

[77] Krenk, S., and H. Schmidt. 1982. "Elastic wave scattering by a circular crack." *Philosophical Transactions of the Royal Society of London* A 308: 167–198. https://doi.org/10.1098/rsta.1982.0158.

[78] Kundu, T. 1986. "Transient response of an interface-crack in a layered plate." *Journal of Applied Mechanics* 53: 579–585. https://doi.org/10.1115/1.3171814.

[79] Kundu, T. 1987. "The transient response of two cracks at the interface of a layered half-space." *International Journal of Engineering Sciences* 25: 1427–1439. https://doi.org/10.1016/0020-7225(87)90021-8.

[80] Kundu, T. 1988. "Dynamic interaction between two interface cracks in a three layered plate." *International Journal of Solids and Structures* 24: 27–39. https://doi.org/10.1016/0020-7683(88)90096-0.

[81] Kundu, T., and A. Boström. 1992. "Elastic wave scattering by a circular crack in a transversely isotropic solid." *Wave Motion* 15: 285–300. https://doi.org/10.1016/0165-2125(92)90012-Q.

[82] Kuo, A. Y. 1984. "Transient stress intensity factors of an interfacial crack between two dissimilar anisotropic half-spaces, part 1: Orthotropic materials." *Journal of Applied Mechanics* 51: 71–76. https://doi.org/10.1115/1.3167600.

[83] Kuo, A. Y. 1984. "Transient stress intensity factors of an interfacial crack between two dissimilar anisotropic half-spaces, part 2: Fully anisotropic half-spaces." *Journal of Applied Mechanics* 51: 780–786. https://doi.org/10.1115/1.3167724.

[84] Langenberg, K.-J., R. Marklein, and K. Mayer. 2012. *Ultrasonic Nondestructive Testing of Materials. Theoretical Foundations.* Boka Raton: CRC Press, Taylor & Francis Group.

[85] Lavrentyev, A. I., and S. I. Rokhlin. 1994. "Models for ultrasonic characterization of environmental degradation of interfaces in adhesive joints." *Journal of Applied Physics* 76: 4643–4650. https://doi.org/10.1063/1.357301.

[86] Lekesiz, H., N. Katsube, S. Rokhlin, and R. R. Seghi. 2013. "Effective spring stiffness for a periodic array of interacting coplanar penny-shaped cracks at an interface between two dissimilar isotropic materials." *International Journal of Solids and Structures* 50: 2817–2828. https://doi.org/10.1016/j.ijsolstr.2013.04.006.

[87] Mal, A. K., and L. Knopoff. 1967 "Elastic wave velocities in two-component systems." *IMA Journal of Applied Mathematics* 3: 376–387. https://doi.org/10.1093/imamat/3.4.376.

[88] Martin, P. A. 2006. *Multiple Scattering: Interaction of Time-Harmonic Waves with N Obstacles*. Cambridge: Cambridge University Press.

[89] Martin, P. A., and J. R. Berger. 2001. "Waves in wood: free vibrations of a wooden pole." *Journal of the Mechanics and Physics of Solids* 49: 1155–1178. https://doi.org/10.1016/S0022-5096(00)00068-5.

[90] Martin, P. A., and R. A. Dalrymple. 1988. "Scattering of long waves by cylindrical obstacles and gratings using matched asymptotic expansions." *Journal of Fluid Mechanics* 188: 465–490. https://doi: 10.1017/S0022112088000801.

[91] Martin, P. A., and F. J. Rizzo. 1989. "On boundary integral equations for crack problems." *Proceedings of the Royal Society* A 421: 341–355. https://doi.org/10.1098/rspa.1989.0014.

[92] Martin, P. A., and F. J. Rizzo. 1996. "Hypersingular integrals: how smooth must the density be?" *International Journal of Numerical Methods in Engineering* 39: 687–704. https://doi.org/10.1002/(SICI)1097-0207(19960229)39:4⟨687::AID-NME876⟩3.0.CO;2-S.

[93] Martin, P. A., F. J. Rizzo, and T. A. Cruse. 1998. "Smoothness-relaxation strategies for singular and hypersingular integral equations." *International Journal of Numerical Methods in Engineering* 42: 885–906. https://doi.org/10.1002/(SICI)1097-0207(19980715)42:5⟨885::AID-NME392⟩3.0.CO;2-X.

[94] Martin, P. A., and G. R. Wickham. 1983. "Diffraction of elastic waves by a penny-shaped crack: analytical and numerical results." *Proceedings of the Royal Society* A 390: 91–129. https://doi.org/10.1098/rspa.1983.0124.

[95] Mattsson, J., and A. J. Niklasson. 1997. "Ultrasonic 2-D SH crack detection in anisotropic solids." *Journal of Nondestructive Evaluation* 16: 31–41. https://doi.org/10.1007/BF03325383.

[96] Mattsson, J., A. J. Niklasson, and A. S. Eriksson. "Three-dimensional ultrasonic crack detection in anisotropic materials." *Research in Nondestructive Evaluation* 9: 59–79. https://doi.org/10.1007/PL00003890.

[97] McNab, A., A. Cochran, and M. A. Campbell. 1990. "The calculation of acoustic fields in solids for transient normal surface forces for arbitrary geometry and apodization." *Journal of the Acoustical Society of America* 87: 1455–1465. https://doi.org/10.1121/1.399442.

[98] Mittleman, J., R. Roberts, and R. B. Thompson. 1995. "Scattering of longitudinal elastic waves from an anisotropic spherical shell." *Journal of Appled Mechanics* 62: 150–158. https://doi.org/10.1115/1.2895896.

[99] Morse, P. M., and H. Feshbach. 1953. *Methods of Theoretical Physics*. New York: McGraw-Hill.

[100] Nayfeh, A. H. 1995. *Wave Propagation in Layered Anisotropic Media with Applications to Composites*. Amsterdam: North-Holland.

[101] Neerhoff, F. L. 1979. "Diffraction of Love waves by a stress-free crack of finite width in the plane interface of a layered composite." *Applied Scientific Research* 35: 237–249. https://doi.org/10.1007/BF00418217.

[102] Niklasson, A. J., and S. K. Datta. 1998. "Scattering by an infinite transversely isotropic cylinder in a transversely isotropic solid." *Wave Motion* 27: 169–185. https://doi.org/10.1016/S0165-2125(97)00038-3.

[103] Norris, A. N. 1986. "Scattering of elastic waves by spherical inclusions with applications to low frequency wave propagation in composites." *International Journal of Engineering Sciences* 24: 1271–1282. https://doi.org/10.1016/0020-7225(86)90056-X.

[104] Norris, A. N., and A. L. Shuvalov. 2012. "Elastodynamics of radially inhomogeneous spherically anisotropic elastic materials in the Stroh formalism." *Proceedings of the Royal Society* A 468: 467–484. https://doi:10.1098/rspa.2011.0463.

[105] Norris, A. N., and Y. Yang. 1991. "Dynamic stress on a partially bonded fiber." *Journal of Applied Mechanics* 58: 404–409. https://doi.org/10.1115/1.2897200.

[106] Oien, M. A., and Y.-H. Pao. 1973. "Scattering of compressional waves by a rigid spheroidal inclusion." *Journal of Applied Mechanics* 40: 1073–1077. https://doi.org/10.1115/1.3423128.

[107] Olsson, P. 1985. "The rigid movable inclusion in elastostatics and elastodynamics." *Wave Motion* 7: 421–445. https://doi.org/10.1016/0165-2125(85)90018-6.

[108] Olsson, S. 1990. "Transmission and reflection of elastic waves by a spherical obstacle in an infinite circular cylindrical rod." *Quarterly Journal of Mechanics and Applied Mathematics* 47: 583–606. https://doi.org/10.1093/qjmam/47.4.583.

[109] Olsson, S. 1994. "Point force excitation of a thick-walled elastic infinite pipe with an embedded inhomogeneity." *Journal of Engineering Mathematics* 28: 311-325. https://doi.org/10.1007/BF00128750.

[110] Olver, F. W. J. 1974. *Asymptotics and Special Functions*. New York: Academic Press.

[111] Pao, Y.-H. 1978. "Betti's indentity and transition matrix for elastic waves." *Journal of the Acoustical Society of America* 64: 302–310. https://doi.org/10.1121/1.381986.

[112] Pao, Y.-H., and C.-C. Mow. 1973. *Diffraction of Elastic Waves and Dynamic Stress Concentrations*. New York: Crane-Russak.

[113] Pao, Y.-H., and V. V. Varatharajulu. 1976. "Huygens' principle, radiation conditions and integral formulas for the scattering of elastic waves." *Journal of the Acoustical Society of America* 59: 1361–1371. https://doi.org/10.1121/1.381022.

[114] Peterson, B., and S. Ström. 1974. "Matrix formulation of acoustic scattering from an arbitrary number of scatterers." *Journal of the Acoustical Society of America* 56: 771–780. https://doi.org/10.1121/1.1903325.

[115] Qu, J. M. 1994. "Scattering of plane waves from an interface crack." *International Journal of Engineering Sciences* 33: 179–194. https://doi.org/10.1016/0020-7225(94)00058-R.

[116] Qu, J. M. 1995. "Interface crack loaded by a time-harmonic plane wave." *International Journal of Solids and Structures* 31: 329–345. https://doi.org/10.1016/0020-7683(94)90110-4.

[117] Robertson, I. A. 1967. Diffraction of plane longitudinal wave by a penny-shaped crack." *Proceedings of the Cambridge Philosophical Society* 63: 229–238. https://doi:10.1017/S0305004100041116.

[118] Ryzy, M., T. Grabec, P. Sedlak, and I. A. Veres. 2018. "Influence of grain morphology on ultrasonic wave attenuation in polycrystalline media with statistically equiaxed grains." *Journal of the Acoustical Society of America* 143: 219–229. https://doi.org/10.1121/1.5020785.

[119] Schmerr, Jr., L. W. 1998. *Fundamentals of Ultrasonic Nondestructive Evaluation—A Modeling Approach*. Switzerland: Springer.

[120] Schmerr, Jr., L. W. and A. Sedov. 1989. "An elastodynamic model for compressional and shear wave transducers." *Journal of the Acoustical Society of America* 86: 1988–2006. https://doi.org/10.1121/1.398578.

[121] Sha, G., M. Huang, M. J. S. Lowe, and S. I. Rokhlin. 2020. "Attenuation and velocity of elastic waves in polycrystals with generally anisotropic grains: Analytic and numerical modeling." *Journal of the Acoustical Society of America* 147: 2442–2465. https://doi.org/10.1121/10.0001087.

[122] Shen, S. P., Z. B. Khuang, and S. Hu. 1999. "On interface crack in laminated anisotropic medium." *International Journal of Solids and Structures* 36: 4251–4268. https://doi.org/10.1016/S0020-7683(98)00199-1.

[123] Stanke, F. E., and G. S. Kino. 1984. "A unified theory for elastic wave propagation in polycrystalline materials." *Journal of the Acoustical Society of America* 75: 665–681. https://doi.org/10.1121/1.390577.

[124] Stein, S. 1961. "Addition theorems for spherical wave functions." *Quarterly of Applied Mathematics* 19: 15–24.

[125] Ström, S. 1991. "Introduction to integral representations and integral equations for time-harmonic acoustic, electromagnetic and elastodynamic wave fields." In *Field Representations and Introduction to Scattering*, edited by V. V. Varadan, A. Lakhtakia, and V. K. Varadan, 37–141. Amsterdam: North-Holland.

[126] Ström, S. 1991. "The scattered field." In *Field Representations and Introduction to Scattering*, edited by V. V. Varadan, A. Lakhtakia, and V. K. Varadan, 143–164. Amsterdam: North-Holland.

[127] Tan, T. H. 1977. "Reciprocity relations for scattering of plane, elastic waves", *Journal of the Acoustical Society of America* 61: 928–931. https://doi.org/10.1121/1.381393.

[128] Thompson, R. B., F. J. Margetan, P. Haldipur, L. Yu, A. Li, P. D. Panetta, and H. Wasan. 2008 "Scattering of elastic waves in simple and complex polycrystals." *Wave Motion* 45: 655–674. https://doi.org/10.1016/j.wavemoti.2007.09.008.

[129] Ting, T. C. T. 1996. *Anisotropic Elasticity*. Oxford: Oxford University Press.

[130] van den Berg, P. M. 1981. "Transition matrix in acoustic scattering by a strip." *Journal of the Acoustical Society of America* 70: 615–619. https://doi.org/10.1121/1.386751.

[131] van den Berg, P. M. 1982. "Scattering of two-dimensional elastodynamic waves by a rigid plane strip or a plane crack of finite width: the transition matrix." *Journal of the Acoustical Society of America* 72: 1038–1045. https://doi.org/10.1121/1.388235.

[132] van der Hijden, J. H. M. T. 1987. *Propagation of Transient Elastic Waves in Stratified Anisotropic Media*. Amsterdam: North-Holland.

[133] van der Hijden, J. H. M. T., and F. L. Neerhoff. 1984. "Scattering of plane waves by a plane crack of finite width." *Journal of Applied Mechanics* 51: 646–651. https://doi.org/10.1115/1.3167687.

[134] Varadan, V. V., and V. K. Varadan. 1981. "Configurations with finite number of scatterers—A self-consistent T-matrix approach." *Journal of the Acoustical Society of America* 70: 213–217. https://doi.org/10.1121/1.386674.

[135] Varatharajulu, V. V. 1977. "Reciprocity relations and forward amplitude theorems for elastic waves." *Journal of Mathematical Physics* 18: 537–543. https://doi.org/10.1063/1.523335.

[136] Wang, G. F. 2008. "Diffraction of shear waves by a nanosized spherical cavity." *Journal of Applied Physics* 103: 053519. https://doi.org/10.1063/1.2895389.

[137] Wang, G. F., X. Q. Feng, and S. W. Yu. 2007. "Interface effects on the diffraction of plane compressional waves by a nanosized spherical inclusion." *Journal of Applied Physics* 102: 043533. https://doi.org/10.1063/1.2775217.

[138] Wang, Y. S., and D. Wang. 1996. "Scattering of elastic waves by a rigid cylindrical inclusion partially debonded from its surrounding matrix. SH case. *International Journal of Solids and Structures* 33: 2789–2815. https://doi.org/10.1016/0020-7683(95)00179-4.

[139] Waterman, P. C. 1976. "Matrix theory of elastic wave scattering." *Journal of the Acoustical Society of America* 60: 567–580. https://doi.org/10.1121/1.381130.

[140] Watson, G. N. 1944. *A Treatise on the Theory of Bessel Functions*, second edition. Cambridge: Cambridge University Press.

[141] Weaver, R. L. 1990. "Diffusivity of ultrasound in polycrystals." *Journal of the Mechanics and Physics of Solids* 38: 55–86. https://doi.org/10.1016/0022-5096(90)90021-U.

[142] J.R. Wait, J. R. 1955. “On the scattering of spherical waves by a cylindrical object.” *Applied Scientific Research* Series B 4, 464–468. https://doi.org/10.1007/BF02920024.

[143] Yang, H. J., and D. B. Bogy. 1985. “Elastic wave scattering from an interface crack in a layered half space.” *Journal of Applied Mechanics* 52: 42–49. https://doi.org/10.1115/1.3169024.

[144] Yang, L., O. T. Lobkis, and S. I. Rokhlin. 2012. “An integrated model for ultrasonic wave propagation and scattering in a polycrystalline medium with elongated hexagonal grains.” *Wave Motion* 49: 544–560. https://doi.org/10.1016/j.wavemoti.2012.03.003.

[145] Yang, Y., and A. N. Norris. 1991. “Shear-wave scattering from a debonded fiber.” *Journal of the Mechanics and Physics of Solids* 39: 273–294. https://doi.org/10.1016/0022-5096(91)90006-A.

[146] Yang, Y., and A. N. Norris. 1992. “Longitudinal-wave scattering from a partially bonded fiber.” *Wave Motion* 15: 43–59. https://doi.org/10.1016/0165-2125(92)90004-L.

[147] Zhang, Ch., and D. Gross. 1998. *On Wave Propagation in Elastic Solids with Cracks*. Southampton: Computational Mechanics Publications.

Index